Polymers for Microelectronics and Nanoelectronics

ACS SYMPOSIUM SERIES **874**

Polymers for Microelectronics and Nanoelectronics

Qinghuang Lin, Editor
IBM T. J. Watson Research Center

Raymond A. Pearson, Editor
Lehigh University

Jeffrey C. Hedrick, Editor
IBM T. J. Watson Research Center

Sponsored by the
**ACS Division of Polymeric Materials: Science and
Engineering, Inc.**

American Chemical Society, Washington, DC

Library of Congress Cataloging-in-Publication Data

Polymers for microelectronics and nanoelectronics / Qinghuang Lin, editor, Raymond A. Pearson, editor Jeffrey C. Hedrick, editor.

    p. cm.—(ACS symposium series ; 874)

Based on selected papers from the ACS symposium entitled Polymers for micro- and nano-electronics," held in August 18–22, 2002 in Boston, Massachusetts.

Includes bibliographical references and index.

ISBN 0–8412–3857–X

    1. Polymers—Congresses. 2. Microelectronics—Materials—Congresses. 3. Molecular electronics—Materials—Congresses.

    I. Lin, Qinghuang, 1963- II. Pearson, R. A., 1958- III. Headrick, Jeffrey C. IV. Series.

TK7871. 15.P6275   2004
621.382—dc22   2003063663

The paper used in this publication meets the minimum requirements of American National Standard for Information Sciences—Permanence of Paper for Printed Library Materials, ANSI Z39.48–1984.

PRINTED IN THE UNITED STATES OF AMERICA

Foreword

The ACS Symposium Series was first published in 1974 to provide a mechanism for publishing symposia quickly in book form. The purpose of the series is to publish timely, comprehensive books developed from ACS sponsored symposia based on current scientific research. Occasionally, books are developed from symposia sponsored by other organizations when the topic is of keen interest to the chemistry audience.

Before agreeing to publish a book, the proposed table of contents is reviewed for appropriate and comprehensive coverage and for interest to the audience. Some papers may be excluded to better focus the book; others may be added to provide comprehensiveness. When appropriate, overview or introductory chapters are added. Drafts of chapters are peer-reviewed prior to final acceptance or rejection, and manuscripts are prepared in camera-ready format.

As a rule, only original research papers and original review papers are included in the volumes. Verbatim reproductions of previously published papers are not accepted.

ACS Books Department

Contents

Preface...xi

1. **Recent Progress in Materials for Organic Electronics**...........................1
   Zhenan Bao

2. **PPV Nanotubes, Nanorods, and Nanofilms as well as
   Carbonized Objects Derived Therefrom**..15
   Kyungkon Kim, Guolun Zhong, Jung-Il Jin, Jung Ho Park,
   Seoung Hyun Lee, Dong Woo Kim, Yung Woo Park,
   and Whikun Yi

3. **Layer-by-Layer Assembly of Molecular Materials
   for Electrooptical Applications**...30
   M. E. van der Boom and T. J. Marks

4. **Oxidative Solid-State Cross-Linking of Polymer Precursors
   to Pattern Intrinsically Conducting Polymers**.....................................44
   Sung-Yeon Jang, Manuel Marquez, and Gregory A. Sotzing

5. **Fluorocarbon Polymer-Based Photoresists for 157-nm
   Lithography**...54
   T. H. Fedynyshyn, W. A. Mowers, R. R. Kunz, R. F. Sinta,
   M. Sworin, A. Cabral, and J. Curtin

6. **157-nm Resist Materials Using Main-Chain
   Fluorinated Polymers**...72
   Toshiro Itani, Hiroyuki Watanabe, Tamio Yamazaki,
   Seiichi Ishikawa, Naomi Shida, and Minoru Toriumi

7. **Correlation of the Reaction Front with Roughness
   in Chemically Amplified Photoresists**..86
   Ronald L. Jones, Vivek M. Prabhu, Darío L. Goldfarb,
   Eric K. Lin, Christopher L. Soles, Joseph L. Lenhart,
   Wen-li Wu, and Marie Angelopoulos

8. Utilizing Near Edge X-ray Absorption Fine Structure
    to Probe Interfacial Issues in Photolithography.......................98
    Joseph L. Lenhart, Daniel A. Fischer, Sharadha Sambasivan,
    Eric K. Lin, Christopher L. Soles, Ronald L. Jones, Wen-li Wu,
    Darío L. Goldfarb, and Marie Angelopoulos

9. Photolabile Ultrathin Polymer Films for Spatially Defined
    Attachment of Nano Elements..118
    B. Voit, F. Braun, Ch. Loppacher, S. Trogisch, L. M. Eng,
    R. Seidel, A. Gorbunoff, W. Pompe, and M. Mertig

10. Soft Lithography on Block Copolymer Films: Generating
    Functionalized Patterns on Block Copolymer Films as a
    Basis to Further Surface Modification...............................129
    Martin Brehmer, Lars Conrad, Lutz Funk, Dirk Allard,
    Patrick Théato, and Anke Helfer

11. Nanoporous, Low-Dielectric Constant Organosilicate
    Materials Derived from Inorganic Polymer Blends....................144
    R. D. Miller, W. Volksen, V. Y. Lee, E. Connor, T. Magbitang,
    R. Zafran, L. Sundberg, C. J. Hawker, J. L. Hedrick, E. Huang,
    M. Toney, Q. R. Huang, C. W. Frank, and H.-C. Kim

12. Porous Low-k Dielectrics: Material Properties....................161
    C. Tyberg, E. Huang, J. Hedrick, E. Simonyi, S. Gates, S. Cohen
    K. Malone, H. Wickland, M. Sankarapandian, M. Toney,
    H.-C. Kim, R. Miller, W. Volksen, P. Rice, and L. Lurio

13. Ultra Low-k Dielectric Films with Ultra Small Pores Using
    Poragens Chemically Bonded to Siloxane Resin...................173
    Bianxiao Zhong and Eric S. Moyer

14. Design of Nanoporous Polyarylene Polymers for Use as
    Low-k Dielectrics in Microelectronic Devices.....................187
    H. Craig Silvis, Kevin J. Bouck, James P. Godschalx,
    Q. Jason Niu, Michael J. Radler, Ted M. Stokich, John W. Lyons,
    Brandon J. Kern, Joan G. Marshall, Karin Syverud, and Mary Leff

15. Molecular Brushes as Templating Agents for Nanoporous
    SiLK* Dielectric Films...199
    Q. Jason Niu, Steven J. Martin, James P. Godschalx,
    and Paul H. Townsend

16. X-ray Reflectivity as a Metrology to Characterize
    Pores in Low-k Dielectric Films..209
    Christopher L. Soles, Hae-Jeong Lee, Ronald C. Hedden,
    Da-wei Liu, Barry J. Bauer, and Wen-li Wu

17. Micro- and Nanoporous Materials Developed Using
    Supercritical CO_2: Novel Synthetic Methods for the
    Development of Micro- and Nanoporous Materials Toward
    Microelectronic Applications..223
    Sara N. Paisner and Joseph M. DeSimone

18. Thermally Degradable Photocross-Linking Polymers......................236
    Masamitsu Shirai, Satoshi Morishita, Akiya Kawaue,
    Haruyuki Okamura, and Masahiro Tsunooka

19. Evaluation of Infrared Spectroscopic Techniques to Assess
    Molecular Interactions...251
    Robert K. Oldak and Raymond A. Pearson

20. Study on Metal Chelates as Catalysts of Epoxy and Anhydride
    Cure Reactions for No-Flow Underfill Applications.......................264
    Zhuqing Zhang and C. P. Wong

21. Benzocyclobutene-Based Polymers for Microelectronic
    Applications...279
    Ying-Hung So, Phil E. Garrou, Jang-Hi Im, and Karou Ohba

22. Formation of Nanocomposites by In Situ Intercalative
    Polymerization of 2-Ethynylpyrdine in Layered
    Aluminosilicates: A Solid-State NMR Study..............................294
    A. L. Cholli, S. K. Sahoo, D. W. Kim, J. Kumar, and A. Blumstein

Indexes

Author Index...309

Subject Index..311

Preface

Semiconductor chips are the brain of the Information Age and the engine of the knowledge-based economy. Sophisticated chips power everything from super computers, personal computers, personal digital assistants, cellular phones to medical devices and household appliances. They also control modern automobiles, airplanes, spaceships, and communication networks. The ubiquitous semiconductor chips have fundamentally altered the way people work, communicate, do business, and receive education and entertainment. Historically, the semiconductor chip industry grew at a compound annual growth rate of more than 15%, and it has grown to become a gigantic business of more than $150 billions (U.S. dollars) per year in less than five decades.

The success of the semiconductor industry has been constantly driven by the relentless pursuit of technological innovations. One key technological innovation is the incredible ability of optical lithography to continuously shrink device features to improve performance and to reduce manufacturing costs. This remarkable feat in optical lithography has been made possible in a large part by a class of photosensitive polymers called resists. The current leading edge semiconductor devices—the so-called 90-nm node devices—in mass productionhave a gate length less than 50 nm.

As the optical lithography and conventional device structures approach their limits, material innovations will become a key driver to maintain the historical pace of performance enhancement of semiconductor devices.

Advanced polymeric electronic materials lie at the heart of these material innovations in the semiconductor industry. These polymeric materials include polymers for patterning, insulating, and packaging of semiconductor circuitries and chips. More recently, electrically conducting polymers have been found useful as the active components of a transistor. The discovery and the maturing of the electrically conducting and semiconducting polymers have led to emerging fields of organic electronics and molecular electronics.

This book is based on selected papers from the American Chemical Society (ACS) symposium entitled *Polymers for Micro- and Nanoelectronics* held August 18–22, 2002 in Boston, Massachusetts. It covers significant recent advances, current status, and future directions in polymeric materials for patterning, insulating, and packaging of semiconductor chips as well as organic electronic materials for organic electronics and optoelectronics.

We hope that this book will prove valuable to the many scientists and engineers working in the fast-moving semiconductor industry. We also hope that it will be useful to faculty members, graduate students, undergraduate students, or anyone who has interests in the exciting fields of microelectronics, nanoelectronics, organic electronics, and optoelectronics. This book will also serve as a valuable reference for those who are interested in nanofabrication, micro- and nano-fluidics, micro- and nano-photonics, Micro-Electro-Mechanical Systems (MEMS), BioMEMS, porous media, high-temperature polymers, as well as nanostructured materials.

We thank the ACS Division of Polymeric Materials: Science and Engineering (PMSE) for sponsoring the symposium in Boston and Jay Dias and Eileen Ernst of PMSE for help in making the symposium a very successful one. We also thank the authors for their valuable contributions to this volume and the referees for careful review and thoughtful comments on the manuscripts. We are also grateful to the Loctite Company and Shipley Company, LLC for their financial support for the symposium.

Finally, we extend our sincere thanks to Stacy VanDerWall, Margaret Brown, and Bob Hauserman of the ACS for their patience and tireless efforts in assembling and publishing this book.

Qinghuang Lin
T. J. Watson Research Center
IBM
P.O. Box 218, Route 34
Yorktown Heights, NY 10598

Jeffrey C. Hedrick
T. J. Watson Research Center
IBM
P.O. Box 218, Route 34
Yorktown Heights, NY 10598

Raymond A. Pearson
Microelectronic Packaging Laboratory
Lehigh University
5 East Packer Avenue
Bethlehem. PA 18015

Polymers for Microelectronics and Nanoelectronics

Chapter 1

Recent Progress in Materials for Organic Electronics

Zhenan Bao

Bell Laboratories, Lucent Technologies, 600 Mountain Avenue,
Murray Hill, NJ 07974

Organic and polymeric materials are active materials used for electronic devices such as thin film field-effect transistors (FETs). These materials are more likely to have practical advantages when coupled with low cost approaches to patterning that are compatible with them. In this chapter, we discuss material requirements and survey different materials used for high performance organic FETs.

Introduction

Organic and polymeric semiconducting materials have been widely pursued in recent years as active materials for low-cost, large area, flexible and lightweight devices. They offer numerous advantages for easy processing (e.g. spin-coating, printing, evaporation), good compatibility with a variety of substrates including flexible plastics, and great opportunities for structural modifications. [1] In addition, these materials can be processed at much lower substrate temperatures (less than 120 °C) and with little or no vacuum involved, as compared to the high substrate temperatures (greater than 900 °C) and high vacuum needed for typical inorganic semiconducting materials. Instead of competing with conventional silicon technologies, printable FETs may find niche applications in low-cost memory devices, such as smart cards and electronic luggage tags, and large area driving circuits for displays. The demonstration of such devices relies heavily on high-performance solution-processable organic and polymeric materials along with novel printing processes. In this article, we will review the current status in transistor material discovery.

Device Operation

A typical FET has the device structure shown in Figure 1. The current flow between the drain and source electrode is low when no voltage is applied between the gate and the drain electrodes as long as the semiconducting material is not highly doped. This state at which the gate voltage is zero is called the "off" state of a transistor. When a voltage is applied to the gate, charges can be induced into the semiconducting layer at the interface between the semiconductor and dielectric layer. As a result, the drain-source current increases due to the increased number of charge carriers, and this is called the "on" state of a transistor. Therefore to construct a FET, materials ranging from insulating (dielectric material), semiconducting, to conducting are required.

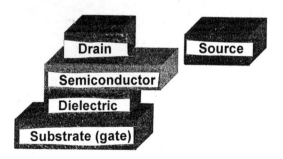

Figure 1. Schematic structure of a thin film field-effect transistor.

The key parameters in characterizing a FET are its field-effect mobility and on/off ratio. Field-effect mobility quantifies the average charge carrier drift velocity per unit electric field, whereas on/off ratio is defined as the drain-source current ratio between the "on" and "off" states. They both depend strongly on the gate voltage used for calculation. For a high performance FET, the field-effect mobility and on/off ratio should be as high as possible. For example, field-effect mobility of greater than 0.1 cm^2/Vs and on/off ratio greater than 10^6 are needed for the driving circuits in liquid crystal displays.

Materials

As mentioned above, materials ranging from conducting, semiconducting, to insulating are required for the construction of organic transistors. In the following sessions, recent developments in the materials aspect are surveyed.

Electrode Materials

Gold has been the most commonly used electrode material for organic electronics so far because it forms good ohmic contact with many organic semiconductors and it is inert to environmental changes. One drawback found with gold electrodes was that some organic semiconducting materials tend to deposit on gold surfaces as small and more rough three-dimensional grains from vapor phase. This results in different film morphology in the channel region above the dielectric layer (typically SiO_2) and the electrode region above gold introducing a significant charge injection barrier. [2,3] The effect of metal work function on organic transistor performance has been studied by Lin, Gundlach, and Jackson. [2] Electrode materials, such as Al, Cu, In, Ni, and Pd were used as the drain and source electrodes. Interestingly, they found that the properties of the source electrode can significantly impact the contact characteristics, indicating that charge injection dominates the contact characteristics. [2]

The metal electrode surface can be modified with self-assembled monolayers (SAM) to improve charge injection as well as the morphology of semiconductor deposited. [3,4] It was found that larger and smoother grains of pentacene are grown on SAM treated gold electrodes and give rise to improved transistor performance. For n-channel materials, a similar improvement in performance was reported attributed mainly to the improved growth of the semiconducting layer. [5]

Conducting polymers, such as polyanilline [6,7] and poly(3,4-ethylene dioxythiophene) (PEDOT) [8,9,10,11] have been used as electrode materials. They generally give good transistor performance. The advantage for using these solution processable polymers is that they can be easily patterned by printing methods with relatively high resolution. Electrodes separated by less than ten μm spacings have been prepared by micromolding in capillaries, [6,11] screen-printing, [12] as well as inkjet printing. [10] Conducting paste made of polymer composites of graphite or silver particles were screen printed as the drain and source electrodes. [12] In this case, only devices with electrodes printed above the semiconducting layer showed good transistor performance while the bottom-contact geometry did not work due to the highly rough electrode surface.

Semiconducting Materials

Vacuum Deposited p-Channel Materials

A number of conjugated oligomers and metallophthalocyanines have been studeied as *p*-channel semiconducting materials (Table 1). These compounds have limited solubility in organic solvents and therefore vacuum evaporation has to be used to fabricate their thin films. The highest field-effect mobility has been reported with pentacene ca. 1.5-2 cm^2/Vs, [13] which is in the same order of magnitude as amorphous Si (α-Si). Its high performance has been attributed to the ability of forming single-crystal-like films upon vacuum deposition onto gently heated (about 80 °C) substrates. [14]

Most p-channel compounds are thiophene-containing oligomers. Mobilities generally in the order of 10^{-2} cm^2/Vs have been reported with these compounds. α-Dihexyltetrathienyl (DH-α-4T) has been found to form single-crystal-like films, similar as pentacene, and high filed-effect mobility ca. 0.2 cm^2/Vs has been reported. [15]

Table 1. Transistor Performance of Some p-Channel Semiconductors Fabricated by Vacuum Deposition

Compound	Field-effect mobility (cm^2/Vs)	Reference
n = 4 - 8	0.002 - 0.02	16,17,18
	0.03 - 0.04	19
n = 4 - 8	0.01 - 0.2	20,21,22

Table 1. *Continued*

Compound	Field-effect mobility (cm^2/Vs)	Reference
M = Cu, Sn, Zn, H2	0.003 - 0.02	23
	0.01 - 0.02	24
	0.001 - 0.01	25
n=1-4	0.01-0.15	26
n=0-1	10^{-4}-0.03	27
	10^{-5}-0.007	28
	0.02-0.04	28
	10^{-4}	29
R = H, alkyl	0.015 - 0.17	30
	0.1- 2.1	13

Metallophthalocyanines (M = Cu,Zn,Pt,Ni,Sn,Fe,H$_2$) is another class of material, which showed promising transistor performance. The best mobility was reported for copper phthalocyanine with mobility of 0.02 cm^2/Vs and on/off ratio greater than 10^5. [23] These materials are attractive since they are commercially available in large quantity and high purity. They are also chemically and thermally stable.

One potential problem with most p-channel semiconductors is that they tend to be photo-oxidized over time and result in decreased mobility and on/off ratio. To solve this problem, fine-tuning the energy levels of the semiconducting material is essential. Recently, several new semiconductors based on fluorenes and thiazoles have shown improved stability and good transistor performance. [26,28] Mobility as high as 0.15 cm^2/Vs was reported for oligo(fluorene bithiophene) oligomers. [26] Most remarkably, the transistor performance was essentially unchanged even after storage in air with exposure to ambient light and without any encapsulation.

It is worth pointing out that most organic semiconductors adapt herringbone packing in their crystal structures and in thin films. Improved mobility may be possible if they can pack face-to-face. Very few such examples are known so far and high mobility from these systems is yet to be demonstrated. [24,31] The grain size and the details of the grain boundaries may be the limiting factor.

Soluble p-Channel Materials.

To truly realize the advantages (i.e. processability and low-cost) of organic materials in device applications, liquid phase processing techniques by spin-coating, casting, or printing are strongly desired. Both organic oligomers and polymers can be deposited from solution to form uniform films. Table 2 summarizes the transistor performance of some organic semiconductors fabricated from solution.

A number of oligomers have been deposited from solution to give reasonable transistor performance. These lower molecular weight materials are easier to purify compared to polymers since they can be chromatographied, sublimed, or recrystallized. On the other hand, the conditions for forming a smooth and uniform film over a large area is much more sensitive to the deposition conditions. Since most of these oligomers have limited solubility (less than 1 mg/ml), they tend to crystallize easily to form disconnected crystals on a surface instead of a continuous film. Carefully experimenting with solvent systems, solvent evaporation rate, concentration, and surface modifications are necessary in order to achieve good results.

Table 2. Transistor Performance of Organic Semiconductors Fabricated from Solution

Compound	Field-effect mobility (cm^2/Vs)	Reference
$n = 4 - 6$	0.01 - 0.2	20,32
n=2	0.005	26
	0.005-0.02	28
X=Cl,Br $\longrightarrow$ pentacene	0.1	33
$\longrightarrow$ pentacene	0.1-0.9	34
	0.22	35
n=4-12	10^{-4}-0.1	12,36,37
	0.008	38
	10^{-4}-0.01	39

The second technique involves the use of a soluble precursor oligomer or polymer which can undergo subsequent chemical reactions to give the desired conjugate oligomer or polymer, such as pentacene [33,34] and poly(thienylene vinylene). [35] High mobilities have been reported for both types of materials. For pentacene, the conversion temperature required is relatively high in some cases for achieving high mobility. [33] In the case of poly(thienylene vinylene), not only a heating temperature of 200 °C is required, but also an acid is needed as the catalyst for the conversion, which may lead to doping of the polymer and result in low on/off ratio.

The third technique utilizes soluble conjugated polymers and they are fabricated by spin-coating, casting, or printing techniques. Different conjugated polymers have been studied. Examples including poly(2,5-dialkylpheneylene-co-phenylene)s, poly(2,5-dialkylphenylene-co-thiophene)s, poly(2,5-dialkylphenylene vinylene)s and the dialkoxyl derivatives of the above polymers. [40] However, very low (less than 10^{-4} cm^2/Vs) or no field-effect mobilities have been found. More extensive effort has been directed towards soluble polythiophene derivatives since they are widely used as conducing and semiconducting materials. [41,42,43] We have studied the electrical characteristics of field-effect transistors using solution cast regioregular poly(3-hexylthiophene) (P3HT). [36] It is demonstrated that both high field-effect mobilities (ca. 0.05 cm^2/Vs in the accumulation-mode and 0.01 cm^2/Vs in the depletion-mode), and relatively high on/off current ratios (greater than 10^3) can be achieved. It was also found that the film quality and field-effect mobility are strongly dependent on the choice of solvents. The field-effect mobility can range from 10^{-4} to 10^{-2} cm^2/Vs when different solvents are used for film preparation. There is also an alkyl chain length dependence of the mobility. For alkyl chain shorter than dodecyl-chain, mobilities greater than 0.01 cm^2/Vs have been found while regioregular poly(3-dodecylthiophene) only had mobility in the order of 10^{-4} cm^2/Vs. [12] Surface treatment of SiO$_2$ surface with HMDS (hexamethyldisilazane) resulted in mobility as high as 0.1 cm^2/Vs for regioregular poly(3-hexylthiophene). [37]

Some rigid-rod conjugated polymers also exhibit thermotropic liquid crystallinity. [44] Therefore, it is possible to align them from the liquid crystalline phase using an alignment layer. Improved mobility using this method was reported for a poly(fluorene bithiophene) copolymer. [39]

Amorphous polymers generally do not give high field effect mobilities. However, a poly(triarylamine)-based system was found to have reasonably high hole mobility (0.008 cm^2/vs) in a field-effect transistor geometry. [38] Most remarkably, the stability of this material is significantly better than polythiophenes due to its much higher band gap.

n-Channel Organic Semiconductors

N-channel semiconductors are important components of *p-n* junction diodes, bipolar transistors, and complementary circuits. A complementary circuit, which consists of both p- and n-channel transistors, is an ideal circuit configuration for organic semiconductors because it has low static power dissipation and the transistors are "on" only during switching. There are only a limited number of high performance n-channel semiconductors discovered so far. Most amorphous and polycrystalline organic materials tend to transport holes better than electrons. The performance of n-channel semiconductors is also easily degraded after exposure to air since oxygen can act as trap to oxidize some negative charge carriers. For example, C_{60} transistor devices are easily degraded upon exposure to air. [50] In addition, some anhydrides exhibited n-channel semiconducting properties, but their mobilities remained relatively low. [45]

The general strategy for air-stable n-channel organic semiconductors is to substitute a conjugated framework with strong electron-withdrawing groups, such as F-, CN-, perfluoroalkyl chain. High mobility and stable operation in air have been reported. Hexadecafluoro- and hexadecachloro-metallophthalocyanines were found to function as *n*-channel semiconductors and copper hexadecafluoro-phthalocyanine showed the best field-effect mobility ca. 0.03 cm^2/Vs. [47] The high mobilities of these compounds are attributed to the highly ordered films upon vacuum deposition. They also possess remarkable stability in air, and their transistors can be operated in both vacuum and air.

Another class of high performance n-channel semiconductors are based on naphthalenetetracarboxylic diimide (NTCDI). [5,48] Mobilities up to 0.06 cm^2/Vs measured in air were reported. In another approach, perfluorinated alkyl chains were substituted onto oligothiophenes and resulted in air-stable n-channel performance. [46] This approach may lead to a variety of new n-channel materials. The above three types of materials are by far the only thin film n-type materials found to achieve longtime stability in air with mobilites greater than 10^{-2} cm^2/Vs (of course, it is desirable to have n-channel materials with higher mobilities in the range of 10^{-1} cm^2/Vs). Suitable n-channel semiconductors, which can resist these environmental effects, are desired or alternatively, proper packaging method has to be implemented.

Dielectric Materials

Identification of solution processable dielectric materials is also crucial to the commercial application of high performance, low cost, organic thin film field effect transistors. The thin dielectric films must be pinhole free and have a high breakdown voltage and good long-term stability. To permit the use of plastic substrates in these devices, it is also desirable that the films can be formed at

Table 3. Transistor Performance of Some n-Channel Organic Semiconductors

Compound	Field-effect mobility (cm^2/Vs)	Operation in air	Reference
	10^{-4} - 0.02	No	46
M=Cu,Co,Fe,H2	10^{-3}-0.03	Yes	47
	0.12	Yes	5,48
	0.3-0.6	No	49
	0.005	Yes	31
C_{60}	0.015 - 0.17	No	50
	10^{-6}-10^{-4}	No	51

lower temperatures than those traditionally used with typical inorganic materials. Other important requirements for the dielectric film are: it must have good adhesion to the substrate; it must be able to be applied by conventional methods, such as spin coating or offset printing; and it must be compatible with the subsequent processing of the device. For some applications it would be advantageous if the film could be patterned using conventional photolithography.

For use with organic transistors, the dielectric material must also be compatible with organic semiconductors. Specifically, the dielectric films must have a low surface trapping density, low surface roughness, and a low impurity concentration and must not degrade the performance of the adjacent, ordered organic semiconducting films.

Previously, a few organic polymers had been used as dielectric materials for with organic transistors. [7,52] Examples are polymethyl methacrylate, [52] polyvinylphenol, [7] polyimides, [12] and BCB. [53] We have found that organosilsesquioxanes meet the stringent requirements for use in these devices. [54] Precursors of these materials, low molecular weight oligomers are commercially available from various sources. They have the empirical formula $RSiO_{1.5}$, where R can be hydrogen, alkyl, alkenyl, alkoxyl and aryl. The Glass Resin oligomers are soluble in common organic solvents which permit the use of conventional liquid coating techniques to produce films of the required thickness, several thousand angstroms. After solvent removal, the precursors are cured at temperatures less than 150°C to yield films that have high dielectric strength, low leakage current, and good hydrolytic and thermal stability. The surface of the cured material can be modified to alter the surface properties. The oligomers having methyl and methyl-phenyl pendant groups on the siloxane backbone were found to be good candidates as they could be cured at 135°C to generate films exhibiting high transistor performance that was similar to that obtained using SiO_2 as the dielectric layer. [54]

Summary and Future Outlook

In this article, organic/polymeric materials for the construction of electrodes, semiconductor, and dielectrics for organic transistors are surveyed. There are now several semiconducting materials, which has performance similar as amorphous Si. For soluble organics and polymers, there are far less candidates, which have mobilities and on/off ratios approaching to those of amorphous Si. Large area uniform coverage still tends to be difficult to achieve for organics while the existing polymers generally have low on/off ratios unless special fabrication conditions are followed. New materials with improved performance and processability should be explored. In addition, better understanding is needed of the chemical structure-performance-processability

12

relationships by systematic material design and characterizations. All the soluble semiconducting materials discussed in this article are *p*-channel materials. Logic families constructed with complementary metal oxide semiconductor (CMOS) circuits require both *p*- and *n*-channel active materials. Only one n-channel semiconductor has been shown to produce a film with useful mobility from solution. [48] There are clear needs for such materials to realize all-printable CMOS circuits.

References

1   (a)Katz, H. E.; Bao, Z. *J. Phys. Chem. B* **2000**, *104*, 671-678; (b)Horowitz, G.; Hajlaoui, M. E. *Adv. Mater.* **2000**, *12*, 1046-1050; (c)Katz, H. E.; Bao, Z.; Gilat, S. *Acct. Chem. Rev.* **2001**, *34*, 359-369; (d)Dimitrakopoulos, C.D.; Mascaro, D.J. *IBM J. Res. & Dev.* **2001**, *45*, 11-28; (e)Bao, Z.; Rogers, J. A.; Katz, H. E. *J. Mater. Chem.* **1999**, *9*, 1895.

2   (a)Lin, Y.Y.; Gundlach, D.J.; Jackson, T.N. *Materials Research Society Symposium Proceedings,* **1996**, *413*, 413-418; (b)Gundlach, D.J.; Nichols, J.A.; Kuo, C.C.; Klauk, H.; Sheraw, C.D.; Schlom, D.G.; Jackson, T.N. *41st Electronic Materials Conference Digest*, **1999**, p. 16; (c)Gundlach, D.J.; Jia, L.L.; Jackson, T.N. *43rd Electronic Materials Conference Digest*, **2001**, p. 31; (d)Gundlach, D.J.; PhD. Thesis, *Small-molecule Organic Thin Film Transistors*, The Pennsylvania State University.

3   Kymissis, I.; Dimitrakopoulos, C.D.; Purushothaman, S. *IEEE T Electron Dev.* **2001**, *48*, 1060-1064.

4   Wang, J.; Gundlach, D.J.; Kuo, C.C.; Jackson, T.N. *41st Electronic Materials Conference Digest*, **1999**, p. 16.

5   Katz, H.E.; Johnson, J.; Lovinger, A.J.; Li, W.J. *J. Am. Chem. Soc.* **2000**, *122*, 7787-7792.

6   Rogers, J.A.; Bao, Z.; Raju, V.R. *Appl. Phys. Lett.* **1998**, *72*, 2716.

7   Gelinck, G.H.; Geuns, T.C.T.; de Leeuw, D.M. *Appl. Phys. Lett.* **2000**, *77*, 1487.

8   Touwslager, F.J.; Willard, N.P.; de Leeuw, D.M. *Appl. Phys. Lett.* **2002**, *81*, 4556.

9   Halik, M.; Klauk, H.; Zschieschang, U.; Kriem, T.; Schmid, G.; Radlik, W.; Wussow, K. *Appl. Phys. Lett.* **2002**, *81*, 289.

10   Sirringhaus, H.; Kawase, T.; Friend, R.H.; Shimoda, T.; Inbasekaran, M.; Wu, W.; Woo, E.P. *Science*, **2000**, *290*, 2123-2126.

11   Beh, W.S.; Kim, I.T.; Qin, D.; Xia, Y.N.; Whitesides, G.M. *Adv. Mater.* **1999**, *11*, 1038-1041.

12   Bao, Z.; Feng, Y.; Dodabalapur, A.; Raju, V.R.; Lovinger, A.J. *Chem. Mater.* **1997**, *9*, 1299.

13   Lin, Y.Y.; Gundlach, D.J.; Nelson, S.F.; Jackson, T.N. *IEEE Trans. Elec. Dev.* **1997**, *44*, 1325.

14 Laquindanum, J.G.; Katz, H.E.; Lovinger, A.J.; Dodabalapur, A.; *Chem. Mater.* **1996**, *8*, 2542-2544.

15 Katz, H.E.; Lovinger, A.J.; Laquindanum, J.G.; *Chem. Mater.* **1998**, *10*, 457-459.

16 Katz, H.E.; Dodabalapur, A.; Bao, Z. in *Oligo- and Polythiophene-based Field-effect Transistors*; D. Fichou Ed.; Wiley-VCH: Weinheim, **1998**.

17 Hajlaoui, R.; Fichou, D.; Horowitz, G.; Nessakh, B.; Constant, M.; Garnier, F. *Adv. Mater.* **1997**, *9*, 557-561.

18 Garnier, F.; Hajlaoui, R.; Yassar, A. *Science* **1994**, *265*, 1684.

19 Laquindanum, J.; Katz, H.E.; Dodabalapur, A.; Lovinger, A.J. *Adv. Mater.* **1997**, *9*, 36.

20 Katz, H.E.; Laquindanum, J.G.; Lovinger, A.J.; *Chem. Mater.* **1998**, *10*, 633-638.

21 Garnier, F.; Yassar, A.; Hajlaoui, R.; Horowitz, G.; Deloffre, F.; Servet, B.; Ries, S.; Alnot, P. *J. Am. Chem. Soc.* **1993**, *115*, 8716-8721.

22 Li, W.; Katz, H.E.; Lovinger, A J.; Laquindanum, J.G.; *Chem. Mater.* **1999**, *11*, 458-465.

23 (a)Bao, Z.; Lovinger, A.J.; Dobabalapur, A. *Appl. Phys. Lett.* **1996**, *69*, 3066; (b)Bao, Z.; Lovinger, A.J.; Dodabalapur, A. *Adv. Mater.* **1997**, *9*, 42-44.

24 Li, X.C.; Sirringhaus, H.; Garnier, F.; Holmes, A.B.; Moratti, S.C.; Feeder, N.; Clegg, W.; Teat, S.J.; Friend, R.H. *J. Am. Chem. Soc.* **1998**, *120*, 2206;

25 Dimitrakopoulos, C.D.; Afzali-Aradakani, A.; Furman, B.; Kymissis, J.; Purushothaman, S. *Syn. Met.* **1997**, *89*, 193.

26 Meng, H.; Bao, Z.; Lovinger, A. J.; Wang, B.-C.; Mujsce, A. M.; *J. Am. Chem. Soc.* **2001**, *123*, 9214-9215.

27 Meng, H.; Bao,Z.; Lovinger, A.J., unpublished results.

28 Hong, X.M.; Katz, H.E.; Lovinger, A.J.; Wang, B.C.; Raghavachari, K.; *Chem. Mater.* **2001**, *13*, 4686-4691.

29 Kunugi, Y.; Takimiya, K.; Yamashita, K.; Aso, Y.; Otsubo, T. *Chem. Lett.* **2002**, *10*, 958-959.

30 Laquindanum, J.; Katz, H.E.; Lovinger, A.J. *J. Amer. Chem. Soc.* **1998**, *120*, 664-672.

31 Pappenfus, T.M.; Chesterfield, R.J.; Frisbie, C.D.; Mann, K.R.; Casado, J.; Raff, J.D.; Miller, L.L. *J. Am. Chem. Soc.* **2002**, *124*, 4184-4185.

32 Garnier, F.; Hajlaoui R.; El Kassmi, A.; Horowitz, G.; Laigre, L.; Porzio, W.; Armanini, M.; Provasoli, F. *Chem.Mater.* **1998**, *10*, 3334-3339.

33 Herwig, P.T.; Müllen, K.; *Adv. Mater.* **1999**, *11*, 480-483.

34 Afzali, A.; Dimitrakopoulos, C.D.; Breen, T.L.; *J. Am. Chem. Soc.* **2002**, *124*, 8812-8813.

14

35 Fuchigami, H.; Tsumura, A.; Koezuka, H. *Appl. Phys. Lett.* **1993**, *63*,1372-1374.
36 Bao, Z.; Dodabalapur, A.; Lovinger, A.J. *Appl. Phys. Lett.* **1996**, *69*, 4108.
37 Sirringhaus, H.; Tessler, N.; Friend, R.H. *Science* **1998**, *280*, 1741.
38 Veres, J.; Ogier, A.; Leeming, S.; Brown, B.; Cupertino, D. *Mat. Res. Soc. Symp. Proc.*, **2002**, 708.
39 Sirringhaus, H.; Wilson, R.J.; Friend, R.H.; Inbasekaran, M.; Wu, W.; Woo, E.P.; Grell, M.; Bradley, D.D.C. *Appl. Phys. Lett.* **2000**, *77*, 406.
40 Bao, Z. unpublished results.
41 Paloheimo, J.; Kuivalainen, P.; Stubb, H.; Vuorimaa, E.; Yli-Lahti, P. *Appl. Phys. Lett.* **1990**, *56*, 1157.
42 Assadi, A.; Svensson, C.; Willander, M.; Inganäs, O. *Appl. Phys. Lett.* **1988**, *53*, 195.
43 Tsumura, A.; Koezuka, H.; Ando, T. *Appl. Phys. Lett.* **1986**, *49*, 1210.
44 Yu, L.; Bao, Z.; Cai, R. *Angew. Chem. Int. Ed.* **1993**, *32*, 1345-1347.
45 (a)Horowitz, G.; Kouki, F.; Spearman, P.; Fichou, D.; Nogues, C.; Pan, X.; Garnier, F. *Adv. Mater.* **1996**, *8*, 242-244; (b)Laquindanum, J.G.; Katz, H.E.; Dodabalapur, A.; Lovinger, A.J. *J. Am. Chem. Soc.* **1996**, *118*, 11331-11332;
46 Facchetti, A.; Deng, Y.; Wang, A.; Koide, Y.; Sirringhaus, H.; Marks, T.J.; Friend, R.H. *Angew. Chem. Int. Ed.* **2000**, *39*, 4547-4551.
47 Bao, Z.; Lovinger, A.J.; Brown, J. *J. Amer. Chem. Soc.* **1998**, *120*, 207-208.
48 Katz, H.E.; Lovinger, A.J.; Johnson, J.; Kloc, C.; Siegrist, T.; Li, W.; Lin, Y.Y.; Dodabalapur, A. *Nature* **2000**, *404*, 478-481.
49 Malenfant, P.R.L.; Dimitrakopoulos, C.D.; Gelorme, J.D.; Kosbar, L.A.; Graham, T.O.; Curioni, A.; Andreoni, W. *Appl. Phys. Lett.* **2002**, *80*, 2517.
50 Haddon, R.C.; Perel, A.C.; Morris, R.C.; Palstra, T.T.M.; Hebard, A.F.; Fleming, R.M. *Appl. Phys. Lett.* **1995**, *67*, 121
51 Babel, A.; Jenekhe, S.A. *Adv Mater.* **2002**, *14*, 371-374.
52 Peng, X.; Horowitz, G.; Fichou, D.; Garnier, F. *Appl. Phys. Lett.* **1990**, *57*, 2013.
53 Klauk, H.; Gundlach, D.J.; Nichols, J.A.; Sheraw, C.D.; Bonse, M.; Jackson, T.N. *Solid State Technology*, **2000**, *43*, 63-77.
54 Bao Z.N.; Kuck, V.; Rogers, J.A.; Paczkowski, M.A. *Adv. Funct. Mater.* **2002**, *12*, 526-531.

Chapter 2

PPV Nanotubes, Nanorods, and Nanofilms as well as Carbonized Objects Derived Therefrom

Kyungkon Kim[1], Guolun Zhong[1], Jung-Il Jin[1,*], Jung Ho Park[2],
Seoung Hyun Lee[2], Dong Woo Kim[2], Yung Woo Park[2],
and Whikun Yi[3]

[1]Division of Chemistry and Molecular Engineering and Center for Electro-
and Photo-Responsive Molecules, Korea University, Seoul 136–701 Korea
(fax: +82 2 921 6901; telephone: +82 2 3290 3123; email: jijin@korea.ac.kr)
[2]School of Physics, Seoul National University, Seoul 151–747, Korea
[3]NCRI, Center for Electron Emission Source, Samsung Advanced Institute
of Technology, P.O. Box 111, Suwon 440–600, Korea

ABSTRACT

Poly(p-phenylenevinylene) (PPV) could be obtained in the form of nanotubes, nanorods and nanofilms by the chemical vapor deposition (CVD) polymerization of α,α'-dichloro-p-xylene followed by thermal dehydrochlorination. The polymerizations were conducted on the inner surface of or inside the nanopores of alumina or polycarbonate membrane filters or on the surface of silicon wafers. The PPVs thus obtained could be thermally converted to the corresponding carbonized tubes, rods, and films. We also could obtain nanopatterns and nanowells of PPV and carbon on silicon wafers by utilizing nanolithographed poly(methyl methacrylate) patterns. The PPV films obtained on the silicon wafers are semicrystalline and produce highly conducting ($\sigma \sim 0.7 \times 10^3$ Scm^{-1}) graphitic films even when treated only at 850 °C. Field-emission properties of some of the graphitic nanotubes are also described in this report.

INTRODUCTION

The present world-wide, explosive interests in nanotechnology is arousing renewed attention to organic conductors and carbonaceous matters because of their potentials in wide variety of applications, especially in nanodevices.

Among many organic conductors, poly(*p*-phenylene vinylene) (PPV),*(1, 2)* is unique in that the polymer can be prepared by many synthetic routes and possesses many interesting electrical and optical properties including photo- and electroluminescence.*(3)* Due to its polyconjugative nature, the polymer can be doped to exhibit electrical conductivity as high as ca. 10^2 Scm^{-1}. The polymer also is known to be photo-conducting. Moreover, structural modification of PPV can be easily conducted to produce many derivatives revealing a wide spectrum of electro-optical properties.

In addition to its many interesting properties potentially very useful in future applications, PPV can be readily converted to graphitic products by simple thermal treatment.*(4)* The degree of graphatization, thus the electrical conductivity, depends highly on the temperature of thermal treatment.

In spite of many attractive properties, their insolubility and infusibility make nanoscacle fabrication or nano wiring of PPV and graphitic carbons on desired substrates almost impossible. Fullerenes,*(5)* however, are known to have enough volatility and solubility, which render them a certain degree of processability. The situation is not the same for carbon nanotubes that are insoluble and unvaporizable. We adopted the CVD*(6-9)* method in the preparation of PPV in various nano shapes in order to circumvent the processability problem. The PPV nano objects thus prepared could be converted to the corresponding carbonaceous products by simple thermal conversion.

This article describes the preparation, structural analyses and properties of nano objects of PPV and graphitic products derived therefrom.

PREPARATION AND PROPERTIES OF NANO PPV OBJECTS

Among the many known methods for the preparation of PPV, there are three different synthetic methods that proceed through soluble precursors before conversion to the final polymer. They are represented below by chemical equations.

$$X-CH_2-\langle\underline{\ }\rangle-CH_2-X \xrightarrow{600\text{-}800\ ^{\circ}C} \left[XCH=\langle\underline{\ }\rangle=CH_2\right] \qquad (3)$$

$$\xrightarrow{\text{cold surface}} \left(\langle\underline{\ }\rangle-CH_2-\underset{X}{CH}\right)_n \xrightarrow[\text{or }\Delta]{\text{base}} \left(\langle\underline{\ }\rangle\diagup\right)_n$$

The first(10) and the second(11) methods are known to be Wessling-Zimmerman method and Gilch-Wheelwright polymerization, respectively. The last reaction route is practically the same as that is utilized in the synthesis of poly(p-xylylene) from p-xylene via the CVD polymerization method. The first two methods have been widely used not only in the synthesis of PPV but also PPV derivatives. Since the methods produce soluble precursor polymers, the final polymer can be obtained in various shapes.

The last method(12) is unique in that the activated species formed at an elevated temperature in the gas phase undergoes self-addition on the cold substrate surface to produce the precursor polymer that can be subjected to thermolysis to the final polymer. This method does not require the use of any solvent and additional chemicals during the polymerization process. Moreover, since synthesis can be performed directly on the surface of desired substrates, we can obtain uncontaminated polymer in various shapes. Taking those advantages into our consideration, we decided to employ the third CVD polymerization method in the preparation of PPV in various nano shapes including nanotubes, nanorods and nanofilms. Recently, Schäfer et al.,(7) Staring and coworkers,(8)

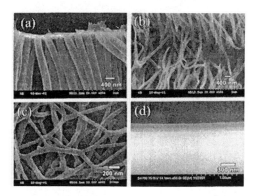

Figure 1. Scanning electron micrographs (SEM) of (a) and (b) nanotubes, (c) nanorods and (d) film of PPV on Si wafer (Reproduced from reference 14. Copyright 2001 American Chemical Society.)

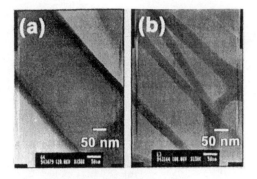

Figure 2. Transmission electron micrographs (TEM) of (a) nanotubes
and (b) nanorods of PPV (Reproduced from reference 14. Copyright
2001 American Chemical Society.)

and Vaeth and Jensen(9, 12) studied the CVD polymerization of α,α'-dihalo-*p*-
xylenes and obtained thin films of PPV.

In the synthesis of PPV in nano shapes we used commercially available nano
porous alumina (nominal pore diameter: 200 nm; Whatman, England) and
polycarbonate (nominal pore diameter: 100 and 10 nm; Osmonics Inc., U. S. A.)
filter membranes(13) and the (001) surface of silicon wafers as substrates. The
filter membranes are removed by dissolution in order to separate the PPV
nanotubes and nanorods. The synthetic details can be found in our earlier
report.(14) Figure 1 shows the scanning electron micrographs of PPV nanotubes
and nanorods obtained after removal of the filter substrates by dissolution either
in 3M NaOH or in dichloromethane. They reveal smooth surface of tubes and
rods. Moreover, no surface defects such as pinholes are detected. Figure 1(d) is
the scanning electron micrograph of a fracture surface of a PPV film (210 nm
thick) deposited on a silicon wafer. Here again, we observe a uniform
morphology. The transmission electron micrographs of the nanotubes and
nanorods given in Figure 1(a) and Figure 1(c) are shown respectively in Figure
2(a) and 2(b). The wall thickness of the nanotubes AL-200 is relatively uniform
and is estimated to be 28 ± 3 nm. Needless to say, the thickness can be controlled
by the reaction condition such as evaporation temperature of the monomer, flow
rate of the carrier gas and reaction time. Figure 2(b) tells us that the nanorods are
completely filled and have a diameter of 31 ± 3 nm.

We also could prepare by the same CVD polymerization method very thin
nanofibers of PPV inside the pores of MCM 41 (mesoporous silica) with pore

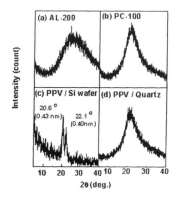

Figure 3. Wide-angle X-ray diffractograms of PPVs prepared (a) in an alumina membrane (nominal diameter: 200 nm), (b) in a polycarbonate membrane (nominal diameter: 10 nm), (c) on a silicon wafer, and (d) on a quartz substrate.

diameters of 2 – 10 nm. Carbonization of the fibers resulted in graphitic nanofibers. In order to facilitate CVD polymerization of PPV it was necessary to modify the inner surface of MCM 41 pores with trimethylsilyl chloride to make it hydrophobic._(15)_

Figure 3 compares the wide-angle X-ray diffractograms of PPVs prepared on different substrates. It is very interesting to note that only the diffractogram (Figure 3(c)) of the PPV film (550 nm thick) obtained on the surface of a Si-wafer reveals crystalline diffraction patterns whereas all the other (Figure 3(a), (b) and (d)) reveal amorphous or less crystalline diffraction patterns. This implies that the nature of surface of and / or the characteristics of a substrate material is important in controlling the morphology of PPV formed. For the sake of comparison we prepared a PPV film (490 nm thick) on the surface of quartz, whose X-ray diffractogram is shown in Figure 3(d). The diffraction peaks of the PPV film at $2\theta = 20.6\,^{\circ}$ (d=0.43 nm), $22.1\,^{\circ}$ (d=0.40 nm) and $28.2\,^{\circ}$ (d=0.32 nm) exactly match those reported by Chen et al._(16)_ for drawing-induced crystalline PPV having a herringbone arrangement of the polymer chains.

The UV-vis absorption spectra and photoluminescence (PL) spectra of the bulk PPV film separately prepared by the Wessling – Zimmerman method and PPV nano objects prepared in this investigation are compared in Figure 4. They are basically the same and maximum absorption was observed at 410 nm for all

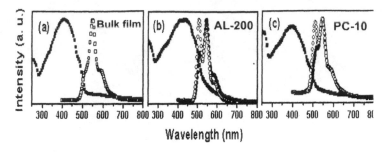

Figure 4. Comparison of UV-vis absorption and photoluminescence (PL) spectra. ■ : UV-vis absorption spectra ((a): bulk PPV film; (b): PPV nanotubes (AL-200) in the methanol suspension; (c): PPV nanorods (PC-10) in methanol suspension), □ : PL spectrum of bulk film, ● : PL spectra of nanotubes and nanorods in filter membranes, ○ : PL spectra of nanotubes and nanorods suspended in methanol (All the PL spectra were obtained at the excitation wavelength of 350 nm) (Reproduced from reference 14. Copyright 2001 American Chemical Society.)

the samples, which correspond to π-π* transition of the PPV backbone.*(17)* Their photoluminescence behavior, however, depends on their preparation method and also whether the PPV nanotubes and nanorods were remaining inside the pores of substrate filters or isolated from them. The range of wavelengths (475 – 625 nm) of luminescence emission is practically the same for the three samples and their PL spectra have emission at three positions (510, 545 and 585 nm). The fine details of the PL spectra, however, are quite different: the major peculiarity lies in the fact that the relative emission intensity at the shortest wavelength (510 nm) grows in the order of the bulk film < nano samples in the filter < nanotubes or nanorods suspended in methanol. This emission peak has been claimed to be from $S_1 \rightarrow S_0$ 0-0 transition, i. e., highest electronic energy transition involving π-electrons. The additionally enhanced $S_1 \rightarrow S_0$ 0-0 emission for the methanol suspensions of nano PPVs is a very intriguing phenomenon that requires further studies. It, however, may be due to reduced reabsorption*(18)* by neighboring chains of the emitted light originated from the 0-0 transition because of complete isolation of the nano particles suspended in the solvent that consist of much fewer polymer chains compared to bulk films. In contrast, when the nanotubes and nanorods remain in the filter pores, the PPV molecules in the interface

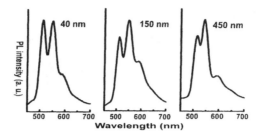

Figure 5. PL spectra of PPV films of varying thickness obtained on the Si-wafer surface.

would interact with the substrate surface causing a reduced emission for the 0-0 transition[19] compared to suspensions. Smith *et al.*[20] observed earlier the same phenomenon for PPV nanocomposites prepared in photochemically polymerized matrix of a lyotropic liquid-crystalline monomers. They[21] also studied photoluminescence detected magnetic resonance (PLDMR) of the composites confirming that higher PL quantum yield was resulted from the isolation of the PPV chains.

Figure 5 compares the PL spectra of PPV films of varying thickness obtained on the Si-wafer surface. Although the difference is not intensive, we note that the emission at the shortest wavelength (510 nm) becomes stronger as the thickness of the film decreases. This observation implies that reabsorption by the neighboring chains of the emitted light originated from the 0-0 transition of a given chain is indeed a very important factor governing details of PL spectra.

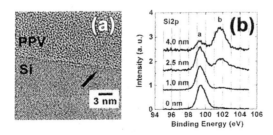

Figure 6. (a) Cross sectional TEM image and (b) XPS Si 2p spectra of PPV film deposited on the Si wafer.

22

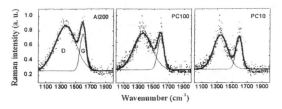

Figure 7. Raman spectra of carbon nanotubes and nanorods obtained from PPV. An Ar laser of 514.5 nm was employed as a light source (Reproduced from reference 14. Copyright 2001 American Chemical Society.)

Since there are earlier reports*(22)* of the possible formation of the carbon-silicon bonds during CVD of olefinic hydrocarbons on silicon, we studied this possibility in our PPV preparation on the surface of Si-wafers. Moreover, the very reactive quinonedimethide type intermediate depicted in equation (3) may be involved in the reaction with the Si atoms on the surface. The TEM image (Figure 6(a)) of the cross-section of PPV-Si-wafer clearly shows the existence of interface of about 0.5 nm. The X-ray photoelectron spectra of the PPV thin films gradually change as we increase its thickness as shown in Figure 6(b). Initially the XPS spectrum shows only a peak at the binding energy of 99.4 eV from Si_{2p} electrons for the Si-Si bonds. As we increase the thickness of the PPV film deposited, a new peak appears at the binding energy of 101.6 eV. And the intensity of this new peak grows with increasing the film thickness. This binding energy of Si_{2p} electrons was observed for the bond of O-Si-C.*(23-25)* Therefore, we may conclude that direct chemical bonds are formed between the Si atoms on the wafer surface and carbon ends of PPV chains constructed on it. It, however, is believed that, due to unavoidable exposure to air during experimentation, the Si atoms in the Si-C bonds are bound to oxygen atoms. The possibility of the existence of SiO_2 could be excluded by the fact that we did not observe any Si_{2p} peak (103.2 eV)*(23-25)* from SiO_2.

We intended to gather further information on the formation of Si-C bonds by examining the C_{1s} XPS peak, but due to much less peak shift, we could not obtain any definitive conclusion.

Although we have not given the data, we could prepare similar nano objects of poly(2,5-thienylenevinylene) by CVD polymerization of 2,5-bis(chloromethyl)-thiophene. They could readily be doped at room temperature with dopants such as I_2 and $FeCl_3$ to conducting materials.

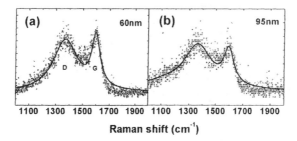

Figure 8. Thickness dependence of Raman spectra of carbonized PPV films deposited on the crystalline Si-wafer surface (a) 60nm thick: I(G) / I(D) = 0.33, (b) 95nm thick: I(G) / I(D) = 0.29 (carbonization temp.: 850 °C).

PREPATATION AND PROPERTIES OF GRAPHITIC NANO OBJECTS OBTAINED FROM PPV

The nano objects of PPV described in the previous section were thermally carbonized at 850 °C for 1 hour under argon atmosphere to produce the corresponding graphitic products.

In order to study the degree of graphitization of the carbonaceous nano products obtained, we obtained their Raman spectra as shown in Figure 7. They commonly show the coexistence of amorphous and crystalline graphitic regions. The peak at 1355 cm^{-1}, which is often called D(26) peak, is originated from a

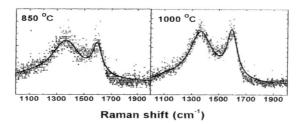

Figure 9. Carbonization temperature dependence of Raman spectra of carbonized PPV films deposited on the crystalline Si-wafer surface   (a) 1000 °C: I(G) / I(D) = 0.38, (b) 850 °C: I(G) / I(D) = 0.29 (film thickness: 95nm).

Figure 10. (a) Cross sectional TEM image of a carbon film on Si wafer and (b) its selected area electron diffraction.

breathing mode of A_{1g} symmetry involving phonons near the K zone boundary of disordered structure. The other peak at 1592 cm^{-1}, usually called G peak,[26] is from the G mode of E_{2g} symmetry and involves the in-plane bond-stretching motion of pares of carbon atoms of sp^2 hybridization. In other words, the G peak represents crystalline graphite region.

As one can see from Figure 7, the ratios of the two Raman scattering peaks, I(G) / I(D), are slightly higher than 0.4. According to the following equation proposed by Tuinstra and Koening,[27] the cluster diameter or in-plane correlation lengths (L_a) in the graphitic clusters are estimated to be slightly larger than 1 nm.

$$I(D) / I(G) = C(\lambda) / L_a \qquad (4)$$

The constant $C(\lambda)$ depends on the wavelength of incident laser light (the value is 44 Å for the wavelength of 514.5 nm that was utilized in this investigation). Moreover, the fact that the I(D) / I(G) ratios is about 2 and the G mode peak appears at 1592 cm^{-1} suggests us that the present carbon nanotubes and nanorods are of nanocrystalline graphite.[27] The perfect graphite exhibits the G mode at 1581 cm^{-1} and the peak moves toward the higher wave numbers as the structure becomes disordered.[26]

We obtained Raman spectra of carbonized films derived from PPV films deposited on the crystalline Si-wafer surface (Figure 8). Carbonization was again performed at 850 °C for 1 hour under argon atmosphere. This figure demonstrates that the degree of carbonization depends on the thickness of the

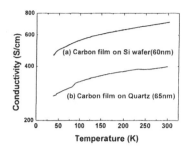

Figure 11. Comparison of Electrical conductivities of the graphitic carbon nanofilms.

original PPV films. Increasing carbonization temperature also enhanced the degree of graphitization as shown by Figure 9 which shows that the higher carbonization temperature augments the $I(G) / I(D)$ ratio.

The formation of graphitic structure by carbonization of PPV is clearly recognizable in the TEM image given in Figure 10(a). This figure exhibits well aligned straight lines that are tilted about 15 ^{O} on the wafer. The electron diffraction pattern given in the inset also supports the well aligned morphology. Figure 10(b) shows a TEM image of a carbon fiber obtained by carbonization at 850 °C of a PPV fiber synthesized in a pore of MCM-41. The fiber diameter is 4.3 nm and the interlayer spacing is estimated to be about 3.5 Å. Althouth the Raman spectrum of this fiber is not given, the $I(G) / I(D)$ ration was rather high, 0.77. This corresponds to the L_a value of about 3.4 nm. This high value is ascribed to the small diameter of the fiber, which would favor more efficient supply of heat flow for carbonization.

Electrical conductivities of the graphitic films prepared on the surface of a Si-wafer and a quartz plate were measured by the four-line probe method and the results are presented in Figure 11. This figure provides us with two important informations: (1) The conductivity of graphitic film obtained on the Si-wafer surface is consistently higher than that on the quartz plate. (2) The room temperature conductivity of the former is fairly high about 0.7×10^3 Scm^{-1}. Electrical conductivity of a thick carbon film obtained by carbonizing at 850 °C a PPV film prepared by Wessling-Zimmerman method was only about 60 Scm^{-1}.[4] We believe that a combination of the crystalline nature of the starting PPV,

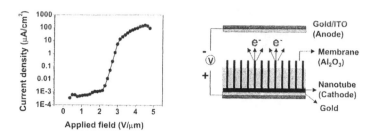

Figure 12. Field emission property of carbonized PPV nanotubes.

thinness of the film, and high thermal conductivity(28) together with low heat capacity(28) of the wafer caused more efficient graphitization leading to a high electrical conductivity for the carbon films prepared on the surface of the Si-wafer.

We examined the field emission properties(29) of the carbon nanotubes prepared in the pores of an alumina membrane whose pore diameter was 230 nm. The wall thickness and the length of the carbon tubes obtained were 15 nm and 60 μm, respectively. The emitting ends of the tubes were open, while the other ends were connected to the same carbonized film formed on the surface of the membrane. As shown in Figure 12, we utilized the gold-coated indium / tin oxide (ITO) as the anode and the carbon tubes as cathode. The gap between the tube ends and the anode was 130 μm. According to the measured results given in Figure 12, the turn on applied electric field is 3 V / μm, which is comparable to

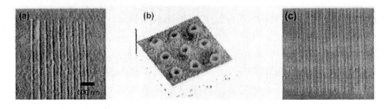

Figure 13. AFM images of patterned PPV in (a) nanowires(width 75 nm, height 4.5 nm), (b) nanowells (width: 230 nm, Height: 16 nm, wall thickness: 75 nm) and (c) carbonized PPV nano patterns (width: 386 nm, height: 3.5 nm).

those reported for carbon nanotubes*(30)* prepared by conventional methods. The value, however, is significantly lower than those for metal cones.*(31)*

NANOPATTERNING OF PPV AND CARBON

The main advantage of the present CVD polymerization of PPV is found in easy nanopatterning on conducting substrates including silicon wafers. For this purpose we utilized lithographed poly(methylmethacrylate) (PMMA) patterns on Si wafers. The CVD polymerization of PPV was performed on a PMMA pattern. The precursor polymer formed was thermally treated at 90 °C for 2 hours to partially thermolyze into insoluble polymer. And then the PMMA pattern was removed by dissolution with acetone. Finally, the insoluble pattern remaining on the wafer was thermally treated at 270 °C for 14 hours converting it to the PPV polymer. As shown in Figure 13, we could obtain nanopatterns of PPV in different shapes. By carbonizing the pattern, one can easily obtain the corresponding nanopatterns of carbon (Figure 13(c)).

Since PPV is known to emit green-light photoluminescence (PL) upon exposure to UV light and also electroluminescence (EL)*(3)* when an electric field is applied, we expect that nano PL and EL devices can be constructed directly on Si wafers by the present CVD method. In addition, the present method enables us to fabricate nano patterns and nanowires of conducting carbon.

ACKNOWLEDGMENT

This work was supported by the Korea Science and Engineering Foundation through the Center for Electro- and Photo-Responsive Molecules, Korea University. K. Kim and G. Zhong were the recipients Brain Korea 21 Fellowship from the Ministry of Education and Human Resources of Korea. The X-ray diffractograms were obtained by the Korea Basic Science Institute – Seoul Branch.

REFERENCES

1.   C. Kvarnstrom and A. Ivaska, in *'Habdbook of Conductive Molecules and Polymers'*, ed. H. S. Nalwa, Chapter 9. Characterization and Application of Poly(p-phentlene) and Poly(p-phenylenevinylene), John Wiley & Sons, New York, 1997.

2.   C. S. Moratti, in *'Handbook of Conducting Polymers'*, ed. T. A. Skothtim,

R. E. Elsenbaumer, and J. R. Reynolds, Chapter 13. The Chemistry and Uses of Polyphenylenevinylenes, Marcel Dekker, New York, 1998.

3.   J. H. Burroughes, D. D. C. Bradley, A. R. Brown, R. N. Marks, K. Mackay, R. H. Friend, P. L. Burn, and A. B. Holmes, *Nature*, 1990, **347**, 539.

4.   H. Ueno and K. Yoshino, *Phys. Rev. B: Condens. Matter*, 1986, **34**, 7158.

5.   H. W. Kroto, J. R. Heath, S. C. O'Brien, R. F. Curl, and R. E. Smalley, *Nature*, 1985, **318**, 162.

6.   S. Iwatsuki, M. Kubo, and T. Kumeuchi, *Chem. Lett.*, 1991, 1071.

7.   O. Schäfer, A. Greiner, J. Pommerehne, W. Guss, H. Vestweber, H. Y. Tak, H. Bassler, C. Schmidt, G. Lussem, B. Schartel, V. Stumpflen, J. H. Wendorff, S. Spiegel, C. Moller, and H. W. Spiess, *Synth. Met.*, 1996, **82**, 1.

8.   E. G. J. Staring, D. Braun, G. L. J. A. Rikken, R. J. C. E. Demandt, Y. A. R. R. Kessener, M. Bauwmans, and D. Broer, *Synth. Met.*, 1994, **67**, 71.

9.   K. M. Vaeth and K. F. Jensen, *Adv. Mater.*, 1997, **9**, 490.

10.   R. A. Wessling and R. G. Zimmerman, U.S. Pat. 3401152, 1968.

11.   H. G. Gilch and W. L. Wheelwright, *J.Polym.Sci., Part A: Polym.Chem.*, 1966, **4**, 1337.

12.   K. M. Vaeth and K. F. Jensen, *Macromolecules*, 1998, **31**, 6789.

13.   C. R. Martin, in *'Handbook of Conducting Polymers'*, ed. T. A. Skothtim, R. E. Elsenbaumer, and J. R. Reynolds, Chapter 9. Template Polymerization of Conductive Polymer Nanostructures, Marcel Dekker, New York, 1998.

14.   K. Kim and J.-I. Jin, *Nano Lett.*, 2001, **1**, 631.

15.   R. Anwander, C. Palm, J. Stelzer, O. Groeger, and G. Engelhardt, *Stud. Surf. Sci. Catal.*, 1998, **117**, 135.

16.   D. Chen, M. J. Winokur, M. A. Masse, and F. E. Karasz, *Polymer*, 1992, **33**, 3116.

17.   A. Kraft, A. C. Grimsdale, and A. B. Holmes, *Angew. Chem., Int. Ed. Engl.*, 1998, **37**, 402.

18.   G. G. Guilbault, *'Practical Fluorescence-Theory, Methods, and Techniques'*, Marcel Dekker, Inc.,New York, 1973.

19.   S. Heun, R. F. Mahrt, A. Greiner, U. Lemmer, H. Bassler, D. A. Halliday, D. D. C. Bradley, P. L. Burn, and A. B. Holmes, *J. Phys.: Condens. Matter*, 1993, 247.

20. R. C. Smith, W. M. Fisher, and D. L. Gin, *J. Am. Chem. Soc.*, 1997, **119**, 4092.

21. E. J. W. List, P. Markart, W. Graupner, G. Leising, J. Partee, J. Shinar, R. Smith, and D. L. Gin, *Opt. Mat.*, 1999, **12**, 369.

22. J. Terry, M. R. Linford, C. Wigren, R. Cao, P. Pianetta, and C. E. D. Chidsey, *J. Appl. Phys.*, 1999, **85**, 213.

23. K. L. Smith and K. M. Black, *J. Vac. Sci. Technol., A*, 1984, **9**, 1351.

24. G. Beamson and D. Briggs, '*High Resolution XPS of Organic Polymers - The Scienta ESCA300 Database*', Wiley Interscience, New York, 1992.

25. W. K. Choi, T. Y. Ong, L. S. Tan, F. C. Loh, and K. L. Tan, *J. Appl. Phys.*, 1998, **83**, 4968.

26. A. C. Ferrari and J. Robertson, *Phys. Rev. B: Condens. Matter*, 2000, **61**, 14095.

27. F. Tuinstra and J. L. Koenig, *J. Chem. Phys.*, 1970, **53**, 1126.

28. D. R. Lide, '*CRC Handbook of Chemistry and Physics*', CRC Press, London, 1994.

29. W. B. Choi, D. S. Chung, J. H. Kang, H. Y. Kim, Y. W. Jin, I. T. Han, Y. H. Lee, J. E. Jung, N. S. Lee, G. S. Park, and J. M. Kim, *Appl. Phys. Lett.*, 1999, **75**, 3129.

30. W. B. Choi, Y. W. Jin, H. Y. Kim, S. J. Lee, M. J. Yun, J. H. Kang, Y. S. Choi, N. S. Park, N. S. Lee, and J. M. Kim, *Appl. Phys. Lett.*, 2001, **78**, 1547.

31. I. Brodie and P. R. Schwoebel, *Proc. IEEE.*, 1994, **82**, 1006.

Chapter 3

Layer-by-Layer Assembly of Molecular Materials for Electrooptical Applications

M. E. van der Boom[1,*] and T. J. Marks[2,*]

[1]Department of Organic Chemistry, Weizmann Institute of Science,
Rehovot, 76100, Israel
[2]Department of Chemistry and Materials Research Center, Northwestern
University, Evanston, IL 60208–3113

Intrinsically acentric organic films are becoming a valuable
alternative to poled-polymers and inorganic materials such a
lithium niobate ($LiNbO_3$) for the formation of electro-optic
devices. The ability to fabricate noncentrosymmetric structures
in a molecular layer-by-layer approach offers nanoscale
control over film dimensions and allows tailoring of
microstructural and optical properties at the molecular level.

Introduction

Formation of "all-organic" electronic and/or photonic devices is at the forefront
of materials science research and offers many intriguing challenges. Device-
quality organic electro-optic materials must meet a number of stringent criteria,
including long-term thermal, chemical, photochemical, and mechanical stability.
Moreover, an excellent optical transparency window to minimize absorption
losses is crucial for optical device performance. Relative few polar organic
materials combine excellent electro-optic responses with long term stability at
elevated temperatures. Moreover, device fabrication often requires organic
materials with a good temporal stability. Electro-optical or Pockel effects result
from a small change in the index of refraction of a noncentrosymmetric material
in an electric field (*1, 2*). Noncentrosymmetric organic films can be produced via
post-deposition poling (*3, 4*) or during an intrinsically acentric film growth
process (*5, 6*). One of the major breakthroughs in the development of organic
devices has been the use of bulky NLO-active chromopheres in poled-polymers

by Dalton and co-workers (7) leading to the formation of "push-pull" electro-optic modulators operating at telecommunication wavelengths (1.3 and 1.55 μm) with switching voltages (V_π) as low as 0.8 V and r_{33} values as high as 60 pm/V. Thermally robust poled-polymers with excellent electro-optic responses (r_{33} = 105 pm/V) have been reported (8) and are almost competitive with conventional, expensive inorganic materials such as LiNbO$_3$. The use of chromophore precursors having sterically demanding groups (i.e., *tert*-butyldimethylsilyl, *n*-hexyl) results in a significant decrease of electrostatic interactions among the chromophores, hence, higher r_{33} values have been achieved. However, the use of poled-polymers requires post-deposition chromophore dipole moment alignment which is obtained by applying electric field poling at high voltages and elevated temperatures on spin-coated films. Subsequently, the polymer is slowly cooled down under the large DC field to maintain and "freeze" the noncentrosymmetric packing of the dyes. Thermal relaxation of the chromophore dipole moment alignment is often suppressed by covalent cross-linking (9, 10). In despite of the considerable recent advancements with poled-polymers, film growth methods which do not require post-deposition steps remain a highly attractive alternative.

Stepwise assembly of organized arrays of functional molecules in noncentrosymmetric solids and thin films offers many attractions including a appreciable level of architectural control. No complex poling steps (or electrically matched buffer layers) are here necessary to achieve a large polar orientation of the chromophore building blocks. Moreover, a modular layer-by-layer approach allows fine-tuning of many physicochemical properties (i.e., r_{33}, n, λ_{abs}, $\chi^{(2)}$, ϵ, morphology, thickness) by ready integration of other functional organic or inorganic components. Langmuir-Blodgett and related film transfer techniques (11), covalent molecular self-assembly (12-19), bipolar amphiphiles and polyelectrolytes (20), metal-coordination (21-24) and many other methods have been used for the formation of intrinsically acentric films (3,5,6,25). There exist a large variation in the kind of functional molecules that can be successfully integrated in highly ordered multilayers, which can be grown on various substrates including gold, silicon, and indium-tin-oxide coated glass. Nevertheless, innovative design of new assembly techniques is still needed since not all layer-by-layer methodologies are suitable for the formation of device-quality functional organic films. In general, the nature of the interlayer bonding is an important factor controlling the stability of multilayer assemblies. Strong (covalent) interlayer bonding seems to be a prerequisite for the formation of applicable intrinsically acentric films. In this concise overview, only a relatively small portion of the many aforementioned promising synthetic approaches can be highlighted. However, the recent developments in molecular layer-by-layer self-assembled zirconium and hafnium phosphonate and siloxane-based multilayer assemblies for electro-optic switching discussed here will provide a sense of some of the fascinating chemistry involved in thin film applications and future synthetic, materials and engineering challenges.

Zirconium and Hafnium Phosphonate Multilayer Assemblies

Katz *et al.* (*21,22*) used layer-by-layer self-assembly to generate noncentro-symmetric zirconium phosphate/phosphonate multilayers of up to 30 chromophore layers with good second-order optical nonlinearity and electro-optic responses comparable in magnitude to that of LiNbO$_3$ (r_{33} = 31 pm/V). Noteably, the relatively low dielectric constant of organic and 'hybrid' organic-inorganic films results often in a favorable figure-of-merit ($n^3 r_{33}/\epsilon$). The three-step sequential deposition process involves (i) surface phosphorylation, (ii) zirconation, and (iii) chromophore deposition. The resulting polar films with an ellipsometric derived thickness of ~50 nm shown in Figure 1 are thermally robust and chemically inert in common organic solvents up to 80°C. This metal-coordination assembly approach has been explored by many others as well (*23,24,26,27*), and is based on a centrosymmetric multilayer synthesis with aqueous ZrOCl$_2$ and 1,10-decanebisphosphonic acid initially reported by Mallouk *et al.* in 1988 (*28*). Intrinsically acentric (siloxane-based) multilayer films were first introduced by Marks (*17*).

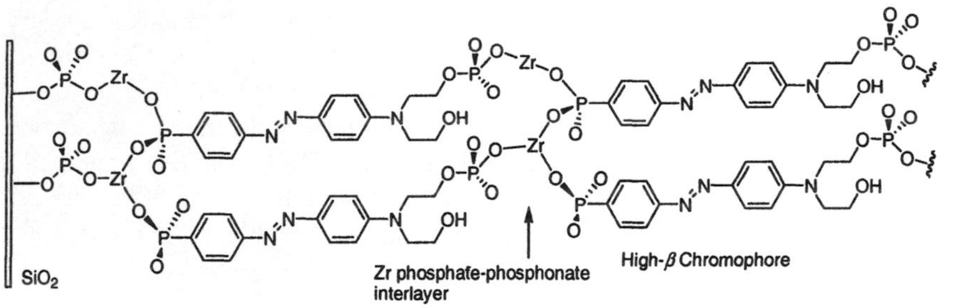

Figure 1. Schematic structure of a polar zirconium phosphate/phosphonate multilayer film.

The phosphate (P–O–C) linkages shown in Figure 1 are robust but under certain reaction conditions are susceptible to hydrolysis. An enhanced deposition technique introduced by Page *et al.* (*24*) uses this fact to build-up an acentric metal-bisphosphonate multilayer film based on relative inert P–C bonds via P–O–C acid hydrolysis. Bis(phosphonate) dyes exhibiting a phosphonic acid group bind to hafnium-primed SiO$_2$-based surfaces (Figure 2). Subsequent hydrolysis of the surface phosphonate ester group with HCl followed by metal (zirconium, hafnium) incorporation generates a fresh zirconium or hafnium surface suitable for sequential phosphonic acid binding. This elegant layer-by-layer hydrolysis method might be extended to a variety of functional molecules, however, the overall process is rather slow. More than two days are necessary for the build-up of a sub-unit of a multilayer structure. Grazing angle X-ray diffraction and ellipsometry measurements indicated a consistent layer thickness of 17-18 Å, and a relative large tilt angle of the molecules of 54° from the surface normal. Noteably, solid state magic angle spinning ^{31}P NMR has been used as well to monitor the phosphonate ester hydrolysis. Solid state NMR is rarely used in analysis of organic thin films, but here it provided a crystal-clear distinction between unreacted phosphonate ester and phosphonic acid groups (29).

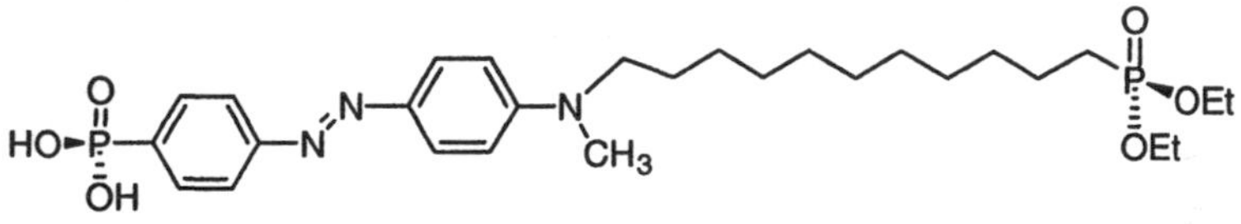

Figure 2. Acentric chromophore building block exhibiting a phosphonic acid for surface anchoring and a labile phosphonate ester group.

A change in the index of refraction Δn of 8.0×10^{-6} at 632.8 nm for zirconium phosphonate-azobenzene multilayers on gold substrates, that have been integrated into airgap capacitors, has been determined by electrochemically modulated surface plasmon resonance (EM-SPR) measurements. Corn *et al.* (23) demonstrated that this method allows accurate determination of the changes in the index of refraction of a noncentrosymmetric organic superlattice that take place in an external electrostatic field. The abovementioned experimentally derived Δn corresponds to an electro-optic coefficient of $r_{33} \approx 11$ pm/V and a $\chi^{(2)}$ value of 41 pm/V (9.7×10^{-8} esu). These values are not enhanced by chromophore absorption. Mixed multilayers on gold substrates have been formed by alternating assembly of either a bis(phosphonate)-derivative or a dye monolayer, allowing systematic variation of the index of refraction within a narrow but potentially useful range ($n = 1.58 \pm 7$ at $\lambda = 632.8$ nm) and allowing accurate phase matching in device configurations. Recent detailed second harmonic generation studies by Blanchard *et al.* (27) on vacancy defect densities in zirconium phosphate/phosphonate multilayer assemblies revealed direct evidence for the role of electric quadrupolar contributions in these systems. The electric dipole term dominates the observed $\chi^{(2)}$ responses in these multilayer assemblies. It is often assumed that the electric dipole term in $\chi^{(2)}$ is larger than the electric quadrupole and higher order terms, but experimental studies are scarce.

Self-Assembled Siloxane-Based Multilayers

Formation of covalently bound siloxane-based multilayers introduced by Sagiv in 1983 (*30*) is considered to be a major step forward in the development of nanoscale precise thin films. The straightforward assembly approach shown in Figure 4 involves solution deposition of a trichlorosilane surfactant having a terminal carbon–carbon double bond. The trichlorosilane unit enables strong, covalent binding to a hydrophilic surface whereas the ethylenic double bond can be chemically oxidized to a terminal, reactive hydroxyl group. Repeating the procedure with the resulting hydrophilic surface resulted in multilayer formation of up to 30 layers (*6*). Other silanes such as 23-trichlorosilyltricosanoate can be used as well by converting the surface methyl ester group into a reactive carboxylic acid.

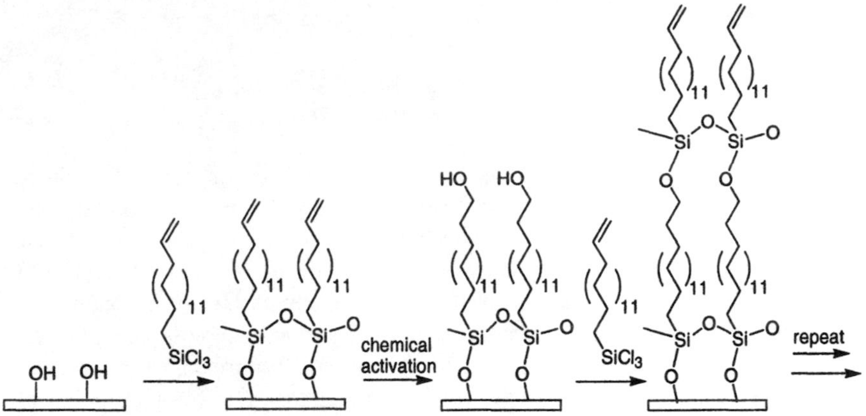

Figure 4. Layer-by-layer formation of a siloxane-based multilayer by chemisorption of 15-hexadecenyltrichlorosilane and chemical regeneration of reactive surface sites.

During the trichlorosilane assembly process only trace amount of moisture is tolerable, and it is known that moisture may induce horizontal polymerization of trichlorosilane surfactants (*31*). For the build-up of structurally regular multilayer assemblies, a quantitave reaction yield is a necessary requirement for each step to maintain the same surface functional density in every monolayer. Several surface functionalization procedures have been reported, but there has been limited application to multilayers (*14, 24, 29*). Despite these technical drawbacks, the resulting siloxane-based films are generally much more stable than those self-assembled by non-covalent interactions such as in the Langmuir-

Blodgett technique. The enhanced materials stability may make functional siloxane-based systems suitable for device integration and operation. It is noteworthy that standard isocyanate/hydrolysis polymerization chemistry can be used as well to build-up structurally regular multilayers.

Since 1990, Marks and co-workers (17) have devised several organosilane-based assembly techniques for the formation of intrinsically acentric self-assembled superlattices (SAS) composed of stilbazolium- or benzothiazole-type chromophores as exemplified in Figure 5 (12-19,32). A "coupling" alkyl or benzyl halide-based monolayer is covalently bound to a surface oxide via a strong siloxane linkage. Subsequent chromophore anchoring by quarternization of a pyridine moiety, followed by treatment with the commercially available "capping" reagent octachlorotrisiloxane, results in a thermally robust, acentric film (~ 2.5 nm thick). The "capping" step deposits a ~0.8 nm thick polysiloxane film, generates a large density of reactive hydroxyl sites necessary for subsequent chromophore layer deposition, and provides orientational stabilization via interchromophore crosslinking.

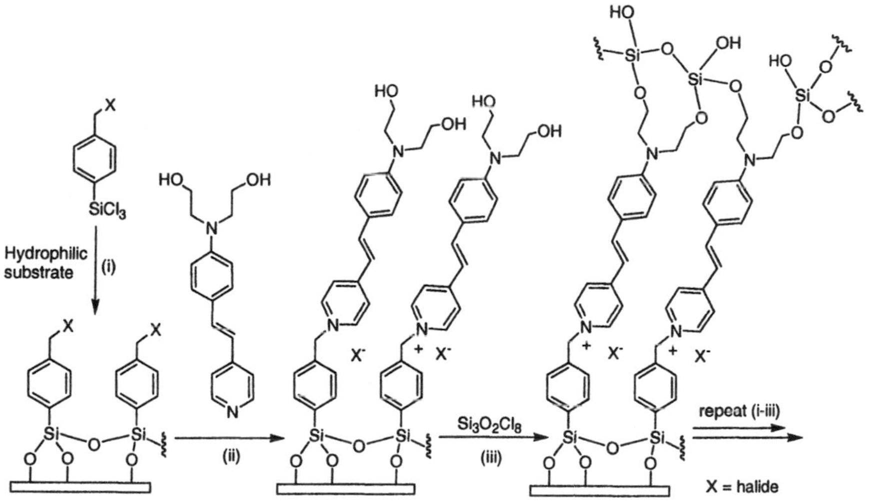

Figure 5. Covalent siloxane-based self-assembly scheme introduced by Marks.

The large chromophore density ($N_s \approx 1$ chromophore/40 Å^2) and high degree of dipole moment aligned results in film exhibiting excellent macroscopic NLO and electro-optic responses (up to $\chi^{(2)} \approx 370$ pm/V and $r_{33} \approx 120$ pm/V at $\lambda = 1064$ nm). Several covalently bound monolayers including zwitterionic dyes (33), phenolthiazine-stilbazole (34), porphyrin (35) and calixarene derivatives (36) have been assembled by Li, Pagani and others in an acentric fashion by applying this general siloxane-based strategy (Figure 6).

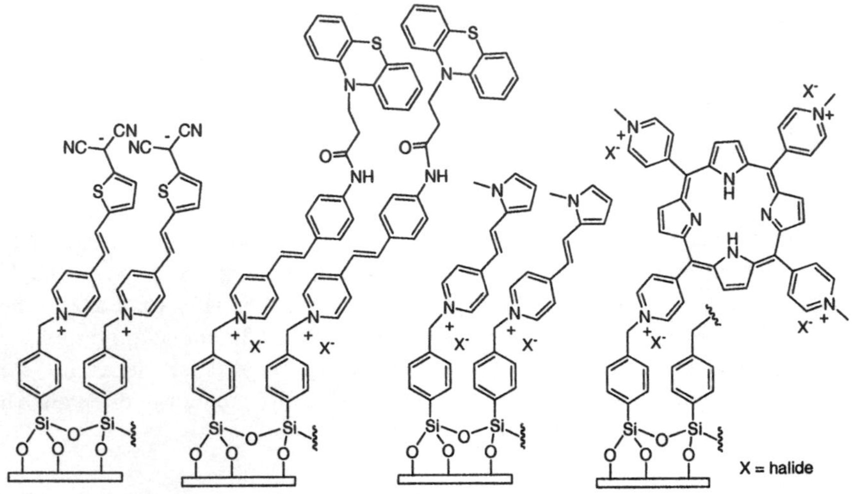

Figure 6. Examples of acentric siloxane-based monolayers.

Recently, a streamlined assembly method has been developed by van der Boom, Marks and co-workers (*13-15,19,32*) to generate highly ordered, intrinsically acentric organic materials which can be integrated into electro-optic and related devices, such as light modulators and switches (*37-40*). This new synthetic approach involves only two rapid alternating solution deposition steps, as shown in Figure 7. First, monolayers of high-β chromophores are covalently bound on hydrophilic substrates (step (i)). The siloxy removal step (ii) renders the surface hydrophilic, thus allowing the rapid build-up of a covalently-bound siloxane-based capping layer from commercially available octachlorotrisiloxane. Each bilayer reaches completion in about 40 min., which much faster than previously reported siloxane-based deposition methodologies. Film deformation due to evaporation of incorporated solvents is not observed. Chromophore and octachlorotrisiloxane solutions can be recycled without compromising the film quality. The film deposition does not require sophisticated tools and can be carried out in a standard organic laboratory in a single reaction vessel using cannula techniques or by dip-coating of the substrates in reaction solutions using a simple nitrogen-filled glovebag. A full battery of physicochemical techniques, including synchrotron X-ray reflectivity, X-ray photoelectron spectroscopy, atomic force microscopy, optical spectroscopy, aqueous contact-angle measurements, and angle dependent second harmonic generation measurements reveal the formation highly ordered materials. Recently, more than 80 chromophore layers (each chromophore + capping layer ≈ 3.2 nm thick) have been assemblied (*19, 40*), resulting in structurally regular sub-μm thick SAS films with good NLO and electro-optic responses ($\chi^{(2)} \approx 180$ pm/V and $r_{33} \approx 65$ pm/V at $\lambda = 1064$ nm).

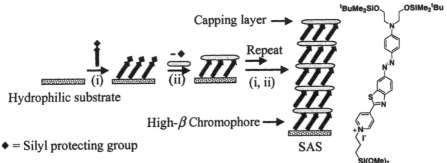

Figure 7. Two-step layer-by-layer SAS self-assembly sequence.

This two-step method is amenable to automation using a single reactor process or a dipping procedure, and allows tailoring of physical properties to meet the needs of specific applications (*15, 32*). As seen in Figure 8, integration of high refractive index inorganic materials by solution deposition of 3 nm thick metal oxide nanolayers (e.g., by hydrolysis of M(i-OC$_3$H$_7$)$_3$ solutions; M = Ga^{3+}, In^{3+}) on the hydrophilic polysiloxane surface results in formation of organic-inorganic superlattices and opens possibilities for modifying the refractive index contrast between electro-optic films and device components (e.g., cladding layers, modulating electrodes). This should result in better light confinement in the active waveguide region and more effective velocity matching of the optical waves and modulating radio frequencies. The intercalation of high-Z oxide components into the microstructure of the superlattice significantly increases the film index of refraction.

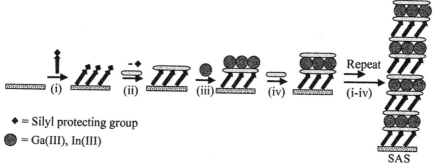

Figure 8. Incorporation of refractive index modifying metal oxide layers.

Refractive indices of $n = 1.76$ (Ga) and $n = 1.84$ (In) at 650 nm are determined from ellipsometry in conjugation with synchrotron X-ray reflectivity thickness measurements (i.e., compare $n = 1.52$ at 650 nm for the metal-free organic films). Importantly, no specially designed ligand systems or metal ion scaffolds are necessary. The high degree of control over film dimensions, electron density, texture, and optical properties has been unambiguously demonstrated using various physicochemical analytical tools, including angle-dependent polarized second harmonic generation, atomic force microscopy, and synchrotron X-ray reflectivity measurements. Refractive indices are tunable via the frequency with which the metal oxide layers are incorporated during the film assembly process.

Typically organic chromophores for NLO and electro-optical applications consist of a conjugated π-electron bridge/spacer unsymmetrically end-capped with strong electron donating and accepting groups (41). The assembly methods allow the integration of many so-called "push-pull" chromophore building blocks as shown in Figure 9. Each dye has a pyridine ring which can be quarternized in high yield with a trichloro- or alkoxysilane functionalized alkyl or benzyl halide to ensure, covalent surface attachement. The resulting pyridinium moiety is electron-deficient, thus, increasing the β value of the chromophore. The anion can be exchanged and appears to have a non-negligible effect on the NLO and electro-optical response properties of the films. The electron-donating amine group (NR_3) contains two alkylhydroxyl functionalities, allowing the formation of a (moderate) hydrophilic surface necessary for subsequent layer formation.

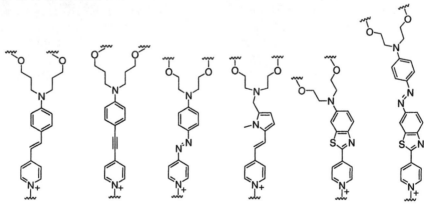

Figure 9. Examples of high-β chromophore building blocks for layer-by-layer formation of siloxane-based superlattices.

Some chromophores might be light-sensitive, especially in the presence of air and/or moisture. Incorporating such chromophore films in polymeric devices, which operate only at specific wavelengths, may still result in valuable systems. For each chromophore, multilayer structures provide the expected dependence of the second harmonic signal intensity on the number of layers. Chromophore building blocks with molecular dipole moment-hyperpolarizability products ($\mu\beta$) as large as $10,000-18,0000 \times 10^{-48}$ esu have been reported, and there is a reliable theoretical basis for understanding the mechanisms involved as well as prediction of new molecular structures with even larger responses (*42*). Computationally aided development of efficient, robust and simple organic chromophores exhibiting large first and second hyperpolarizabilities (β) is expected to dramatically improve nonlinear optical (NLO) responses. Indeed, single molecules with responses as large as $\mu\beta = 800,000 \times 10^{-48}$ esu have been predicted, while solvent/environment effects and interchromophore interactions have recently been modeled (*42*). Modelling studies have also revealed that molecular chromophores with a stereochemically enforced zwitterionic behavior in the ground state and high oscillator strengths may exhibit unprecedentedly large NLO responses (Fig. 10). These classes of never synthesized, "twisted" organic compounds may be chemically modified and applied as potential building blocks for the formation of novel multilayer assemblies. Trends in β values as a result of substituent pattern, chromophore properties, and spacer lengths between donor and acceptor units have been established for organic compounds (*43*). Nevertheless, innovative design of suitable chromophores having excellent NLO responses remains a formidable synthetic task.

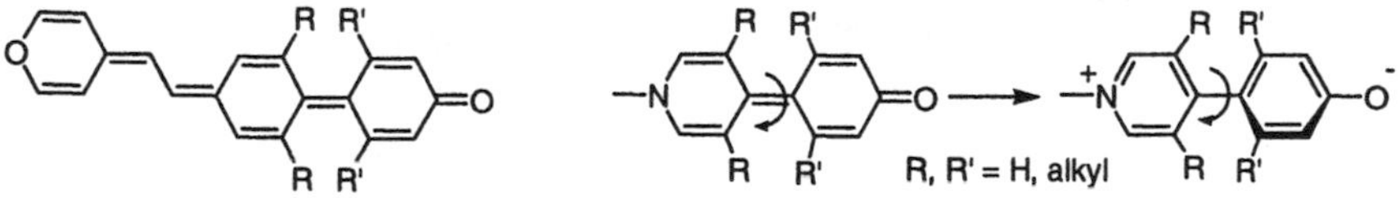

Figure 10. 'Twisted' chromophores with outstanding NLO responses.

"All-Organic" Self-Assembled Electro-Optical Devices

Various frequency doubling devices (39), a frequency-selective, ultrafast optical switch (38) and siloxane-based prototype modulators have been demonstrated (37, 40). The electro-optic phase modulator shown in Figure 11 consists of a two-component electro-optic SAS/waveguiding structure with Cytop™ as cladding layer. The glassy polymers Cytop™ and Cyclotene™ are used as low-loss waveguiding and cladding materials because of excellent optical transparency over the ultraviolet to infrared wavelength range ($\sim 95\%$) with a substantial refractive index contrast, $\Delta n \approx 0.21$. The straightforward architecture using well-established fabrication processes described below allows formation of a wide variety of optical waveguide building-block structures and at the same time ensures manufacturability. For instance, the modulator was constructed on a gold-coated silicon wafer, which acts as the ground-modulating electrode. A chemical vapor-deposited 2.0 μm thick SiO$_2$ film was used as: (i) a lower cladding layer to separate the guided light from the lower electrode, and (ii) as a hydrophilic substrate for siloxane-based film assembly. Multilayers with varying thickness (100–200 nm) were grown on the Si/Au/SiO$_2$ substrates. Cyclotene™ and Cytop™ solutions were then spin-coated onto the self-assembled thin film, and thermally cured at 130 °C for 30 min., resulting in smooth polymeric films with ellipsometry-derived thicknesses of 2.0 μm and 1.6 μm, respectively.

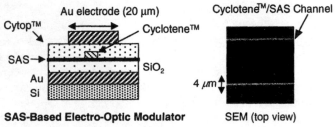

Figure 11. Self-assembled electro-optic modulator design and scanning electron microscope image showing the 'hybrid' Cyclotene™/SAS waveguides.

Channel-type ridged waveguide patterns (4.0 μm × 2.0 μm × 4.0 mm = w × h × l) were then patterned by photolithographic and reactive ion etching techniques. Subsequently, a Cytop™ cladding layer was spin-coated onto the device structure and thermally cured, followed by vapor phase deposition of the gold modulating top-electrode (20 μm width). The Cyclotene™ refractive index, n ~1.56, is almost identical to that of the ellipsometry-derived refractive index of the electro-optic self-assembled film at $\lambda = 1064$ nm, permitting formation of a two-component electro-optic/waveguiding structure with Cytop™ (n ~1.34) as cladding layer. The prototype spatial light modulator presented here has an operating voltage $V_\pi × L \approx 175$ V·cm and an effective $r_{33} \approx 22$ pm/V, which results in an appreciable figure-of-merit of $n^3 r_{33}/\epsilon \approx 14$ pm/V (at $\lambda = 1064$ nm). This is truly remarkable considering the fact that only a relatively thin electro-optic film (40 chromophore layers) was used with a model azobenzene type chromophore as the electro-optical component. Utilization of new, ultra-high β-chromophores (*42*), increasing the refractive index contrast between the cladding and guiding layers (*15, 32*), and thicker electro-optic films (*19*) are only three factors which can be varied to reduce the modulator voltage for practical applications.

Conclusions and Future Outlook

The examples of molecular layer-by-layer self-assembly highlighted here show only some of the significant progress made by many groups during the last decade in terms of deposition methods, improved electro-optic activity, and prototype device formation. However, there are still many remaining issues to be addressed before applicable self-assembled electro-optic devices become reality. For future advancement, a thorough fundamental understanding of factors affecting adhesion between self-assembled films and device materials under operating conditions is necessary. Recent adhesion studies led to an appreciable enhancement of organic light emitting diode performance (*44*). Likewise, little is know about factors controlling optical (absorption and coupling) losses in self-assembled optical devices. Of course, replacing C–H moieties with C–F and/or C–D groups will reduce absorption at telecommunication wavelengths but no experimental studies have been carried out for layered assemblies. Many reported multilayer assemblies are too thin for direct, conventional electro-optic measurements, and responses are often estimated from second-harmonic generation data (using the known relationship: $r_{33} \approx -2\chi^{(2)}/(n_z)^4$. At present it is difficult to qualitatively compare properties of different multilayer assemblies especially when different techniques have been used to determine $\chi^{(2)}$ and/or r_{33} values. Note that films significantly absorbing the second-harmonic light may exhibit non-negligible resonance enhancement. Film growth is frequently demonstrated with model chromophores for practical reasons and therefore many methods do not necessarily reflect the maximum possible electro-optical responses. Numerous ultrahigh-β chromophores are available but as of yet have not been integrated into thin polar films. Moreover, series of new compounds exhibiting superb NLO-characteristics have been predicted by computational chemistry (*42*) and organic chemists are challenged to synthesize these sometimes unusual molecules. Formation of microdomains and grains is a common phenomenon in layered thin films and is undoubtfully hampering the formation of high-quality sub-μm thick films. New layer-by-layer film growth techniques which prevent and/or suppress domain formation, or regenerate smooth, reactive surfaces are essential and this is perhaps the greatest remaining challenge. The use of thin, robust, covalently bound planarization layers seems a promising direction. New methods must be automable and based on highly reliable, fast chemical transformations. A "third-level self-assembly" approach resulted in exponential layered film growth (*45*), which would be extremely useful for the ultrafast deposition of hundreds of chromophore layers allowing more accurate electro-optic measurements and real device formation. Fundamental studies dealing with factors controlling the electro-optical properties of μm-scale films will be of equally importance.

Acknowledgments

Research supported by the NSF MRSEC program (Grant DMR0076077), by ARO/DARPA (DAAD19-00-1-0368), and by a research grant from the Henri Gutwirth fund for research. M. B. is incumbent of the Dewey David Stone and Harry Levine career development chair.

References

1. Prasad, P. N.; Williams, D. J. *Introduction to Nonlinear Optical Effects in Molecules and Polymers*; Wiley: New York, NY, 1990.
2. *Molecular Nonlinear Optics - Materials, Physics and Devices*; Zyss, J., Ed.; Academic Press: San Diego, CA, 1994.
3. *Functional Organic and Polymeric Materials*; Richardson, T. H., Ed.; Wiley: Chichester, England, 2000.
4. van der Boom, M. E. *Angew. Chem. Int. Ed.* **2002**, *41*, 3363.
5. Ulman, A. *An Introduction to Ultrathin Organic Films: From Langmuir-Blodgett to Self-Assembly*; Academic: San Diego, 1991.
6. Ulman, A. *Chem. Rev.* **1996**, *96*, 1533.
7. Shi, Y.; Zhang, C.; Zhang, H.; Bechtel, J. H.; Dalton, L. R.; Robinson, B. H.; Steier, W. H. *Science* **2000**, *288*, 119.
8. Dalton, L. R. *Opt. Eng.* **2000**, *39*, 589.
9. Davey, M. H.; Lee, V. Y.; Wu, L.-M.; Moylan, C. R.; Volkson, W.; Knoesen, A.; Miller, R. D.; Marks, T. J. *Chem. Mater.* **2000**, *12*, 1679.
10. Ma, H.; Chen, B.; Sassa, T.; Dalton, L. R.; Jen, A. K.-Y. *J. Am. Chem. Soc.* **2001**, *123*, 986.
11. Schwartz, H.; Mazor, R.; Khodorkovsky, V.; Shapiro, L.; Klug, J. T.; Kovalev, E.; Meshulam; Berkovic, G.; Kotler, Z.; Efrima, S. *J. Phys. Chem. B* **2001**, *105*, 5914.
12. Yitzchaik, S.; Marks, T. J. *Acc. Chem. Res.* **1996**, *29*, 197.
13. van der Boom, M. E.; Zhu, P.; Evmenenko, G.; Malinsky, J. E.; Lin, W.; Dutta, P.; Marks, T. J. *Langmuir* **2002**, *18*, 3704.
14. van der Boom, M. E.; Richter, A. G.; Malinsky, J. E.; Lee, P.; Armstrong, N. A.; Marks, T. J. *Chem. Mater.* **2001**, *13*, 15.
15. van der Boom, M. E.; Evmenenko, G.; Dutta, P.; Marks, T. J. *Adv. Funct. Mater.* **2001**, *11*, 393.
16. Marks, T. J.; Ratner, M. *Angew. Chem., Int. Ed. Engl.* **1995**, *34*, 155.
17. Li, D.; Ratner, M.; Marks, T. J.; Zhang, C.; Yang, J.; Wong, G. *J. Am. Chem. Soc.* **1990**, *112*, 7389.
18. Facchetti, A.; Abbotto, A.; Beverina, L.; van der Boom, M. E.; Dutta, P.; Evmenenko, G.; Pagani, G. A.; Marks, T. J. *Chem. Mater.* **2002**, *14*, 4996.
19. Zhu, P.; van der Boom, M. E.; Kang, H.; Evmenenko, G.; Dutta, P.; Marks, T. J. *Chem. Mater.* **2002**, *14*, 4982.
20. Roberts, M. J.; Lindsay, G. A.; Herman, W. N.; Wynne, K. J. *J. Am. Chem. Soc.* **1998**, *120*, 11202.

21. Katz, H. E.; Scheller, G.; Putvinski, T. M.; Schilling, M. L.; Wilson, W. L.; Chidsey, C. E. D. *Science* **1991**, *254*, 1485.
22. Katz, H. E.; Wilson, W. L.; Scheller, G. *J. Am. Chem. Soc.* **1994**, *116*, 6636.
23. Hanken, D. G.; Naujok, R. R.; Gray, J. M.; Corn, R. M. *Anal. Chem.* **1997**, *69*, 240.
24. Neff, G. A.; Helfrich, M. R.; Clifton, M. C.; Page, C. J. *Chem. Mater.* **2000**, *12*, 2363.
25. Maoz, R.; Yam, R.; Berkovic, G.; Sagiv, J. *Thin Films*; Academic Press: San Diego, CA, 1995.
26. Doron-Mor, H.; Hatzor, A.; Vaskevich, A.; van der Boom-Moav, T.; Shanzer, A.; Rubinstein, I.; Cohen, H. A. *Nature* **2000**, *406*, 382.
27. Flory, W. C.; Mehrens, S. M.; Blanchard, G. J. *J. Am. Chem. Soc.* **2000**, *122*, 7976.
28. Lee, H.; Kepley, L. J.; Hong, H.-G.; Mallouk, T. E. *J. Am. Chem. Soc.* **1988**, *110*, 618.
29. Neff, G. A.; Page, C. J.; Meintjes, E.; Tsuda, T.; Pilgrim, W.-C.; Roberts, N.; Warren Jr., W. W. *Langmuir* **1996**, *12*, 238.
30. Netzer, L.; Sagiv, J. *J. Am.Chem. Soc.* **1983**, *105*, 674.
31. Fadeev, A. Y.; McCarthy, Y. J. *Langmuir* **2000**, *16*, 7268.
32. Evmenenko, G.; van der Boom, M. E.; Kmetko, J.; Dugans, S. W.; Marks, T. J.; Dutta, P. *J. Chem. Phys.* **2001**, *115*, 6722.
33. Facchetti, A.; van der Boom, M. E.; Abbotto, A.; Beverina, L.; Marks, T. J.; Pagani, G. A. *Langmuir* **2001**, *17*, 5939.
34. Huang, W.; Helvenston, M.; Cason, J. L.; Wang, R.; Bardeasu, J.-F.; Lee, Y.; Johal, M. S.; Swanson, B. I.; Robinson, J. M.; D., L. *Langmuir* **1999**, *15*, 6510.
35. Li, D.-Q.; Swanson, B. I.; Robinson, J. M.; Hoffbauer, M. A. *J. Am. Chem. Soc.* **1993**, *115*, 6975.
36. Yang, X.; McBranch, D. W.; Swanson, B. I.; Li, D. Q. *Angew. Chem. Int. Ed. Engl.* **1996**, *35*, 538.
37. Zhao, Y. G.; Wu, A.; Lu, H. L.; Chang, S.; Lu, W. K.; Ho, S. T.; van der Boom, M. E. *Appl. Phys. Lett.* **2001**, *115*, 6722.
38. Wang, G.; Zhu, P.; Marks, T. J.; Ketterson, J. B. *Appl. Phys. Lett.* **2002**, *81*, 2169.
39. Lundquist, P. M.; Lin, W.; Zhou, H.; Hahn, D. N.; Yitzchaik, S.; Marks, T. J.; Wong, G. K. *Appl. Phys. Lett.* **1997**, *70*, 1941.
40. Zhao, Y. G.; Chang, S.; Wu, A.; Lu, H. L.; Ho, S. T.; van der Boom, M. E.; Marks, T. J. *Opt. Eng. Lett.* **2003**, *42*, 298.
41. Marder, S. R.; Kippelen, B.; Jen, A. K.-Y.; Peyghambarian *Nature* **1997**, *388*, 845.
42. Albert, I. D. L.; Marks, T. J.; Ratner, M. A. *J. Am. Chem. Soc.* **1998**, *120*, 11174.
43. Würthner, F.; Wortmann, R.; Meerholz, K. *ChemPhysChem* **2002**, *3*, 17.
44. Cui, J.; Huang, Q.; Wang, Q.; Marks, T. J. *Langmuir* **2001**, *17*, 2051.
45. Moaz, R.; Matlis, S.; DiMasi, E.; Ocko, B. M.; Sagiv, J. *Nature* **1996**, *384*, 150.

Chapter 4

Oxidative Solid-State Cross-Linking of Polymer Precursors to Pattern Intrinsically Conducting Polymers

Sung-Yeon Jang[1], Manuel Marquez[2], and Gregory A. Sotzing[1]

[1]Department of Chemistry and Polymer Program, Institute of Materials Sciences, University of Connecticut, Storrs, CT 06269
[2]Chemical Science and Technology Division, Los Alamos National Laboratory, Los Alamos, NM 87545

Herein we describe the patterning of an intrinsically conducting polymer (ICP) utilizing electrochemical oxidative solid-state crosslinking, what we have termed electrochemically induced pattern transfer (ECIPT) of a precursor polymer. The precursor polymer was a polynorbornylene prepared via ring opening metathesis polymerization (ROMP) of a norbornylene consisting of two terthiophene groups. An ICP network of poly(terthiophene) was prepared from this precursor polymer utilizing solid-state oxidative crosslinking of the pendent terthiophene moieties via both electrochemical and chemical means. Both chemical and electrochemical solid-state crosslinking were confirmed using cyclic voltammetry and UV-Vis-NIR. Pattern generation via ECIPT was demonstrated using a glass substrate consisting of two sets of 10 μm width interdigitated gold leads (50 leads per set) having 10 μm gaps. A film of 0.5 microns thickness of the precursor polynorbornylene was spun coat onto the substrate and application of 0.9 V for 0.1 seconds to one set of the gold leads afforded 50 lines of the poly(terthiophene) ICP network of 16 micron width within the precursor polynorbornylene film. From this, the oxidative propagation speed of the poly(terthiophene) ICP network from the gold line was calculated to be 30 microns/second.

Intrinsically conducting polymers (ICP) are receiving great attention[1] due to their potential use in applications such as electrochromic devices,[2] protective coatings,[3] organic light emitting diodes (LED),[4] energy storage batteries,[5] volatile gas sensors,[6,7] nonlinear optics,[8] and charge dissipating films.[9] Typically, the synthesis of ICPs takes place either electrochemically or chemically in monomer solutions resulting in infusible and insoluble polymers. These characteristics are attributed to the rigid character of the conjugated ICP backbone and often lead to significant difficulties in their processing. Many recent efforts have been focused on increasing processibility. Much of these have focused on derivatized aromatics to render the polymer water soluble,[10] organic soluble.[11,12] Preparation of main chain alkyl substituted ICPs from the polymerization of alkyl substituted monomer have received the most attention.

A recent strategy that has emerged for the preparation of ICPs is the preparation of precursor polymers that are able to form conducting polymers upon oxidative crosslinking. Within presently reported procedures, precursor polymers with backbones consisting of either poly(siloxane),[13,14] poly(acrylate),[15] poly(ethylene),[16,17,18] or poly(norbornylene),[19] having appended heterocycles are either crosslinked in solution via application of an appropriate oxidative potential or are crosslinked via introduction of a chemical oxidant. Oxidative electrochemical coupling of the precursor polymer in solution through the appended heterocycles results in the precipitation of a presumably crosslinked thin film of conducting polymer on the electrode surface.

Patterning of ICPs has been actively studied due to its potential impact on the fabrication of both electronic and optoelectronic devices.[20,21,22,23] Several different techniques have been reported which include soft photolithography,[24] chemically amplified soft lithography,[31,25,26] photochemical lithography,[27] and electrochemical Dip-Pen Nanolithography (e-DPN).[28,29,30] Presently, these pattern generating techniques require multiple steps and/or slow development times. The most promising technique for patterning conducting polymers with nanometer width scales is e-DPN. However, the present slow scan rates (1 to 10 nm/s), serial addressment, and the limited amount of monomer that can be adsorbed to the atomic force microscope (AFM) tip renders this technique infeasible for large scale pattern production. More recently, Mirkin et al. has reported a very promising technique for depositing conductive polymer onto a charged surface utilizing DPN involving an electrostatic driving force.[31]

Herein, we describe our technique of solid-state oxidation of a precursor insulating polymer film to produce conducting polymer networks using both electrochemical and chemical methods. Furthermore, we describe our electrochemically induced pattern transfer (ECIPT) technique by which we can rapidly pattern intrinsically conducting polymer into an insulating matrix.

Precursor Polymer Preparation

The precursor polymer, poly(5-norbornene-endo-2,3-bis(methylene-3'-[2,2':5',2"]-terthiophene acetate)), poly(1a), was prepared via ring opening metathesis polymerization (ROMP) of 1 utilizing Grubbs' alkylidene catalyst in accordance to Figure 1.[32] Polymerization of the monomer, 1, was complete within 30 minutes to give poly(1a) in a 75% yield. The structure of poly(1a) was confirmed by ^{1}H and ^{13}C NMR and both the molecular weight and polydispersity index were determined by GPC using monodisperse polystyrene standards to be 30,000g/mol and 1.08, respectively. Poly(1a) was soluble in common organic solvents such as chloroform ($CHCl_3$), methylene chloride (CH_2Cl_2), tetrahydrofuran (THF) etc., however not soluble in highly polar aprotic solvents or polar protic solvents such as acetonitrile, water, alcohol, etc.

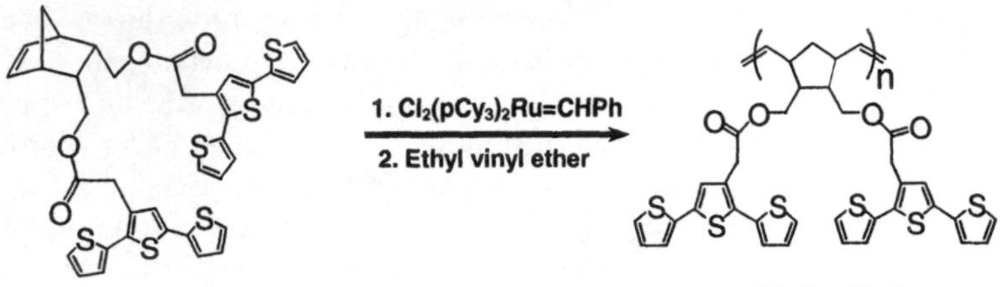

Figure 1. Preparation of precursor polymer, poly(1a), via ring-opening metathesis polymerization (ROMP) of monomer, 1.

Oxidative Solid-State Crosslinking via Oxidative Electrochemistry

In previously reported polymeric precursor systems,[15-21] the precursor polymers were originally dissolved in solvents, then crosslinked either by chemical reagents[17-21] or electrochemical methods.[15,16] In our procedure, the precursor polymer, is crosslinked to the ICP network in the solid-state using either electrochemistry or chemical oxidants. The precursor polymer, poly(1a) is soluble in several common solvents and thereby is easily processed. Thus, the precursor polymer can be coated onto many different types of surfaces or processed into various forms such as films and fibers before the crosslinking process. Since poly(1a) was soluble in solvents such as chloroform ($CHCl_3$), methylene chloride (CH_2Cl_2), and tetrahydrofuran (THF), and was insoluble in highly polar solvents such as water, alcohol, acetonitrile, a film of precursor polymer, poly(1a), can be easily cast onto substrates such as glass from the solution in organic solvents, then crosslinked in acetonitrile. Although insoluble, the precursor polymer swells in acetonitrile increasing main chain mobility. We have found that swelling is a very important factor in performing solid-state oxidative crosslinking. For example, attempts at electrochemical oxidative crosslinking in ethanol and water were unsuccessful.

Oxidative solid-state electrochemical crosslinking of poly(1a) is depicted in Figure 2. In a typical experiment, poly(1a) is coated onto the working electrode via either dip or spin coating from a 1 wt% poly(1a) CH_2Cl_2 solution, and then placed into 0.1M tetrabutylammonium perchlorate/acetonitrile electrolyte solution. The electrochemical crosslinking process was monitored via cyclic voltammetry (Figure 3) by scanning the potential between 0 V and 0.85V at the scan rate of 50mV/s. It should be noted that for the electrochemically induced crosslinking experiments, the electrolyte solution contains no dissolved precursor polymer, poly(1a), within the sensitivity limits of UV spectroscopy. The oxidation of pendant terthiophene moieties occurs at an onset potential of 0.65V with a peak at 0.75V. On the reverse scan, a reduction peak is observed at 0.6V. Crosslinking via oxidative coupling of pendant terthiophene moieties is complete after the first scan as indicated by the absence of current for terthiophene oxidation in the second scan at a potential of 0.75V. The resulting polymer after crosslinking exhibits quite different properties than the precursor in that it is no longer soluble.

Oxidative Solid-State Crosslinking via Chemical Oxidants

Preparation of poly(1b) in the solid swollen state can be effected utilizing chemical oxidants such as iron(III) chloride. A 0.5 µm thick film of poly(1a), prepared by spin coating from a methylene chloride solution of poly(1a) at the rate of 1500 rpm onto a glass substrate, was dipped into 0.005 M iron (III) chloride/ACN solution in order to achieve oxidative coupling of terthiophene moieties. In a few seconds, the color of the film changed from transparent colorless to deep blue indicating that solid-state coupling of terthiophene moieties to form polythiophene was successful. The crosslinking reaction was confirmed using UV-Vis spectroscopy as shown in Figure 4. Terthiophene moieties in poly(1a) (Figure 4A) exhibit a λ_{max} at 350nm. After exposure to iron (III) chloride followed by reduction with hydrazine hydrate, the λ_{max} red shifts to 440 nm, a wavelength commensurate with neutral polythiophene. Disappearance of the absorption at 350 nm in Figure 4B indicates the complete consumption of terthiophene units within the sensitivity limit of UV-Vis spectroscopy. The conductivity of poly(1b) films, chemically prepared using iron(III) chloride, measured via the four-point probe collinear array technique were found to be 0.04 S/cm, which is approximately 1 order higher than conductivities reported for poly(terthiophene)s prepared by conventional oxidative polymerizations.[19,33,34,35]

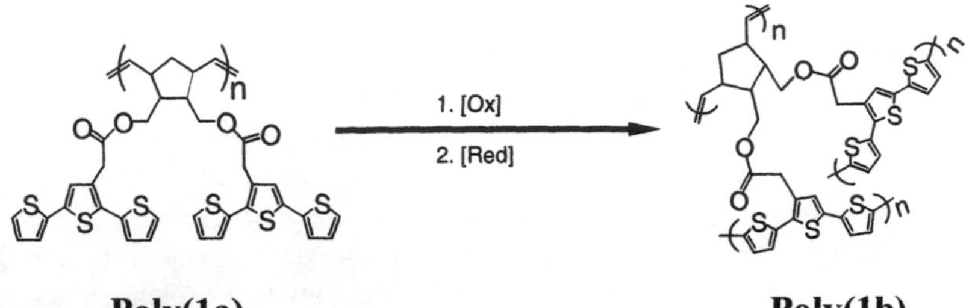

48

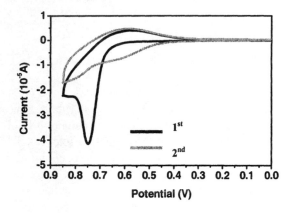

Poly(1a)   **Poly(1b)**

Figure 2. Oxidative solid-state crosslinking of poly(1a)

Figure 3. Electrochemical oxidative crosslinking of a poly(1a) coating on a Pt button working electrode via cyclic voltammetry at a scan rate of 50mV/s in 0.1M tetrabutylammonium perchlorate (TBAP)/ACN. Potentials reported vs. Ag/Ag+ nonaquous reference electrode (0.455V vs NHE). Black and grey lines depict 1st and 2nd scan, respectively.

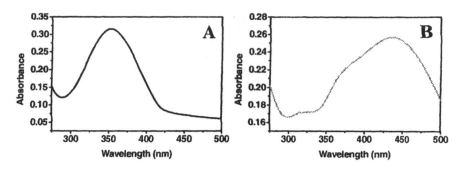

Figure 4. UV-Vis spectra of poly(1a) before (A) and after chemical oxidative crosslinking (B)

Patterning via Solid-State Oxidative Electrochemistry

Direct pattern generation of intrinsically conductive polythiophene into the precursor polymeric film, poly(1a), was accomplished via electrochemical solid-state crosslinking. The precursor polymer was spun coat onto a glass substrate consisting of two sets of interdigitated gold leads. Each set of interdigitated leads consisted of 50 fingers having ten micron width with ten micron finger/finger separation. This precursor polymer coated substrate was then placed into 0.1 M tetrabutylammonium perchlorate/acetonitrile solution. To one set of interdigitated leads was applied a potential of 0.9V vs. Ag/Ag$^+$ in order to induce oxidative solid-state electrochemical crosslinking. In Figure 5 the gold leads are shown in black, the precursor polymer is colorless transparent and the regions by which oxidative solid-state crosslinking occurred are orange. It should be noted that crosslinking only occurred above the gold leads in which the oxidative potential was applied. Since the crosslinking process produces intrinsically conductive polymer in the conducting form, the crosslinked material serves as the new electrode surface by which further propagation can occur. After a 0.1 second pulse of 0.9V, the crosslinked conductive front of poly(1b) had propagated approximately three microns away from the gold surface indicating a propagation speed of *ca.* 30 microns/second. Our electrochemically induced pattern transfer (ECIPT) via solid-state oxidative coupling is limited to the patterning of the conductive surface used as the working electrode for the procedure. However, it should be noted that a substrate may be reused. The width of the conducting polymer, poly(2a), lines that are transferred can be controlled by manipulating the time the potential is applied.

— 10 μm

Figure 5. Conducting pattern of poly(1b) generated by electrochemically induced pattern transfer (ECIPT) via application of 0.9 V for 0.1 sec using 0.1 M TBAP/ACN electrolyte solution. Potential is referenced vs. Ag/Ag⁺. As observed by transmission optical microscopy, orange lines are lines of reduced poly(1b), black lines are gold leads located under the polymer film, and colorless regions consist of the precursor polymer, poly(1a).

Surface Roughness

Surface roughness of both the poly(1a) and poly(1b) films could be important for several different application purposes. For example, film smoothness has implications toward advancing the use of conductive polymers in various types of devices where very thin films of <50 nm are required. Surface roughness was studied using atomic force microscopy (AFM) in the tapping-mode of three films; one prepared by conventional electrochemical polymerization of **1** from solution onto indium doped tin oxide (ITO) coated glass, one from poly(1a) prepared by spin-coating onto an ITO glass substrate, one of poly(1b) prepared via solid state electrochemical polymerization of poly(1a) carried out at 0.9V on ITO glass. Tapping mode AFM images of these three different films are depicted in Figure 6A, 6B and 6C, respectively. Conductive polythiophene films prepared via conventional electrochemical polymerization of **1** were found to have film roughnesses of approximately 15 nm over a twenty five square micron surface area. 0.5 micron thick films of poly(1a) before and after solid-state oxidative crosslinking were found to have film roughnesses of 1.8 and 1.9 nm, respectively, averaged over twenty five square microns. Conventional electrochemical polymerization of **1** from electrolyte solution involves a coupling process in solution within close proximity to the electrode surface. Once higher order oligomers form, they eventually become insoluble in the electrolyte solution thereby electroprecipitating onto the working electrode surface. This nucleation and growth mechanism for film formation inherently makes these films more rougher than those of poly(**1a**) or poly(**1b**). In the conversion of poly(**1a**) to poly(**1b**), surface smoothness is retained. Our oxidative solid-state crosslinking process does not produce conductive polymer via nucleation and growth. Instead, the monomer is covalently attached to the precursor polymer backbone and hence is mobile when the precursor polymer chains are mobile. The monomers are already in place on the substrate surface to produce conductive polymer. Since conductivity, a bulk property, is strongly influenced by film morphology the higher conductivity (0.04 S/cm) of poly(**1b**) prepared by solid-state crosslinking compared to other poly(terthiophene)s most likely results from a smooth film surface .

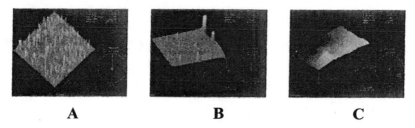

A   **B**   **C**

Figure 6. Tapping mode AFM images of conducting polymer prepared via conventional electroprecipitation from monomer ,1, solution (A), poly(1a) (B), and poly(1b) (C).

Conclusion

Electrochemically induced pattern transfer (ECIPT) is a novel technique by which to rapidly pattern intrinsically conducting polymer (ICP) into an insulating precursor polymer. This technique allows for easy processing of the precursor polymer before oxidative solid-state crosslinking. After crosslinking, the intrinsically conductive pattern is generated and retains film smoothness. Furthermore, these polymers are more resilient than intrinsically conductive polymers prepared conventionally by oxidative electrochemistry from solution. ICP formation via this technique was confirmed via cyclic voltammetry and UV-vis spectroscopy. As an application the ECIPT process, patterning of ICP was successfully performed on a gold patterned substrate consisting of 100 interdigitated leads by oxidative electrochemistry in the solid state. This patterning process is not limited to transfer of gold lines.

References

1. Handbook of Conducting Polymers, 2nd ed.; Skotheim, T. A.; Elsenbaumer, R. L.; Reynolds, L. R. Marcel Dekker: New York, NY, **1998.**
2. Sapp, S. A.; Sotzing, G. A.; Reynolds, J. R. *Chem.Mater.***1998**, *10*, 2101-2108.
3. Perucki, M.; Chandrasekhar, P. *Synth. Met.* **2001**, *119*, 385-386.
4. Yu, G.; Heeger, A. J. *Synth. Met.* **1997**, 85, 1183.
5. Ferraris, J. P.; Eissa, M. M.; Brotherston, I. D.; Loveday, D. C. *Chem. Mater.* **1998**, *11*, 3528-3535.
6. MacQuade, D. T.; Pullen, A., E.; Swager, T. M. *Chem. Rev.* **2000**, *100*, 2537-2574.
7. Sotzing G. A.; Briglin, S.; Grubbs, R. H.; Lewis, N. S. *Anal. Chem.* **2000**, *72*, 3181.
8. Ma, H.; Chen, B.; Sassa, T.; Dalton, L. R.; Jen, A. K.-Y. *J. Am. Chem. Soc.* **2001**, *123*, 986-987.
9. Lerch, K.; Jonas, F.; Linke, M., J.; *Chim. Phys. Phys.-Chim. Biol.* **1998**, *95*, 1506
10. Chayer, M.; Faid, K.; Leclerc, M. *Chem. Mater.* **1997**, *9*, 2902-2905.
11. Jen, K. Y.; Miller, G. G.; Elsenbaumer, R. L. *J. Chem. Soc. Chem. Commun.* **1986**, *17*, 1346-7.
12. Elsenbaumer, R. L.; Jen, K. Y.; Oboodi, R. *Synth. Met.* **1986**, 15(2-3), 169-174.
13. Xia, C.; Fan, X.; Park, M.; Advincula, R. C. *Langmuir,* **2001**, *17(25)*, 7893-7898.
14. Xia, C.; Advincula, R. C. *Chem. Mater.* **2001**, *13*, 1682-1691.
15. Khanna, R. K.; Bhingare, N. *Chem. Mater.* **1993**, *5(7)*, 899-901.
16. Khanna, R. K.; Cui,H. *Macromolecules,* **1993**, *26*, 7076-7078.
17. Nawa, K.; Imae, I.; Noma, N.; Shirota, Y. *Macromolecules,* **1995**, *28*, 723-729.
18. Imae, I.; Nawa, K.; Ohsedo, Y.; Noma, N.; Shirota, Y. *Macromolecules,* **1997**, *30*, 380-386.
19. Watson, K. J.; Wolfe, P. S.; Nguyen, S. T.; Zhu, J.; Mirkin, C. A. *Macromolecules,* **2000**, *33(13),* 4628-4633.
20. Holdcroft, S. *Adv. Mater.* **2001**, *13(23)*, 1753-1765.
21. Yu, J.; Holdcroft, S. *Chem. Commun.* **2001**, 1274-1275.
22. Sirringhaus, H.; Tessler, N.; Friend, R. H. *Science* **1998**, *280*, 1741-1744.
23. Granlund, T.; Nyberg, T.; Roman, L. S.; Svensson, M.; Inganas, O. *Chem.Mater.,* **2000**, *12*, 269-273.
24. Beh, W. S.; Kim, I. T.; Qin, D.; Xia, Y.; Whitesides, G. M. *Adv. Mater.,* **1999**, *11*, 1038-1041.
25. Yu, J.; Abley, M.; Yang, C.; Holdcroft, S. *Chem. Commun.* **1998**, *73*, 775.
26. Lowe, J.; Holdcroft, S. *Synth. Met.,* **1997**, *85*, 1427-1430.

27. Dai, D.; Griesser, H. J.; Hong, X.; Mau, A. W. H.; Spurling, T. H.; Yang, Y. *Macromolecules*, **1996**, *29*, 282-287.
28. Maynor, B. W.; Filocamo, S. F.; Grinstaff, M. W.; Liu, J. *J. Am. Chem. Soc.* **2002**, *124*, 522-523.
29. Li, Y.; Maynor, B. W.; Liu, J. *J. Am. Chem. Soc.* **2001**, *123*, 2105-2106.
30. Maynor, B. W.; Li, Y.;Liu, J. *Langmuir* **2001**, *17*, 2575-2578.
31. Lim, J.; Mirkin, C. A. *Adv. Mater.* **2002**, *14*, 1474-1477.
32. Jang, S-Y.; Sotzing, G. A.; Marquez, M. *Macromolecules*, **2002**, *35*, 293-7300.
33. Roncali, J. *Chem. Rev.* **1992**, *92*, 711-738.
34. Roncali, J.; Lemaire, M.; Garreau, R.; Garnier, F. *Synth. Met.* **1987**, *18*, 139-144.
35. Roncali, J.; Garnier, F. *Synth. Met.* **1986**, *15*, 323-331.

Chapter 5

Fluorocarbon Polymer-Based Photoresists for 157-nm Lithography

T. H. Fedynyshyn, W. A. Mowers, R. R. Kunz, R. F. Sinta,
M. Sworin, A. Cabral, and J. Curtin

Lincoln Laboratory, Massachusetts Institute of Technology,
Lexington, MA 02420

A series of fluorinated polymers derived from 4-hexafluoro-isopropanol styrene were prepared and evaluated for potential use in 157-nm photoresists. Physical properties such as MW, T_g, thermal stability, and VUV absorbance were measured and contrasted. The nature of the blocking group has a large effect on all of the aforementioned properties. Several candidate polymers were formulated into resists and imaged at 157 nm. The imaging results as well as the etch resistance are reported.

Introduction

As photolithography has progressed toward shorter and shorter wavelengths as a means of decreasing feature size, the need for more transparent polymers at these wavelengths has become more critical. The move to excimer (KrF) laser-based imaging at 248 nm necessitated the shift from novolak polymers, which were used extensively at 435 and 365 nm, to poly(hydroxystyrene) (PHS) derivatives. Fortunately, PHS exhibits a transmission window at this wavelength while retaining some of properties of novolak such as base solubility and etch resistance.

As the industry next moved to 193 nm (ArF), phenolic polymers became unusable due to the high absorbance of the aromatic moiety. Multicyclic hydrocarbons (1, 2) (e.g. norbornene) and acrylic monomers (3, 4) were then introduced into resist systems for 193-nm imaging. Increased transparency was achieved at a cost of decreased etch resistance and some compromise to the base solubility of the resist. These matters are currently being addressed and improved upon.

Now the use of 157-nm (F$_2$) radiation is seriously being considered as the next and possibly last optical wavelength to be employed for the manufacture of IC's. At this wavelength most organic moieties absorb strongly. Since the use of ultra-thin (<100 nm) resists is not desirable, the need for polymer binders with minimized 157-nm absorbance is critical.

A survey of various polymer classes indicated that two classes, fluorinated polymers and siloxanes, exhibit fairly good transparency at 157 nm (5). Since this observation was reported, several groups of researchers have been exploiting both of these chemistries in the design of new polymers for 157-nm lithography. Fluorinated materials have become the more popular of the approaches and to this end several types of fluorinated monomers have been explored to achieve the required polymer transparency (6-8).

The purpose of this work is to explore the utility of the 4-hexafluoroisopropanol styrene (HFIP) moiety as a base soluble, aromatic component in polymer backbones for 157-nm lithography (9). The homopolymer exhibits an absorbance of 3.44/μ at 157 nm (5). Copolymers of HFIP with t-butyl acrylate were prepared as well as OH blocked versions of the homopolymer. Various blocking groups were used including carbonates and acetals. The thermal stability of the blocked polymers is compared and contrasted. Properties such as the 157-nm absorbance and thermal degradation are shown to be a strong function of the amount of blocking group present in the polymer.

Finally photoresist formulations were prepared from representative polymers and imaging results are presented which indicate that these systems are capable of sub-100 nm resolution. Etch studies show that the absolute polysilicon and oxide etch rates are greater than those of phenolic analogs. However because the HFIP polymers exhibit reduced absorbance at 157 nm, greater film thicknesses can be employed which in turn more than compensates for the increased rates.

Experimental

Synthesis

The 4-hexafluoroisopropanol styrene (HFIP) and its 4-t-BOC derivative were prepared by a modification of the procedure of Przybilla et al (10).

tert-Butyl[2,2,2-Trifluoro-1-trifluoromethyl-1-(4-vinyl-phenyl)ethoxy]-
acetate (4-t-BuAc). To a solution of 1,1,1,3,3,3-hexafluoro-2-(4-vinylphenyl)-
2-propanol (5.48 g, 20.30 mmol) in THF (25 mL) was slowly added a 1M
solution of t-BuOK in THF (25.00 mL, 25.00 mmol). The reaction mixture was
allowed to stir at room temperature for 30 min, then cooled to -78°C
using a dry ice /acetone bath followed by the addition of t-butyl bromoacetate
(3.75 mL, 25.40 mmol). The reaction mixture was allowed to slowly warm to
room temperature overnight, concentrated, diluted with hexane, filtered through
celite, and concentrated to provide 7.96 g of a reddish-yellow oil. This material
was purified by column chromatography on silica gel using 50:1 hexane/ethyl
acetate to provide 6.44 g (82%) of a colorless oil; ^{1}H NMR (CDCl$_3$) δ 7.60 (d,
2ArH), 7.50 (d, 2ArH), 6.75 (dd, CH=), 5.85 (dd, =CHH), 5.40 (dd, =CHH),
4.05 (s, CH$_2$), 1.50 (s, t-Bu).

**1-(2,2,2-Trifluoro-1-methoxymethoxy-1-trifluoromethyl-ethyl)-4-vinyl
benzene (4-MOM).** To a mixture of NaH (60% dispersion, 968 mg, 24.20
mmol), washed with pentane, and THF (50 mL) was slowly added a solution of
1,1,1,3,3,3-hexafluoro-2-(4-vinylphenyl)-2-propanol (5.00 g, 18.51 mmol) in
THF (10 mL). The reaction mixture was allowed to stir at room temperature for
1 h, then cooled to −78°C using a dry ice/acetone bath followed by the addition
of chloromethyl methyl ether (1.50 mL, 19.75 mmol) in THF (5 mL). The
reaction mixture was allowed to slowly warm to room temperature overnight,
quenched with H$_2$O, concentrated, and partitioned between ether/ H$_2$O. The
combined extracts were washed with saturated NaHCO$_3$, brine, dried (Na$_2$SO$_4$),
and concentrated to provide a crude oil. This material was purified by
Kugelrohr distillation (bp 65-9°C, 1.3 mm), followed by column
chromatography on silica gel using 50:1 hexane/ethyl acetate to provide 1.96 g
(34%) of a colorless oil; ^{1}H NMR (CDCl$_3$) δ 7.60 (d, 2ArH), 7.50 (d, 2ArH),
6.75 (dd, CH=), 5.85 (dd, =CHH), 5.40 (dd, =CHH), 4.85 (s, CH$_2$), 3.60 (s, Me).
*Chloromethyl methyl ether (highly toxic/cancer suspect agent) should be
handled only in well ventilated areas and the user outfitted with high quality
nitrile gloves.*

**1-[1-(*tert*-Butoxymethoxy)-2,2,2-trifluoro-1-trifluoro-methylethyl]-4-
vinylbenzene (4-BOM).** A solution of 1,1,1,3,3,3-hexafluoro-2-(4-
vinylphenyl)-2-propanol (9.00 g, 33.31 mmol) in THF (10 mL)was slowly added
to a mixture of NaH (60% dispersion, 1.11 g, 27.75 mmol), washed with
pentane, and THF (50 mL). The reaction mixture was allowed to stir at room
temperature for 1 h, then cooled to −78°C using a dry ice /acetone bath followed
by the addition of *tert*-butyl chloromethyl ether (22.75 mmol, generated in situ in
hexane, 12.50 mL). The *tert*-butyl chloromethyl ether was prepared by a
modification of the procedure of Goff et al (*11*) The reaction mixture was

allowed to slowly warm to room temperature overnight, quenched with H_2O, concentrated, and partitioned between ether/H_2O. The combined extracts were washed with saturated $NaHCO_3$, brine, dried (Na_2SO_4), and concentrated to provide a crude oil. This material was purified by column chromatography on silica gel using 50:1 hexane/ethyl acetate to provide 3.25 g (41%) of a colorless oil; ^1H NMR (CDCl$_3$) δ 7.60 (d, 2ArH), 7.50 (d, 2ArH), 6.75 (dd, CH=), 5.80 (dd, =CHH), 5.40 (dd, =CHH), 4.90 (s, CH$_2$), 1.30 (s, tBu).

Polymer Synthesis

Polymers were synthesized in THF using AIBN as the initiator (1-3%). Typical polymer yields were between 60 and 80%.

Characterization

Transmission measurements were made by coating a 75 to 150 nm polymer film on a calcium fluoride substrate followed by baking in a convection oven for 30 minutes at 150°C. Complete details of the determining the 157-nm absorbance (A157) have been previously reported (5). Thermal analysis was carried out using a TA Instruments 2920 DSC and 2950 TGA under nitrogen at 20°C/min. T_g's are reported as second heats. Molecular weights determined using a Waters Alliance Chromatograph (Model 2690) equipped with a 2410 refractive index detector. Molecular weights were calculated using poly(styrene) standards.

Lithography

Projection lithography at 157 nm was performed on HMDS treated 8-inch silicon wafers with the Exitech 0.60-NA small field stepper using either binary or phase shift masks. The resist thickness was 100 nm, the post apply bake (PAB) was for 140°C for 60 seconds on a hot plate and the post exposure bake (PEB) was at 140°C for 90 seconds on a hot plate. Development was by single puddle for 45 seconds with 2.38% TMAH (0.26 N) developer.

Etch Conditions

Plasma etching was performed on polymer films coated on 6-inch silicon wafers to a thickness of 150 to 200 nm. Thermal oxide etching was performed on a Lam Rainbow etcher for 60 seconds employing an etch gas mixture of 8 sccm of CF_4, 12 sccm of CHF_3, and 200 sccm of Ar, at a pressure of 350 mTorr, with an electrode power of 700 Watts, and a lower electrode temperature of $-20°C$. Selectivity to oxide is calculated by dividing the material etch rate by the thermal silicon dioxide etch rate of 6.05 nm/second. Polysilicon etching was performed on a Lam 9400 TCP etcher for 60 seconds employing an etch gas mixture of 37.5 sccm of Cl_2, 13.5 sccm of HBr, 6 sccm of He, and 1.5 sccm of O_2, at a pressure of 10 mTorr, with a source power of 200 Watts, a plate power of 100 Watts, and a lower electrode temperature of 45°C. Selectivity to polysilicon is calculated by dividing the material etch rate by the amorphous silicon etch rate of 2.85 nm/second.

Table I. Polymer Properties

Polymer(Ratio)	$M_w(x10^3)$	$T_g(°C)$	A157/μ
HFIP-co-TBA 60/40	17.6	124	3.74
t-BuAc 100	14.5	55	4.29
HFIP/t-BuAc 70/30	45.7	107	3.71
HFIP/t-BuAc 60/40	61.6	93	3.80
t-BOC 100	6.70	62	2.95
HFIP/t-BOC 70/30	25.8	73	3.57
HFIP/t-BOC 60/40	21.8	73	3.44
HFIP/t-BOC 50/50	16.9	69	3.39
MOM 100	16.2	69	2.60
HFIP/MOM 70/30	26.9	117	3.27
HFIP/MOM 60/40	25.5	107	3.08
BOM 100	16.6	63	-
HFIP/BOM 70/30	26.3	106	3.16
HFIP/BOM 60/40	26.3	97	2.82

Results and Discussion

Polymer Properties

Modestly high yields were obtained for the polymerizations in most cases and no attempts were made to either maximize the yield or control the molecular weights (an M_w target of around 25,000 was chosen) and in the case of copolymers no effort was made to finely tune the composition. The polymer structures are shown in Figure 1. In general the molecular weights range from about 15 to 50K with the exception of the 4-t-BOC homopolymer. Polydispersities are approximately 2 for all of the entries. The 4-t-BOC result is somewhat surprising although steric factors may be playing a role in reducing the molecular weight. Copolymers of this material with 4-HFIP behaved more predictably with the HFIP/t-BOC 70/30 and HFIP/t-BOC 60/40 giving M_w's around 25 K. Further evidence of possible steric effects was obtained when the methyl carbonate homopolymer was prepared high molecular weights were achieved (12). Polymer properties including M_w, T_g and VUV absorbance at 157 nm are collected in Table I.

Many groups are working hard to increase the transparency of polymer binders for 157-nm photoresists. In most cases this is being accomplished by incorporating fluorine into the polymer chain either in the backbone or in pendant groups. The absorbance at 157 nm (A157) has been shown to decrease as the amount of fluorine in the polymer increases.

In the all aromatic-based systems described here the introduction of the HFIP group has a dramatic effect on A157. The absorbance of the homopolymer of HFIP styrene has been reported (5) as 3.44/μ and when it is copolymerized with t-butyl acrylate (TBA) a modest increase of 0.3/μ is observed. The increase, which is due to the acrylate (A157 = 5.5/μ), follows a linear relationship. The same is true for the HFIP blocked with the t-butyl acetate (t-BuAc) group. The introduction of the ester group into this polymer causes the A157 to rise as shown in Figure 2. Here again the copolymer exhibits a linear relationship between copolymer composition and A157. The opposite case is observed for both the acetal and carbonate blocking groups. When these are substituted on the OH of the HFIP moiety, a decrease in optical density is observed. The decrease is caused by the reduction of the major absorption band (via volume dilution) at 190 nm which in turn, lowers the absorption minimum located near 157 nm. No spectral shifts were seen for any of these blocking groups. The plots for t-BOC and MOM are also included in Figure 2.

The t-butyl acetal, BOM group (not shown in Figure 2), also lowers the A157. The homopolymer of HFIP-BOM was not able to be coated uniformly on CaF_2 to get a direct A157 measurement. The hydrophobic nature of the polymer prevented it from wetting and adhering to the CaF_2 surface. However the copolymers HFIP/BOM 60/40 and 70/30 coated well. From the linear plot of composition vs. A157 an extrapolated value of 2.06/μ for the BOM-homopolymer was obtained. This is the lowest A157 reported to date for an all-aromatic, 157-nm resist polymer candidate.

The T_g's of the blocked, candidate polymers that were synthesized tend to be lower than those obtained for the t-butyl acrylate copolymers especially for the fully substituted polymers. Ideally the polymers only need to be sufficiently blocked so as to lower the unexposed dissolution rates to an acceptably low

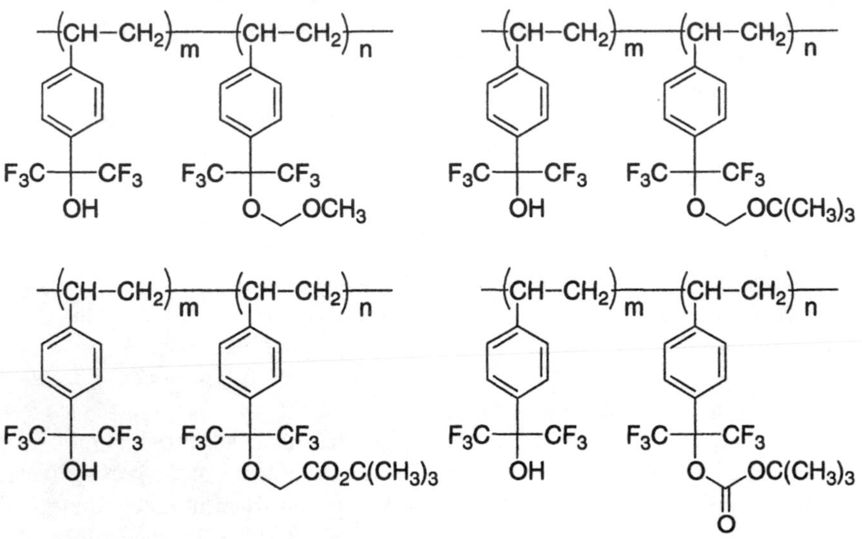

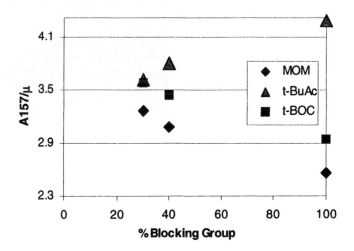

Figure 1. Polymer structures.

Figure 2. Polymer absorbance at 157 nm.

level. Blocking levels of 30-40% are typical levels in order to achieve this. The T_g's of the copolymers exhibit a linear dependence on composition (Figure 3) and can be predicted from the Fox equation The T_g's of these polymers fall within a usable range.

The thermal decomposition temperatures as a function of the %-blocking are plotted in Figure 4 for the t-BuAc and t-BOC substituted polymers. In both cases a linear decline in T_d as the blocking level decreases is observed. A decrease in T_d is also exhibited when these groups are substituted onto the phenolic OH in poly(4-hydroxysytrene) (PHS) (13). In the case of PHS the thermal stability of the t-BuAc group is compromised by a greater extent as the blocking level decreases compared with t-BOC and it is less thermally stable at all blocking levels. This situation is reversed in the case of the partially protected HFIP polymers. Both T_d's fall at about the same rate but clearly the t-BuAc is the more thermally stable moiety by about 60°C at all substitution levels. Decomposition in the blocked-PHS polymers occurs via a nucleophilic displacement reaction involving the OH moiety. This is not the case with the HFIP which is a poor nucleophile. Increased acidity of the HFIP hydroxyl is most likely the cause of the depressed decomposition temperatures.

For both of the acetals (MOM and BOM). the percent substitution had little effect on the polymer decomposition temperature (~330°C). However these groups are readily cleaved off in the presence of strong acids. Figure 5 shows the TGA traces for the homopolymers BOM and MOM. The traces were recorded for the neat polymers and for the polymers plus 5 weight-% camphorsulfonic acid (CSA). In the absence of any acid, the blocking groups are not cleaved until temperatures greater than 320°C. When the acid is added, this temperature is lowered to below 100°C. From the shape of the TGA trace and the weight loss, it can be concluded that the deblocking reaction occurs without side reactions and is quantitative. In fact CSA doping can be used to calculate the %-substitution in the acetal copolymers by TGA. These results indicate that both MOM and BOM are attractive blocking groups in so far as these will be very stable in resist solutions (e.g. long shelf life) but in the presence of acid should yield very sensitive resists.

Lithography Results

Several of the synthesized polymers were formulated with a photoacid generator (di-t-butylphenyl iodonium nonaflate) and a base (tetrabutyl ammonium lactate) in ethyl lactate. Projection lithography was performed on a 0.60 NA 157-nm micro-stepper with both a binary and phase shift mask. A critical assumption in pursuing the fluoroaromatic approach toward the development of 157-nm resists is the ability of these resists to show imaging

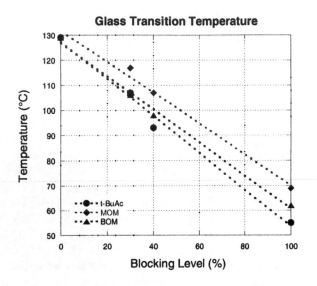

Figure 3. Relationship of Tg and blocking level.

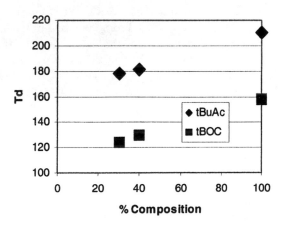

Figure 4. Decomposition temperature as a function of blocking level.

characteristics appropriate to sub-100 nm lithography at resist thicknesses of 100 nm or greater. Several experimental resists were evaluated and compared including the basic ESCAP polymer, poly(HOST-co-TBA), along with resists formulated with the HFIP-TBA and HFIP-acetal copolymers. The resist names and the base polymers are reported in Table II. Also listed in this table is the exposure dose to size 120-nm lines with a 300-nm pitch (1:1.5) employing a binary mask.

It can be seen that all resists have high sensitivity to 157-nm exposure. The hydrocarbon-based resist has about twice the sensitivity of its HFIP analog, which can be explained by noting that the HOST-based polymer has almost twice the absorbance of the fluoroaromatic polymer. It is also observed that the HFIPS/BOM based resist has an higher sensitivity, 14.6 mJ/cm^2 versus 22.7 mJ/cm^2. This is consistent with the prediction of higher sensitivity made by observing the thermal decomposition of the two polymers in the presence of camphorsulfonic acid where the BOM polymer showed thermal decomposition at a lower temperature than the MOM polymer (vide supra).

Table II. Sizing dose (E_S) with 157-nm exposure

Resist	Polymer	E_S (mJ/cm²)
LUVR99071	60:40 HOST/TBA	5.0
LUVR99099	60:40 HOST/TBA	-
LUVR20045	60:40 HFIP/TBA	10.0
LUVR20039	70:30 HFIP/MOM	22.7
LUVR20048	70:30 HFIP/BOM	14.6

Table III. Summary of 157-nm exposure using a phase shift mask

Resist	Phase Shift Mask		
	1:1	1:1.5	1:3
LUVR99071	100	70	50
LUVR99099	100	70	70
LUVR20045	-	70	60
LUVR20039	150	100	90
LUVR20048	100	100	70

A summary of highest resolution with 157-nm exposure with phase shift mask is presented in Table III. The bold numbers in the second row refer to the pitch of the line and space pair with the smallest resolved linewidth given below in nm. All six resists are capable of sub-100-nm resolution with a 1:3 pitch. The resolution decreases with decreasing pitch, although several resists show 70-nm resolution with a semi-isolated line pitch of 1:1.5. In general these resists were capable of printing features of down to 110 nm using a binary mask.

The resolution capability of the hydrocarbon resists, LUVR99071 and LUVR99099, in 60-nm thick films, is shown in Figure 6. The LUVR99071 exhibits slightly higher resolution, 50 versus 60 nm, than LUVR99099. All subsequent resists were based on the LUVR99071 formulation. The resolution capability of the HFIP/TBA-based resists are shown in Figure 7 and that of the HFIP-acetal resists are shown in Figure 8.

LUVR20045, exhibited 60-nm imaging capability for the phase shift mask (PSM) on the 157-nm projection system with a resist thickness of 100 nm. The resist LUVR-20039, containing the MOM copolymer, gave 90-nm lines with a 1:3 pitch and LUVR20047, containing the BOM copolymer, gave 70-nm lines with the same 1:3 pitch. All the resists were able to print features over several doses and focus fields implying that potentially excellent overall latitude are possible with these resists.

Plasma Etch Studies

An ongoing concern with using thin resists, even with hard mask technology, is the ability of the thin film to resist the harsh environments found during plasma etching. This concern may be magnified when fluoropolymers are employed in the resist, since fluoropolymers are expected to have very high plasma etch rates. It is difficult, if not impossible, to test resists under the variety of different plasma etch conditions found throughout the semiconductor fabrication process. Therefore the ability of the fluoropolymers to withstand a plasma etch was tested under two representative plasma etch processes used in industry today. These processes are a thermal oxide etch process based on a fluorine plasma that has a thermal oxide etch rate of 6.05 nm/second and a polysilicon etch process based on a chlorine/bromine plasma that has an amorphous silicon etch rate of 2.85 nm/second .

The evaluation of an HFIP/TBA copolymer and acetal functionalized HFIP polymers was compared against a hydroxystyrene/t-butyl acrylate copolymer (HOST/TBA) under the same conditions. The choice to look at polymers and not resists was based on removing the effects of other resist components on the plasma etch rate and focusing only on the polymer etch characteristics. The rates and selectivities of both plasma etches are presented in Table IV. It can be seen that the HFIP/TBA polymer has a higher etch rate than the HOST/TBA polymer in both the oxide and poly etch. The use of acetal functionalized HFIP gave the

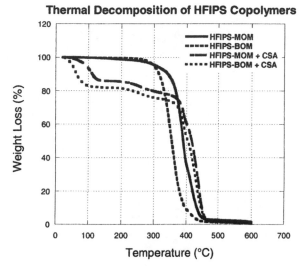

Figure 5. Thermal stability of partially blocked polymers.

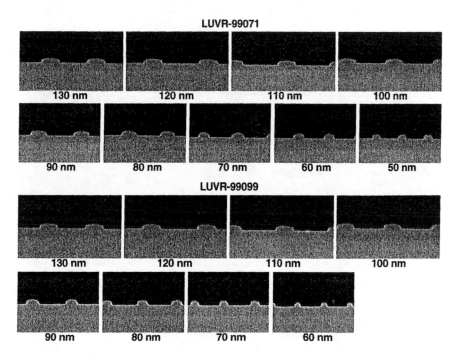

Figure 6. Comparison of HOST/TBA resists showing 1:3 lines at 60 nm.

66

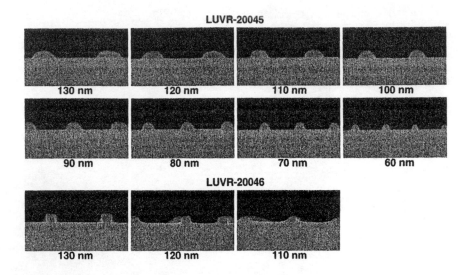

Figure 7. Comparison of HFIP/acrylate resists showing 1:3 lines at a 100 nm.

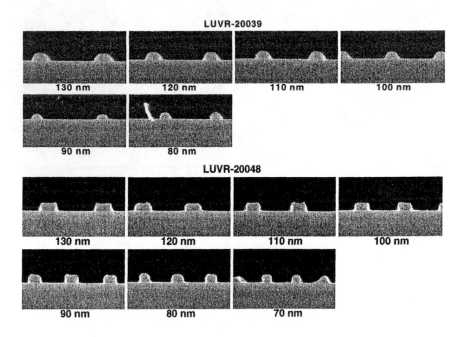

Figure 8. Comparison of HFIP/acetal resists showing 1:3 lines at a 100 nm.

most dramatic improvement in plasma etch selectivity toward both oxide and poly. The HFIP/MOM polymer shows an etch rate similar to that of the phenolic-based copolymer in the poly etch, while the HFIP/BOM polymer not only shows comparable etch selectivity in the oxide etch but superior etch selectivity in the poly etch.

Table IV. Plasma etch rates and selectivities

Polymer	Oxide Etch (nm/sec)	Oxide Selectivity
60:40 HOST/TBA	0.86	7.1
60:40 HFIP/TBA	1.34	4.5
70:30 HFIP/MOM	1.11	5.5
70:30 HFIP/BOM	0.89	6.8
Thermal oxide	6.05	1.0
Amorphous silicon	0.60	10.2

Polymer	Poly Etch (nm/sec)	Poly Selectivity
60:40 HOST/TBA	0.71	4.0
60:40 HFIP/TBA	1.01	2.6
70:30 HFIP/MOM	0.72	4.0
70:30 HFIP/BOM	0.62	4.6
Thermal oxide	0.13	22.6
Amorphous silicon	2.85	1.0

The raw plasma etch rate in different plasmas is a useful number to compare, but even more useful is the plasma etch selectivity toward the substrate of interest. Both of these criteria are important, however there may even be more value in determining the time to strip the resist in a given plasma as a measure of plasma etch resistance for the thin resists that are required for 157-nm lithography. The allowable resist thickness is dependent on the absorbance of the resist, defined as some value of optical density per unit thickness. The time to strip the resist is therefore a function of both the resist thickness allowed at a defined optical density and the resist etch rate.

The time to strip the fluoroaromatic polymers and the HOST/TBA polymer, was determined based on the 157 nm absorbance of the polymer and the plasma etch rate. It is shown graphically in Figures 9 and 10. The y-intercept of the two graphs represents the thickness of a polymer film with an optical density (O.D.) of 0.4, and the slope of the graphs is the plasma etch rate of the polymer. An O.D. of 0.4 is often used as the maximum absorbance that is tolerable for a

68

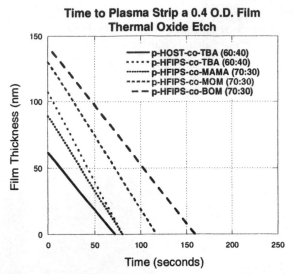

Figure 9.   Performance of 0.4 O.D. fluorocarbon and hydrocarbon polymers during oxide plasma etching.

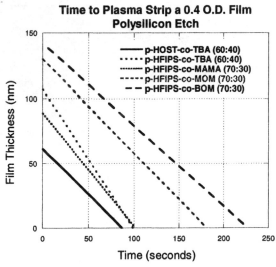

Figure 10.   Performance of 0.4 O.D. fluorocarbon and hydrocarbon polymers during poly plasma etching.

photoresist film to give good imaging characteristics. It can been seen that the higher plasma etch rate of the HFIP-based polymers is more than compensated for by the higher film thickness (i.e. lower 157 nm absorbance) leading to a net gain in plasma etch resistance as measured by film removal time. The greatest gain in plasma etch resistance is shown by the HFIP-acetal polymers, where the combination of low absorbance coupled with plasma etch rates comparable to those of the copolymer shows a more than 2 fold increase in the time to strip, as shown in both of the plots. This is compared to the HFIP/TBA polymer that provides about a 25% increase in plasma etch resistance over the HOST/TBA polymer when film removal time is used as the figure of merit.

Conclusions

An alternative approach to 157-nm resist development was presented, employing fluoroaromatic polymers, that represent an evolutionary step in resist materials development. Both partially blocked and copolymers of 4-hexafluoro-isopropanol styrene have been synthesized and characterized. Both absorb less than 4/□ at 157 nm. Both the acetals and the t-BOC blocking groups lower the absorbance further and the 100% blocked HFIP/BOM has an extrapolated O.D. of 2.06/□. Thermal properties were also measured and shown to be dependent upon the degree of blocking as is the case in PHS-based resists.

Acetal blocked hexafluoroisopropanolstyrene-based copolymers have been prepared and compared with both ESCAP based resists as well as hexafluoroisopropanolstyrene analogs to the hydroxystyrene containing ESCAP polymers. The HFIP-acetal platform offers the ability to tailor lithographic performance from relatively simple synthetic schemes that allow facile manipulation of the incorporated monomer ratios while building on the knowledge base derived from previous platforms. Absorbance of the HFIP polymer platform can be tuned such that 100-150 nm resist film thicknesses are possible. Thicknesses that, when used with a hard mask, could enable the early introduction of 157-nm lithography into device manufacturing and effectively negate the need for heavily fluorinated resins that maximize transparency at the expense of manufacturing cost and development effort.

The preliminary resist imaging of the HFIP-based polymers is very encouraging. Examples of the high performance imaging capability of this resist design are shown to have imaging capability of 60 nm for isolated lines with a 0.60 NA microstepper. These resists are not based on optimized formulations and represent only a first generation resist system. It can be expected that further work on HFIP-based polymers can lead to improved resist performance. The imaging capability observed so far, coupled with the relatively straightforward polymer synthesis, makes insertion of this class of resists into the International

Technology Roadmap's 95 nm node in 2004 and the 65 nm node in 2007 very possible if coupled with hard mask resist processing technology.

The plasma etch rates for the fluorocarbon polymers were determined under two representative plasma etch processes, an oxide etch and a poly etch. The fluoroaromatic acetals are a significant improvement over the fluoroaromatic ester etch rates. The most promising fluoroaromatic acetal is the HFIP/BOM copolymer in which plasma etch rates are comparable or superior to those of the ESCAP polymer under typical oxide or poly etch conditions. When film removal time is used as the figure of merit in evaluating plasma etch resistance, the HFIP/BOM polymer provides over a two-fold improvement in plasma etch resistance over the ESCAP polymer.

High resolution images in 70-80 nm range were obtained with non-optimized formulations on a 157-nm microstepper. These results coupled with the increased transparency and potential etch resistance due to the aromatic backbone, make these polymers attractive candidates for 157-nm photoresists.

Acknowledgements

We acknowledge Sue Cann for SEM support, and Jeff Meute and Georgia Rich of SEMATECH for 157 nm exposures. A special thank you is given to Dr. Maeda of Central Glass Co. Ltd. for the generous gifts of fluorinated monomers.

This work was performed under the Advanced Lithography Program of the Defense Advanced Research Projects Agency under Air Force Contract #F19628-00-C-0002. Opinions, interpretations, conclusions, and recommendations are those of the authors, and do not necessarily represent the view of the Department of Defense.

References

1. Kaimoto, Y.; Nozaki, K.; Takechi, S.; Abe, N. *Proc. SPIE – Advances in Resist Technology and Processing IX*, **1992**, 1672, 66.
2. Kunz, R.R.; Allen, R.D.; Hinsberg, W.D.; Wallraff, G.M. *Proc. SPIE – Advances in Resist Technology and Processing X*, **1993**, 1925, 167.
3. Allen, R.D.; Sooriyakumaran, R.; Opitz, J.; Wallraff, G.M.; DiPietro, R.A.; Breyta, G.; Hofer, D.C.; Kunz, R.R.; Jayaraman, S.; Shick, R.; Goodall, B.; Okoroanyanwu, U.; Willson, C.G. *Proc. SPIE – Advances in Resist Technology and Processing* **1996**, 2724, 334.
4. Wallow, T.I.; Houlihan, F.M.; Nalamasu, O.; Chandross, E.A.; Neenan, T.X.; Reichmanis, E. *Proc. SPIE – Advances in Resist Technology and Processing XIII*, **1996**, 2724, 355.

5. Kunz, R.R.; Sinta, R.; Sworin, M.; Mowers, W. A.; Fedynyshyn, T.H.; Liberman, V.; Curtin, J. E. *Proc. SPIE – Advances in Resist Technology and Processing XVIII*, **2001**, 4345, 285.

6. Ito, H.; Wallraff, G.M.; Brock, P.J.; Fender, N.; Troung, H.D.; Breyta, G.; Miller, D.C.; Sherwood, M.H.; Allen, R.D. *Proc. SPIE – Advances in Resist Technology and Processing XVIII,* **2001**, 4345, 273.

7. Dammel, R.R.; Sakamuri, R.; Romano, A.R.; Vicari, R.; Hacker, C.; Conley, W.; Miller, D.A. *Proc. SPIE – Advances in Resist Technology and Processing XVIII*, **2001**, 4345, 350.

8. Hung, R.J.; Tran, H.V.; Trinque, B.C.; Chiba, T.; Yamada, S.; Sanders, D.P.; Conner, E.F.; Grubbs, R.H.; Klopp, J.M.; Frechet, J.M.J.; Thomas, B.H.; Shafer, G.J.; DesMarteau, D.D.; Conley, W.; Willson, C.G. *Proc. SPIE – Advances in Resist Technology and Processing XVIII*, **2001**, 4345, 385.

9. Fedynyshyn, T.H.; Kunz, R.R.; Sinta, R. F..; Sworin, M.; Mowers, W. A.; Goodman, R. B..; Doran, S. P. *Proc. SPIE – Advances in Resist Technology and Processing XVIII*, **2001**, 4345, 296.

10. Przybilla, K.-J.; Dammel, R.; Pawlowski, G. *German Patent* DE4207261A1, **1993**.

11. Goff, D. A.; Harris III, R.N.; Bottaro J.C.; Bedford, C.D. *J. Org. Chem.*, **1986**, *51*, 4711.

12. Results to be published.

13. Sinta, R.; Barclay, G.G.; Adams, T. G.; Medeiros, D.R. *Proc. SPIE– Advances in Resist Technology and Processing XIII*, **1996**, 2724, 238.

Chapter 6

157-nm Resist Materials Using Main-Chain Fluorinated Polymers

**Toshiro Itani, Hiroyuki Watanabe, Tamio Yamazaki,
Seiichi Ishikawa, Naomi Shida, and Minoru Toriumi**

Semiconductor Leading Edge Technologies, Inc. (Selete), 16–1 Onogawa,
Tsukuba, Ibaraki 305–8569, Japan

157-nm lithography is one of the most promising technologies for next-generation lithography. Due to the high absorbance at 157 nm, existing photoresist materials can not be used for 157-nm lithography. Fluoropolymers show good optical transparency and fluorinated resist materials have been intensively studied. Side-chain-fluorinated polymers have been shown a possibility of use for multi-layer resists. However they can not be applied to a single-layer resist, which is the most favored approach among potential industry. We have studied the main-chain-fluorinated polymers that meet the transparency requirement for use as single layer resists. They are two families composed of tetrafluoroethylene and monocyclic fluorocarbons. They showed good absorption coefficients of less than 1.5 μm^{-1} and fine dry etching durability. Positive-working resists were developed using these fluoropolymers and showed good sensitivity less than 10 mJ/cm^2. The resolution of the 80-nm dense pattern was determined at the film thickness thicker than 200 nm.

Introduction

157-nm lithography is one of the most promising technologies for next-generation lithography. Conventional organic materials used in the resist have fatal disadvantages of opaqueness at the exposure wavelength. Resist materials must meet the transparency requirement for use as single layer resists. High fluorine or silicon content can lead to a substantial increase in transparency at 157 nm. Therefore fluoropolymers are extensively studied. (1-6) Side-chain-fluorinated polymers based on alicyclic and aromatic structures have been studied. (1-4) These polymers inherited good dry-etching resistance in addition to good spinnability, adhesion, developability and so on. Their absorption coefficients are ca. 2 to 3 μm^{-1}. They are better than conventional materials and may be used as multi-layer resists. However they can not be applied to a single-layer resist.

Main-chain-fluorinated polymers were reported to have an absorption coefficient lower than 1.5 μm^{-1}. (5,6) They can be potentially used in a single-layer resist. In this paper we focus on the characteristics of the resist materials for 157-nm lithography: main-chain-fluorinated base resins containing tetrafluoroethylene (TFE) and monocyclic fluorocarbons.

Experimental

Materials

Fluoropolymers studied here were synthesized either at Asahi Glass Co., Ltd. or Daikin Industries, Ltd. through procedures described in the literatures (7,8). They are many kinds of fluoropolymers fluorinated at main-chain and side chain. As mentioned in the following sections, main-chain-fluorinated polymers show the better results. The fluoropolymers can be classified into two groups as shown in Scheme. One group is characterized by α-fluorination of polymer backbones and the presence of dry-etching-resistant norbornene (NB) as poly(TFE/NB/α-fluoroolefin) fluoropolymers (FP1). The other group includes new monocyclic fluoropolymers (FP2). The monocyclic rings are considered as resistant to dry etching. The synthesis procedures and the details of the preparation and chemistry are described.

Measurements

Polymer samples were dissolved in casting solvent of propylene glycol methyl ether acetate (PGMEA), or ethyl lactate (EL) with a solids content of 5-10%. Sample processing was carried out on a Si, calcium fluoride or quartz crystal at a film thickness of 100-300 nm with softbake of 130-150 °C. The vacuum ultra-violet (VUV) absorption properties of the sample films were recorded on a calcium fluoride substrate by using a single-beam VUV spectrophotometer, VU-201, of Bunkoh-Keiki Co., Ltd. Dry-etch measurements were done using an IEM etcher of Tokyo Electron, Ltd. The process parameters were under contact-hole etching conditions with a total pressure of 30 mTorr, a C_4F_8 flux of 11 sccm, an O_2 flux of 8 sccm, and an Ar flux of 400 sccm. Quartz crystal microbalance (QCM) method was used to study the dissolution characteristics of fluoropolymers in an aqueous tetramethylammonium hydroxide (TMAH) solution in a fashion similar to the reported procedure (9). Acidities of alkaline soluble functional groups in fluoropolymers were measured by potentiometric titration in dimethylsulfoxide non-aqueous solution (9). Resists used triphenylsulfonium triflate as sensitizer at approx. 5% of solids. Resist processing was carried out on organic bottom antireflection coating on Si at a film thickness of 100-300 nm with softbake of 130-150 °C. Exposures to resist films were performed in an open-frame F_2 laser exposure system, VUVES-4500 of Litho Tech Japan Co. on 1-cm^2 areas with doses ranging from 0.1 to 100 mJ/cm^2 to obtain the photospeed and contrast curve. To evaluate the resolution, the resists were patterned by using a Levenson-type strong phase-shift mask on a 157-nm laser microstepper (NA=0.6 and 0.85, σ =0.3). For lithographic evaluation, micrographs of fine patterns were obtained using a Hitachi S-5000 scanning electron microscope.

Results and Discussions

Characterization of Fundamental Properties of Fluoropolymers

We have measured the absorption coefficients at 157 nm of many polymers including more than 400 fluoropolymers. Figure 1 shows the histograms of fluoropolymers and non-fluoropolymers. The average absorption coefficient of fluoropolymers is ca. 2.3 μm^{-1}, which is smaller than 6.3 μm^{-1} of non-fluoropolymers. The fluorination makes polymers more transparent by 4 μm^{-1}.

Figure 2 shows the details of fluoropolymer bar chat in Figure 1. They are classified into fluoropolymers incorporating fluorine at their backbones and side-chains. The main-chain-fluorinated polymers have the good optical property transparent than side-chain-fluorinated polymers. Some fluoropolymers have the good absorption coefficient less than 1 μm^{-1}, which can not be achieved by fluorination at side-chains. Tetrafluoroethylene (TFE) is one of good monomers for backbone-fluorinated polymers. Figure 3 shows the TFE-content dependence of absorption coefficient of TFE-norbornene (NB) copolymers. The more TFE-content increases, the more transparent the copolymer becomes. Note that the TFE polymers have the potential of the absorption coefficient less than 1 μm^{-1}.

Figure 4 shows the dependence of optical absorption coefficients of the fluoropolymers on the fluorine concentration. Fluoropolymers generally become more transparent with increasing fluorine content. The second ordinate, or the "50%T thickness", shows the film thickness at the bottom of which the light intensity decreases to half the incident intensity of 157-nm light. Some FP1

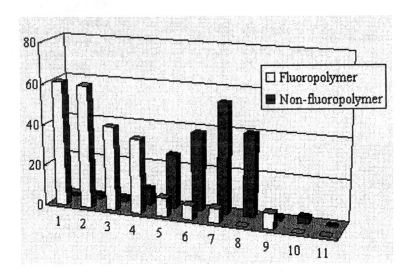

Scheme. Molecular structures of FP1 and FP2.

Figure 1. Absorption coefficient histogram of polymers.

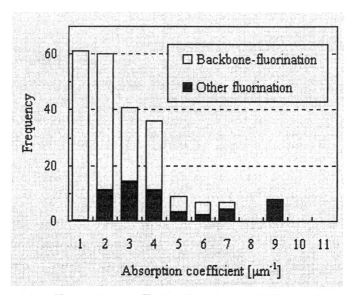

Figure 2. Absorption coefficient histogram of fluoropolymers.

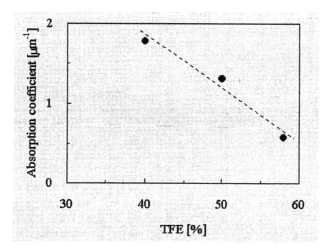

Figure 3. The dependence of absorption coefficient on TFE contents in
TFE-NB-polymers.

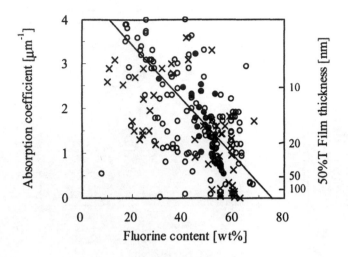

Figure 4. The dependence of absorption coefficient on fluorine contents in fluoropolymers. FP1: open dots, FP2: closed dots, Others: x.

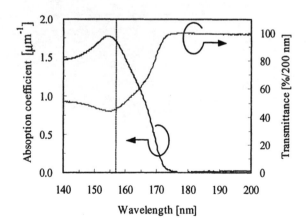

Figure 5. VUV spectrum of a perfluoropolymer, Cytop[©].

polymers achieve good transparency even at a thickness of more than 200 nm. The dashed line was determined by a least-squared fit using all the data. This figure shows that fluorine content of more than 40% is needed for 200-nm-thick films. Data are widely scattered in the figure. Figure 2 indicates also that some fluoropolymers have the larger absorption coefficients larger than 3 μm^{-1} in spite of its backbone fluorination. Note that main-chain fluorination not always make sufficiently polymers more transparent at 157 nm. Typical example is shown in figure 5, which is the Vacuum ultraviolet (VUV) spectrum of Cytop[©]. It is a fluoropolymer of perfluoroether with high fluorine content of 68 %. However it has the absorption coefficient of 1.7 μm^{-1}, which is not so good as the estimated value. The absorption spectrum has the maximum at 157nm and it increases the absorption coefficient at 157nm. In order to obtain the transparent fluoropolymers we have to optimize the molecular structure such as fluorinated positions in addition to the high fluorine content. We have studied two platforms of main-chain-fluorinated polymers for resist base resins. One family is a TFE-type polymer, FP1. The other family is a non-TFE-type polymer, FP2 as shown in Scheme. Open dots indicate FP1, closed dots FP2 and x others in Figure 4. FP1 shows the wide range of absorption coefficients, which is depending upon the molecular structures. FP2 shows the narrower range and has absorption coefficient smaller than the fitting line.

Figure 6 shows relative etching rates to the acrylate-base ArF resist. The etching rates were measured under the conditions of the fluorocarbon-based plasma oxide etching. FP1 polymers contain NB derivatives and some FP1 polymers achieve etching resistance higher than that of ArF resists. Poly(TFE/NB)s have a dry-etching rate 0.8-1.1 times larger than that of ArF resists, and some FP1 polymers are comparable in this respect to KrF resists. The etching rates of monocyclic FP2 polymers are scattered more widely than those of FP1 polymers. It indicates the weak dry-etching durability of the main structure in itself. Some FP2s have slightly lower etching selectivity but it is still comparable to that of acrylate-based ArF resists. The fluorine incorporation to the backbones such as TFE increases the durability (8). This is another advantage of main-chain fluorination.

Figure 7 shows dissolution behavior of FP1 and FP2, both of which dissolved away in the standard alkaline developer of 0.26N aqueous tetramethylammonium hydoxide (TMAH). The FP1 polymer dissolves as fast due to the larger acidic functional group than that of FP2. However FP1 shows the interface layer where it dissolves slower than FP2. This is one of the causes of difference of their resist resolving powers. Most of monomers of FP2 polymer contain the functional group soluble in the alkaline solution. The high contents of alkaline soluble groups may enable its good solubility in the standard developer.

Lithography Performances

To formulate a positive tone resist, fluoropolymers with protective functional groups were evaluated. These polymers underwent a chemically amplified change in solubility through a deprotection reaction. Figure 8 shows the contrast curves of the positive-working resists using FP1 and FP2. The photo-acid generator (PAG) was triphenylsulfonium (TPSTf). They showed no film-thickness loss in the unexposed areas and good clearing doses of 1.7 and 3.2 mJ/cm^2 for FP1 and FP2, respectively. They also had good contrast curves over two dose ranges and did not display negative tone behavior at an exposure dose of 100 mJ/cm^2.

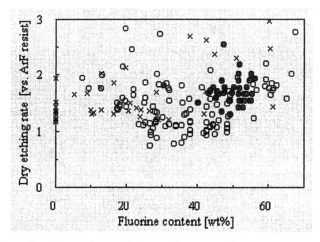

Figure 6. Dependence of dry-etching rate of fluoropolymers on fluorine atom content. FP1: open dots, FP2: closed dots, Others: x.

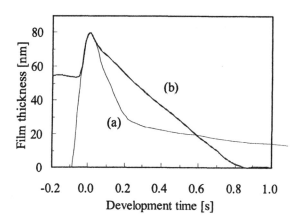

Figure 7. QCM traces of (a) FP1 and (b) FP2 .

Imaging Performances

Figure 9 shows lithographic results obtained using an FP1- and FP2-based resists. FP-1 has *tert*-butyl group as the protection group of 65%, and FP-2 has methoxymethyl group as the protection group of 15%. This imaging experiment was performed on a 0.6-NA microstepper, using a phase-shift mask at a 150-nm film thickness. The resist showed good imaging performance for dense patterns with 1-to-1 lines and spaces. Projection imaging yielded 90- and 80-nm patterns respectively. FP1-based resists lose resolutions. This footing in FP-1 patterns may be the result of the dissolution-rate retardation in the substrate region as shown in Figure 7. We are now improving the dissolution behavior or FP1 polymers to accomplish the higher resolution.

The FP2-based resist shows the delineated patterns with a round top. To improve the resist profile we increased the content of the protective functional group from 20% to 30%. The 30%-protected FP2-based resist achieves the better rectangular shape of fine patterns as shown in Figure 10.

These main-chain fluorinated have high transmittance at the exposure wavelength and can be used to obtain resist films thicker than 200 nm. For instance, Figure 11 shows the 95-nm 1:1 patterns delineated by a 0.6-NA microstepper using a phase-shift mask at a 300-nm film thickness. This a great advantage of main-chain-fluorinated polymers with high transmittance.

Conclusions

Main-chain-fluorinated polymers were investigated and these polymers have the great advantages of high transmittance with absorption coefficients of less than 1 μm^{-1} at 157 nm and good solubility without swelling during the development and fine dry-etching resistant comparable to ArF resist. These main-chain-fluorinated polymers can thus be used as single-layer resists to obtain films larger than 200 nm.

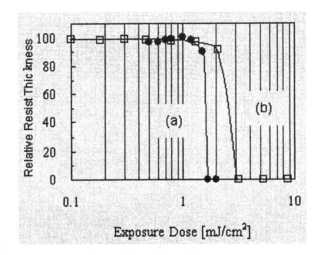

Figure 8. Sensitivity curves of (a) FP1 and (b) FP2-TPSTf systems.

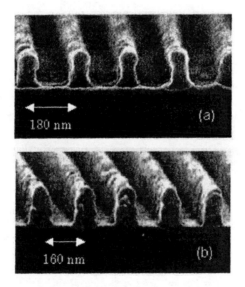

Figure 9. (a) 90-nm patterns in FP1 resist and (b) 80-nm in FP2 resist after development on a F_2 laser microstepper, NA 0.6, using a phase-shift mask. Film thickness is 150 nm.

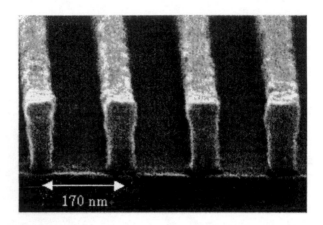

Figure 10. 85-nm features in FP2 resist with high content protective-group.

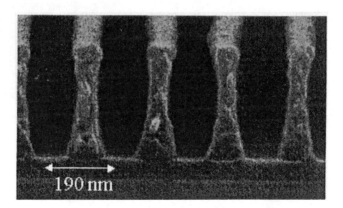

Figure 11. 95-nm patterns in FP2 resist with the film thickness of 300 nm.

References

1. Fedynyshyn, T. H.; Kunz, R. R.; Sinta, R. F.; Sworin, M.; Mowers, W. A.; Goodman, R. B.; Doran, S. P., *Proc. SPIE* **2001**, *4345*, p.296 and references therein.

2. Ito, H.; Wallraff, G. M.; Brock, P.; Fender, N.; Truong, H.; Breyta, G.; Miller, D. C.; Sherwood M. H.; Allen, R. D., *Proc. SPIE* **2001**, *4345*, p.273.

3. Hung, R. J.; Tran, H. V.; Trinque, B. C.; Chiba, T.; Yamada, S.; Sanders, D. P.; Connor, E. F.; Grubbs, R. H.; Klopp, J.; Frechet, J. M. J.; Thomas, B. H.; Shafer, G. J.; DesMarteau, D. D.; Conley, W.; Willson, C. G.; *Proc. SPIE* **2001**, *4345*, p.385 and references therein.

4. Schmaljohann, D.; Bae, Y. C.; Weibel, G. L.; Hamad, A. H.; Ober, C. K., *Proc. SPIE* **2000**, *3999*, p.330.

5. Crawford, M. K.; Feiring, A. E.; Feldman, J.; French, R. H.; Petrov, V. A.; Schadt III, F. L.; Smalley, R. J.; Zumsteg, F. C., *Proc. SPIE* **2001**, *4345*, p.428 and references therein.

6. Toriumi, M.; Ishikawa, S.; Miyoshi, S.; Naito, T.; Yamazaki, T.; Watanabe, M.; Itani, T., *Proc. SPIE* **2001**, *4345*, p.371 and references therein.

7. Shun-ichi Kodama, S.; Kaneko, I.; Takebe, Y.; Okada, S.; Kawaguchi, Y,; Shida, N.; Ishikawa, S.; Toriumi, M.; Itani, T., *Proc. SPIE* **2002**, *4690*, p.76.

8. Koh, M.; Ishikawa, T.; Araki, T.; Aoyama, H.; Yamashita, T.; Yamazaki, T.; Watanabe, H.; Toriumi, M.; Itani, T., *Proc. SPIE* **2002**, *4690*, p486.

9. Toriumi, M.; Itani, T.; Yamashita, J.; Sekine, T.; Nakatani, K., *Proc. SPIE* **2002**, *4690*, p191 and references therein.

Chapter 7

Correlation of the Reaction Front with Roughness in Chemically Amplified Photoresists

Ronald L. Jones[1], Vivek M. Prabhu[1], Darío L. Goldfarb[2],
Eric K. Lin[1], Christopher L. Soles[1], Joseph L. Lenhart[1,3],
Wen-li Wu[1], and Marie Angelopoulos[2]

[1]Polymers Division, National Institute of Standards and Technology,
Gaithersburg, MD 20899–8541
[2]IBM T. J. Watson Research Center, Yorktown Heights, NY 10598
[3]Current address: Sandia National Laboratories, Box 1411, MS 1411,
Albuquerque, NM 87185

Abstract

A model bilayer geometry is used to correlate the reaction front profile width with roughness after development in 0.26 N tetramethylammonium hydroxide aqueous base developer. The bilayer geometry utilizes a bottom layer of protected photoresist polymer with a top layer of deprotected photoresist loaded with photoacid generator. Neutron reflectivity measurements show that the reaction front profile broadens during post-exposure bake (PEB) times between 15 s and 90 s to a width approaching 150 Å. The subsequent development and atomic force microscopy experiments reveal an increase in nominal root-mean-squared (RMS) roughness as well as increased lateral length scale features with PEB time. While the form and size of the deprotection profile have been proposed as an important factor in line edge roughness (LER) formation, this study shows the connection of sidewall morphology to a measured deprotection profile.

Introduction

The influence of photogenerated acid diffusion on side-wall or line-edge roughness (LER) is an increasingly important problem for photoresist imaging. The push to reduce feature widths to dimensions on the order of 30 nm in the next decade, where tolerances are typically on the order of (1 to 5) %, dictates a reduction of LER tolerance to sub-nanometer levels (1). In addition to effects arising from optical blurring in the image projection (2-5), material factors contributing to LER include acid diffusion (6-8), photoresist chemistry (9,10), and developer characteristics (11). Formation of the line edge occurs through a process that includes projection of an optical mask image on a polymer-based thin film containing a photosensitive small molecule, termed a photoacid generator (PAG). Upon exposure, the photogenerated acid deprotects the polymer matrix, forming a base-soluble matrix that is selectively removed by a developer solution. The line edge is therefore created at an internal interface between protected and deprotected species. The factors contributing to LER can therefore be grouped into factors that define the internal deprotection interface (i.e. image blur, acid diffusion) and factors defining the selectivity of the dissolution process (i.e. developer concentration, photoresist/developer interaction). Recent work by our group has demonstrated an ability to measure this interface directly under normal processing conditions (12), along with ongoing studies of the early time dependence of the root mean square (RMS) roughness (13). The RMS roughness is observed to increase during early times, reaching a plateau. In this work, we probe the limit of small image blur to provide data directly connecting the breadth of the deprotection profile interface to the final surface morphology. The morphology is characterized by the lateral correlations of RMS, or the scan size dependence of RMS, rather than the total value of roughness.

To facilitate measurements of the deprotection profile, we follow a procedure outlined previously to produce a model pattern "sidewall" as a top surface (12-14). Here, the line edge of a chemically amplified resist is modeled using a bilayer prepared with a bottom layer of protected polymer and a top "feeder" layer of deprotected polymer loaded with PAG. Upon blanket exposure, the photogenerated acid diffuses across the interface, generating an interfacial profile of deprotected species that increases in width with post exposure bake (PEB) time. This study represents a direct experimental connection of the deprotection profile caused by acid diffusion and reaction to the final surface morphology. The profile is measured using high resolution neutron (NR) and x-ray (XR) reflectivity, while the surface morphology is characterized using atomic force microscopy (AFM).

88

Experimental

Materials

Bilayer structures were prepared on cleaned silicon wafers (approximately 3 mm thick and 75 mm diameter) as follows: 5 min exposure to oxygen plasma, followed by removal of native oxide layer by immersion into a solution of $(10 \pm 2)\,\%$ volume fraction HF and $(5 \pm 2)\,\%$ volume fraction NH_3F in ultra pure water for (15 ± 5) s. An oxide layer was regrown in a UV/Ozone chamber for (120 ± 1) s followed by priming with hexamethyldisilazane vapor (HMDS). The lower layer consisting of the deutero-poly(butoxycarboxy styrene) (d PBOCSt) ($M_{r,n} = 21000$, $M_{r,w}/M_{r,n} = 2.1$) was spin-coated from a propylene glycol methyl ether acetate (PGMEA) solution and post-apply baked (PAB) for 90 s on a 130 °C hotplate to remove residual solvent. The corresponding deprotected polymer, poly(hydroxystyrene) (PHOSt) ($M_{r,n} = 5260$, $M_{r,w}/M_{r,n} = 1.12$), was spin-coated from a 1-butanol solution directly onto the lower layer. The PHOSt layer is loaded with a 5 % mass fraction of the photoacid generator, di(*tert*-butylphenyl) iodonium perfluorooctanesulfonate. The bilayer is subjected to another PAB for 90 s at 130 °C. The model bilayer stack was exposed with a broadband UV dose of ≈ 1000 mJ/cm^2 to generate acid within the top PHOSt layer followed by PEB at 110 °C for varying times of 15 s, 20 s, 30 s, and 90 s. The original PHOSt layer and the soluble deprotected d-PBOCSt reaction products were removed (developed) by immersion in a 0.26 N tetramethylammonium hydroxide (TMAH) solution for 30 s followed by a rinse with deionized water. The use of deuterated PBOCSt facilitates the measurement of the deprotection profile using neutron reflectometry, described below. The approximate deprotection reaction of d-PBOCSt into PHOSt is shown schematically in figure 1. **Neutron Reflectivity**

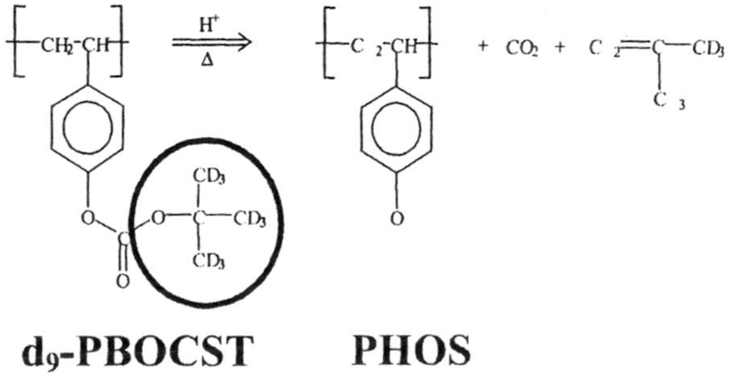

d$_9$-PBOCST   PHOS

Figure 1. Schematic of the deprotection reaction showing the protected polymer, PBOCSt, and deprotected analog, PHOSt. Shown encircled is the protecting group cleaved by the photogenerated acid.

The bilayer samples were measured both before and after aqueous base development by specular neutron reflectivity (NR) on the NG7 reflectometer at the National Institute of Standards and Technology Center for Neutron Research. The NR experiments measure the specular reflected intensity, such that the angle of incidence equals the angle of reflectance that defines a scattering wavevector q ($q = 4\pi\lambda^{-1}\sin(\theta/2)$), where λ is the neutron wavelength of 4.75 Å and θ is the angle of reflectance. The deprotection profile is then extracted from the data using a common modeling procedure. The details of these measurements are provided in a prior publication (12).

Figure 2. AFM tapping-mode images as function of scan size: (a) 10 x 10 μm, (b) (5 x 5) μm , (c) (2 x 2) μm, (d) (1 x 1) μm , (e) (0.5 x 0.5) μm , for a random copolymer o ƒPIIOSt with 20 % mass fraction PBOCSt (No PAG) No PAB, 0.14 N TMAH for 30 s.

Atomic Force Microscopy

The surface image of all samples was measured using a Digital Dimension 3000 atomic force microscope (AFM) in tapping mode. The acquired images were corrected with a plane-fit. RMS roughnesses were obtained using the DI software. As in prior reports (15), the RMS roughness was found to be scan-size dependent. Therefore, the morphology is characterized here using the Fourier components of the image. Fourier transformations of multiple topographic images were found to be isotropic in 2-dimensions and subsequently circularly averaged into a 1-D spectrum. These spectra were averaged to provide a statistical average over a large area of the sample. The final power spectrum provides the lateral structure, in which the image is understood in terms of the probability amplitude of lateral-length scale correlations versus wavevector, q (= $2\pi/d$, where d is a real space length scale). A complete power spectrum covering two-decades of length scale information, was prepared by superposition of Fourier-transform images of different AFM scan sizes ranging from (0.5 μm x 0.5 μm) to (10.0 μm x 10.0 μm). Discrete Fourier transforms often result in large uncertainty near summation limits due to sampling errors, finite size, and the pixel dimensions. In an effort to determine the appropriate range of

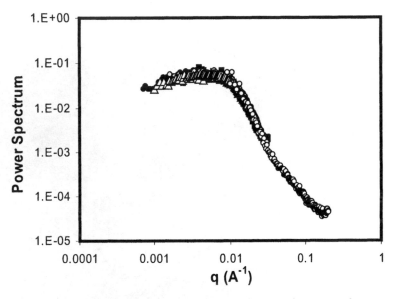

Figure 3. Fourier transform of images shown in figure 2 as function of scan size. Data shown outline the limits of q used for each scan size in this study.

wavevector for each scan size, a test film of a random copolymer of PHOSt with 20 % mass fraction PBOCSt was partially developed in 0.14 N TMAH and imaged (see figure 2). As shown in figure 3, Fourier transforms from 5 different scan sizes were then overlaid. By selecting ranges of Q where two overlapping data sets agreed to ± 20 % of the power spectral intensity, limits of minimum and maximum Q were established for each scan size. In this report, quantitative analysis of the power spectral intensity is not utilized, and therefore the intensity is arbitrarily shifted for clarity in the figures. The relative intensities within a scan have a maximum relative uncertainty of ± 20 % at low q values, with substantially smaller errors as q increases.

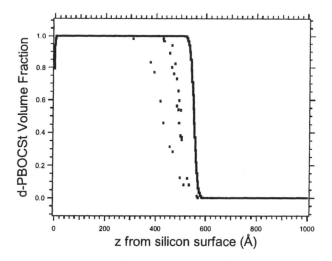

Figure 4. Neutron reflectivity results of reaction front profile Shown are the volume fraction of protected polymer (PBOCSt) as a function of distance from the substrate, z, for varying PEB times. Curves progress left with increasing PEB time, with the earliest (30 s) shown as a solid line, followed by data from 30 s, 60 s, and 90 s (dotted lines).

Results and Discussion

Reaction Front Profile

The reflectivity experiments are sensitive to gradients in the scattering length density of the thin film. For the case of NR, the deuterated protecting group has a different scattering length density than the top layer of protonated PHOSt. This scattering length density difference will lead to the ability to probe the deuterium labeled species composition profile. This is contrasted with XR, which measures differences in electron density, for which the density difference is insufficient to measure the bilayer structure. From model fitting of scattering length density profiles, we are able to measure the fraction of protected species throughout the film thickness for the case of NR, while for XR a single layer of average electron density adequately fits the data (see figure 4). In this paper, we restrict our discussion to the effect of the internal, or pre-developed, interfacial deprotection profile on the resulting roughness observed after development. The details of the internal and developed ᵔᵔ'ᴿ and XR experimental data and reaction kinetics are outside the scope of this ᴊer and will be presented elsewhere.

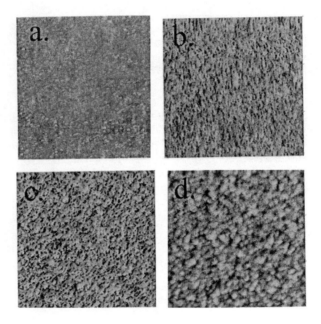

Figure 5. Tapping-mode AFM images from samples measured in figure 4. Shown are samples with PEB times of: (a) 0 s, (b) 15 s, (c) 30 s, (d) 90 s

Developed Surface Morphology

After the development of the bilayers of different PEB times, we measured the surface morphology using AFM. The surface images are presented in Fig. 5. The topographic images illustrate increased lateral structure during the initial stages of reaction front propagation. The result is a continuous progression from an intially featureless surface, with an RMS roughness less than 1 nm, to a nodular structure with an RMS roughness of [3.1 ± 0.5] nm. The nodular morphology is found to persist without significant change for all PEB times greater than 90 s. Similar morphologies have been reported at the line edge of patterned photoresists (2,13). With only a limited number of points, the time evolution of the RMS roughness is not addressed further here. Instead, the morphology is characterized through lateral correlations of height.

The Fourier transform imaging quantifies the increase in lateral inhomogeneity through a characteristic wave vector, labeled here as Q^*. The form of the data in figure 5 is similar to that found in studies of homopolymer surfaces, where a power law dependence at large Q transitions to a plateau at low Q vectors. Using a two regime representation, the transition between power law and plateau behavior occurs over a region centered at Q^*. The value of Q^* then signifies a characteristic length scale of the nodular structure observed in figure 4. In figure 6, a shift of Q^*, signified by a the characteristic "knee" in the logarithmic power spectral density, to lower values of Q is observed with increasing PEB time. Further increases in PEB time did not result in significant changes in Q^*.

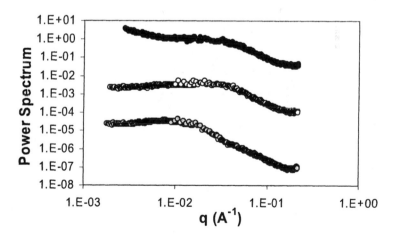

Figure 6. Fourier transform of topographic data from figure 5. Shown are data from PEB of 15 s (top), 30 s (middle), and 90 s (bottom).

Reaction-Front Width and RMS roughness

A composite plot of the reaction front profile width and RMS roughness as functions of the PEB time are shown in Fig. 7. This provides the opportunity to correlate the increased observed surface roughness with the development of an increasingly broadened reaction front profile. The deprotection profile widths range from an intially sharp interface of 20 Å to a broadened 140 Å. However, the development of this internal interface leads to RMS roughness of only (10 to 25) Å. Thus, design criteria for PEB processing that minimizes compositional broadening could serve as a controllable goal to minimize LER.

We should emphasize that the origin of the observed increase in surface roughness reflects a result consistent with true line-edge roughness. Fig. 8 provides a schematic of the types of interface geometries used to understand LER. The first is the typical single layer method in which either a blend or resist formulation is examined for roughness after typical wafer processing. Partially developing a uniform film would not include the interaction of the developer with the reaction front gradient. The second geometry is of the type used in this

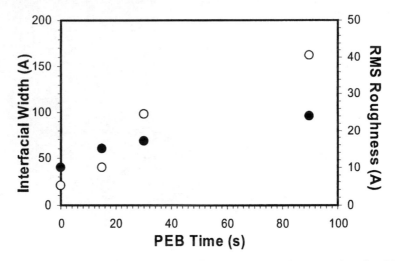

Figure 7. Composite plot of neutron reflectivity reaction front interfacial width (open circles) and AFM RMS roughness (filled circles) versus post-exposure bake time. PEB temperature is 110 °C. AFM roughness values have a relative uncertainty of ± 2 nm, while the deprotection profile width has a relative uncertainty of ± 3 nm.

work. The use of a bilayer to create a gradient is similar in scope to efforts to create gradients using an exposure wavelength that cannot fully penetrate the resist film. The advantage of the bilayer is the ability to create well defined starting points suitable to studies where image blur is considered negligible compared to the effects of acid diffusion. The last experimental method involves the creation of a true line edge. While this case is the most relevant to semiconductor processing, it is the least convenient geometry to measure. Only in the last two geometries are composition gradients present.

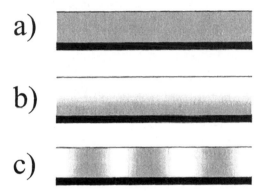

Figure 8. Model film geometries to study photoresist roughness. (a) Single layer with resist and/or blend of protected and deprotected components with additives. (b) This work, bilayer with deprotected bottom layer and top deprotected PAG feeder layer. (c) True profile prepared with mask to study resist performance.

The movement of the lateral length scale observed by AFM with the deprotection profile width is consistent with the model proposed by Schmid et al. (16). Here, the spatial distribution of deprotection is in great part a determining factor of the final line edge morphology. As the deprotection profile width increases, the level of inhomogeneity experienced by the developer may also increase, resulting in a larger length scale of line edge roughness. Other workers have attributed similar forms of line edge morphology to polymer aggregates present in chain scission resists during the developing step (17, 18). In our experiments, the evolution of the surface morphology follows the development of the reaction front. This evolution is not entirely consistent with preformed aggregates. However, in our case the reaction front width of 140 Å does not exceed the lateral scale morphology, thus we can not rule out that such aggregation could exist, or be enhanced by the development process.

Conclusions

The magnitude of lateral correlations in average line edge, or sidewall, roughness was measured as a function of the size of the deprotection profile width. As the deprotection profile broadens during the initial stages of post exposure baking, both the overall size of the RMS roughness and the characteristical lateral length scale defining the morphology were found to increase. For PEB times longer than 90 s at 110 °C, the characteristic length reaches a plateau and becomes invariant with PEB time (11). The dependence on deprotection profile width complements prior reports of the PEB time dependence of RMS roughness, suggesting a mechanism of lateral correlation development and RMS roughness.

Acknowledgements

This work was supported by the Defense Advanced Research Projects Agency under grant N66001-00-C-8803 and the NIST Office of Microelectronics Programs. J.L.L and V.M.P. acknowledge support through the National Research Council-National Institute of Standards and Technology Postdoctoral Fellowship Program. The authors would like to thank David R. Medeiros for the synthesis of h-PBOCST, B.C. Trinque, S. D. Burns, and C. G. Willson for the synthesis of d_9-PBOCST and R.W. Kwong for the GPC measurements.

References

1. *2001 National Technology Roadmap for Semiconductors*, The Semiconductor Industry Association: San Jose, CA (2001).
2. G. W. Reynolds, J. W. Taylor, J. Vac. Sci. Technol. B. **17**, 334 (1999).
3. G. P. Patsis, N. Glezos, I. Raptis, E. S. Valamontes, J. Vac. Sci. Technol. B. **17**, 3367 (1999).
4. W. D. Hinsberg, F. A. Houle, M. I. Sanchez, G. M. Walraff, IBM J. Res. & Dev. **45**, 667 (2001)
5. F. A. Houle, W. D. Hinsberg, M. I. Sanchez, Macromolecules **35**, 8591 (2002).
6. G. M. Schmid, M. D. Smith, C. A. Mack, V. K. Singh, S. D. Burns, C. G. Willson, Proc. SPIE **3999**, 675 (2000).
7. Y.-S. Kim, Y.-H. Kim, S. H. Lee, Y.-G. Yim, D. G. Kim, J.-H. Kim, Proc. SPIE **4690**, 829 (2002).
8. M. Yoshizawa, S. Moriya, J. Vac. Sci. Technol. B. **20**, 1342 (2002).
9. Q. Lin, R. Sooriyakumaran, W.-S. Huang, Proc. SPIE **3999**, 230 (2000).
10. Q. Lin, D. L. Goldfarb, M. Angelopoulos, S. R. Sriram, J. S. Moore, Proc. SPIE **4345**, 78 (2001).

11. L. W. Flanagin, V. K. Singh, C. G. Willson, J. Vac. Sci. Technol. B. **17**, 1371 (1999).
12. E. K. Lin, C. L. Soles, D. L. Goldfarb, B. C. Trinque, S. D Burns, R. L. Jones, J. L. Lenhart, M. Angelopoulos, C. G. Willson, S. K. Satija, W.-L. Wu, Science, **297**, 372 (2002).
13. D. L. Goldfarb et al., in preparation (2002).
14. D. L. Goldfarb, M. Angelopoulos, E. K. Lin, R. L. Jones, C. L. Soles, J. L. Lenhart, W.-L. Wu, J. Vac. Sci. Technol. B. **19(6)**, 2699 (2001).
15. D. He, F. Cerrina, J. Vac. Sci. Technol. B **16**, 3748 (1998)
16. G. M. Schmid, M. D. Smith, C. A. Mack, V. K. Singh, S. D. Burns, C. G. Willson, Proc. of the SPIE **4690**, 381 (2002).
17. T. Yamaguchi, H. Namatsu, M. Nagase, K. Yamazaki, K. Kurihara, Appl. Phys. Lett. **71**, 2388 (1997).
18. T. Yoshimura, H. Shiraishi, J. Yamamoto, S. Okazaki, Appl. Phys. Lett. **63**, 764 (1993).

Chapter 8

Utilizing Near Edge X-ray Absorption Fine Structure to Probe Interfacial Issues in Photolithography

Joseph L. Lenhart[1], Daniel A. Fischer[2], Sharadha Sambasivan[2],
Eric K. Lin[2], Christopher L. Soles[2], Ronald L. Jones[2], Wen-li Wu[2],
Darío L. Goldfarb[3], and Marie Angelopoulos[3]

[1]Sandia National Laboratories, Box 5800, MS 1411,
Albuquerque, NM, 87185 (jllenha@sandia.gov)
[2]National Institute of Standards and Technology,
Gaithersburg, MD 20899–8541
[3]IBM T. J. Watson Research Center, Yorktown Heights, NY 10598

Control of the shape, critical dimension (CD), and roughness is critical for the fabrication of sub 100 nm features, where the CD and roughness budget are approaching the molecular dimension of the resist polymers. Here we utilize near edge X-ray absorption fine structure (NEXAFS) to provide detailed chemical insight into two interfacial problems facing sub 100 nm patterning. First, chemically amplified photo-resists are sensitive to surface phenomenon, which causes deviations in the pattern profile near the interface. Striking examples include T-topping, closure, footing, and undercutting. NEXAFS was used to illustrate that the surface extent of deprotection in a model resist film can be different than the bulk deprotection. Second, line edge roughness becomes increasingly critical with shrinking patterns, and may be intimately related to the line edge deprotection profile. A NEXAFS technique to surface depth profile for compositional gradients is described with the potential to provide chemical information about the resist line edge.

Introduction

Control of the shape, critical dimension (CD), and line edge roughness (LER) is essential for the fabrication of sub-100 nm features, where the CD and roughness budget are approaching the molecular dimension of the resist polymers (1). With shrinking pattern sizes the performance of chemically amplified photo-resists will become increasingly prone to interfacial or surface phenomena, which cause deviations in the pattern profile near the interface. Striking examples include T-topping, closure, footing, and undercutting. In T-topping or closure, the developed lithographic pattern contains an insoluble skin at the resist – air interface (2). In footing and undercutting the developed pattern broadens or shrinks respectively near the substrate surface, typically an anti-reflective coating (3). These types of pattern deviations are extremely undesirable and may result in poor pattern transfer to the substrate during the reactive ion etch. In addition, line edge roughness that is acceptable for current patterning dimensions will be unacceptable in smaller patterns. It is therefore important to develop and utilize new tools to probe the interfacial composition and structure of photo-resist films. Here we demonstrate the utility of NEXAFS for providing information about lithographic interfaces, focusing initially on the t-topping / closure issue and probing the extent of deprotection at the resist surface. Second, a NEXAFS technique is described to surface depth profile in a model line edge region offering the potential to provide detailed chemical information about surface compositional gradients.

Experimental

Materials and Methods. The model resist solution was composed of 0.7 g of protected polymer poly(tertbutyloxy- carbonyloxy- styrene, $M_{n,r}$=15,000) (PBOCSt) mixed with 0.035 g (0.05 mass fraction of PFOS relative to the polymer) of the photo acid generator, bis(p-tert-butylphenyl) iodonium perfluorooctanesulfonate (PFOS). This mixture was dissolved in 20 mL of propylene glycol methyl ether acetate (PGMEA). The resist solution was spun cast onto silicon wafers at 1500 rpm for 60 s and then post apply baked (PAB) for 60 s at 100 °C. The PBOCSt / PFOS films were blanket exposed to ultra-violet radiation from a broadband source with wavelengths ranging between (220 and 260) nm with a total dose of 500 mJ/cm^2. After exposure the films were post exposure baked (PEB) at 100 °C. Poly(hydroxystyrene), $M_{n,r}$=5,000,

(PHS) / PFOS films were made according to the same procedures described above.

NEXAFS. NEXAFS measurements were conducted at the U7A beam line of the National Synchrotron Light Source at Brookhaven National Laboratory. A monochromator with 600 line/mm grating, providing ± 0.15 eV resolution, was used for all the NEXAFS spectra. The monochromator energy scale was calibrated by the carbon K-edge π^* transition of graphite at 285.35 eV. All the spectra were recorded at room temperature in the NIST – Dow material characterization chamber (4) at 10^{-6} Pa. The spectra were normalized to the incident beam intensity, I_0, by collecting the total electron yield intensity from a gold coated 90% transmitting grid placed in the incoming X-ray beam path. The carbon fluorescence-yield intensity was measured utilizing a differentially pumped, UHV compatible proportional counter filled with 200 Torr of P-90 (90% methane, 10% argon) in an energy dispersive mode (5) to reduce background fluorescence from other elements. Surface sensitive partial electron yield measurements were made (probe depth of approximately 1 to 6 nm) by applying a negative bias on the entrance grid of the channeltron electron detector. For the carbon K-edge spectra (260 to 330) eV, the electron yield detector was set with a negative bias of 150 eV. The spectra were collected with the incident beam at the magic angle (54.7°) relative to the sample in order to remove any polarization dependence. For the NEXAFS spectra in this paper the experimental standard uncertainty in the peak position is similar to the grating resolution of ± 0.15 eV. The relative uncertainty in the NEXAFS intensity is less than ± 5% and was determined by multiple scans on a sample.

NEXAFS at the Resist-Air Interface

Figure 1 shows a schematic depicting the principles of NEXAFS. The sample is exposed to tunable plane polarized, monochromatic X-ray radiation from a synchrotron light source. In these experiments, the incident radiation is scanned over the carbon K-edge region, an energy range from (260 to 330) eV. X-rays are preferentially absorbed by the sample when the incident radiation is at the appropriate energy to allow the excitation of a core shell electron to an unoccupied molecular orbital. During electronic relaxation Auger electrons and characteristic fluorescence photons are released. The electronic relaxation processes may release more than one electron. These electrons can only escape from the top surface of the sample (1 to 10) nm. The photons are detected from approximately 200 nm within the sample. Because the characteristic binding

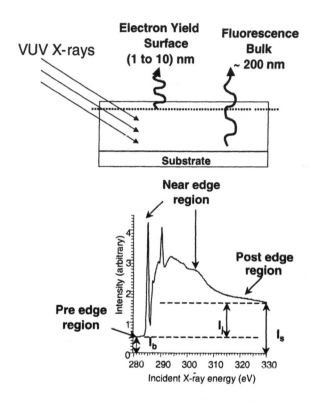

Figure 1. A schematic is shown of the NEXAFS experiment and typical spectra.

energies, carbon, nitrogen, oxygen, and fluorine core electrons are well separated in energy, giving NEXAFS elemental sensitivity. In addition, due to the well-defined energy gap associated with a core shell / unoccupied orbital transition, NEXAFS is also sensitive to the bonding characteristics of the atom (6).

Figure 1 also shows a carbon K-edge electron yield NEXAFS spectrum for PBOCSt. In the pre-edge region, the incident radiation is weakly absorbed by the sample. The intensity in this region, I_b, is the background signal (often from the substrate and sample, lower energy absorption edges, orbital transitions other than core level transitions, etc.). Above the carbon K-edge (285 eV), the signal intensity (electron or fluorescence yield) increases when the incident radiation is strongly adsorbed by the sample. In the near edge region, the peaks represent chemical bonding structure in the sample because the emission signal increases when the incident energy is the appropriate energy to cause an electron transition from the core 1s orbital to an unoccupied molecular orbital. The absorption edge represents the ionization of the core shell electron to the continuum. The edge jump, I_j, is defined as I_s-I_b. In the post edge region the signal intensity, I_j, represents total amount of carbon (since the scan is over the carbon K-edge energy range) in the sampling volume. All the NEXAFS spectra in this paper are pre-edge jump normalized to zero, by subtracting I_b from the spectrum.

It is appropriate to discuss the advantages and disadvantages of NEXAFS in the context of analysis tools more conventional to the microelectronics industry. X-ray photoelectron Spectroscopy (XPS) can provide similar surface information to NEXAFS, since the binding energy of the photo-ejected electron is dependent on the atomic number of the element as well as the type of chemical bonds that are present. However, in NEXAFS we directly monitor the transition of a core level electron into an unoccupied molecular orbital. This transition energy is well defined, and so NEXAFS will typically provide better bonding sensitivity. In addition, NEXAFS utilizes a high intensity synchrotron light source, allowing for monochromatic incident radiation. The monochromatic excitation provides excellent peak resolution. In XPS a standard source has a comparably broad incident wavelength range and the bonding peaks are usually extremely convoluted making data interpretation difficult. An additional advantage of NEXAFS is that the fluorescence yield (sensitive to bulk of film) can be collected simultaneously allowing a direct comparison between surface and bulk chemistry on the same sample. The disadvantage of NEXAFS is that the technique requires a synchrotron light source and thus will not be "in house".

Dynamic secondary ion mass spectrometry (DSIMS) and Rutherford back scattering (RBS) are techniques that can provide information about composition profiles in polymer films. Both techniques provide elemental sensitivity, but neither will provide chemical bonding information as NEXAFS does, since both techniques rely on mass differences as a contrast mechanism. The depth profile is obtained on the basis of elemental composition profiles in the film. In addition, the resolution of these techniques is ~ 50 nm so the surface sensitivity is poor. The advantage of DSIMS and RBS is that the elemental depth profile can be obtained throughout the film thickness, while for NEXAFS a surface depth profile can only be achieved over the first 10 nm. The fluorescence yield in NEXAFS provides bulk sensitivity but does not provide a depth profile through the film.

Ellipsometry is another common technique in the semiconductor industry. It provides an accurate measure of the film thickness. Since many resist shrink upon deprotection due to a loss of volatile components, ellipsometry has been used do monitor the extent of deprotection in the resist. However, the technique does not provide direct information about chemical bonding. Fourier transform infrared (FTIR) spectroscopy provides the best detail about chemical bonding, however it is a bulk technique with little surface sensitivity. Moderate surface sensitivity can be achieved by utilizing total internal reflection FTIR, however the evanescent field will still propagate ~ 500 nm away from the surface making it difficult to detect true surface chemistry changes in a resist film.

In this work we present NEXAFS as a tool that will compliment (not replace) the array of techniques common to the microelectronics industry. The combined advantages of simultaneous surface and bulk sensitivity, bonding and chemical sensitivity, and peak resolution are unique to NEXAFS and difficult to accomplish with other techniques.

Figure 2 shows the carbon edge NEXAFS spectra for the neat components used in our model resist system. The spectra are vertically offset for clarity. The top spectrum is for the protected polymer, PBOCSt. The peak at 285.0 eV reflects the π^* transition from the carbon-carbon aromatic bonds in the styrene ring. At 290.3 eV is a peak associated with the protective group, specifically the π^* transition of the carbon-oxygen double bond from the carbonyl group. The middle spectrum, for the PFOS photo-acid generator, also displays a sharp carbon-carbon π^* transition similar to PBOCSt. However, the broad peaks between (292.0 and 298.0) eV are due to σ^* transitions for carbon-fluorine bonds (292.0 and 298.0) eV and carbon-carbon bonds, 295.0 eV, on PFOS. The bottom spectrum is for the deprotected polymer, PHS, which also contains the strong π^* transition at 285.0 eV. Distinct peaks can be used to detect the individual resist components. For example, the peak at 290.3 eV in PBOCSt, associated with the protective group, is not present in PHS or PFOS, allowing the direct monitoring of the de-protection reaction. Also the carbon-fluorine peaks in PFOS are not present in the other two spectra, although they may partially overlap with the carbon-carbon σ^* transition. Since the spectra are measured over the carbon K-edge, the peaks in Figure 2 represent core level transitions from the C1s orbital to the unoccupied orbital.

Figure 3 compares the electron and fluorescence yield for the PBOCSt / PFOS films after various processing conditions. By monitoring the C=O π^*

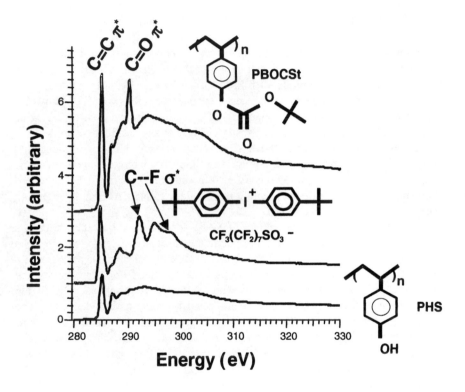

Figure 2. The electron yield NEXAFS spectra are shown for the pure resist components. The top is for PBOCSt. The middle is for PFOS. The bottom is for PHS.

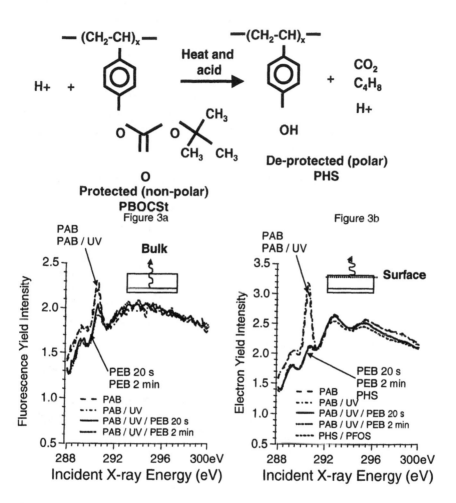

Figure 3. a) The fluorescence yield spectrum (bulk) is shown for the PBOCSt / PFOS films after different processing. b) The electron yield spectrum (surface) is shown for the same PBOCSt / PFOS films. Comparison of the electron and fluorescence yield shows that the surface reaction rate is faster than the bulk. The top illustrates the deprotection reaction of PBOCSt to PHS.

transition from the carbonyl on PBOCSt, the extent of deprotection can be followed during processing. A schematic of the deprotection reaction is shown in Figure 3. In Figure 3 the spectra are both pre and post edge jump normalized. The post edge jump normalization involves dividing the pre-edge jump normalized spectra by the edge jump intensity (I_j from Figure 1). This eliminates the spectral dependence on total carbon content in the sampling volume, thus changes in the NEXAFS peak intensity are due to chemical changes in the system. The films in Figure 3 were treated with typical resist processing conditions including; post apply bake (PAB) for 60 s at 100 °C; UV exposure 500 mJ/cm^2 from a broadband source 220 nm to 260 nm; post exposure bake (PEB) at 100 °C for various times. Figure 3a shows the fluorescence yield (bulk) spectrum from the samples. A strong carbonyl peak is present in the PAB, and PAB / UV treated samples, which overlap each other. After a 20 s PEB at 100 °C, the carbonyl peak at 290.3 eV decreases, but not completely. Even after 2 min PEB, the peak at 290.3 eV is not completely gone.

Figure 3b shows the electron yield (surface) spectra for the same PBOCSt / PFOS films. After both the PAB and a PAB + UV, the carbonyl π^* transition at 290.3 eV is large in the electron yield (Figure 3b) indicating the polymer is still protected. However, after a short 20 s PEB at 100 °C, the carbonyl peak completely disappears in the electron yield indicating complete deprotection at the film surface. Also shown in Figure 3b is the curve for the PBOCSt / PFOS film after 2 min PEB, and a PHS / PFOS film after a PAB. These curves overlap with the PBOCSt / PFOS film after the 20 s PEB in the carbonyl region near 290.3 eV, verifying that complete surface deprotection occurs in the first 20 s of post exposure baking. Comparison of the electron yield with the fluorescence yield spectrum clearly illustrates that the surface reaction rate is faster than the bulk. In addition, the electron yield spectra exhibit strong carbon-fluorine σ^* peaks between (292 and 298) eV from the PFOS. Since these peaks are not observed in the fluorescence yield spectra this illustrates PFOS segregation to the film surface. This surface segregation of the PFOS leads to a higher acid content near the air interface upon UV exposure and increases the deprotection reaction rate. The PFOS is present in small quantities in the bulk of the film (0.05 mass fraction of PFOS relative to polymer, 0.013 mole fraction of PFOS relative to PBOCSt monomers). At small PFOS concentration, the large carbon background from the PBOCSt polymer will dampen the contribution of the C-F σ^* peaks from PFOS. Since the C-F peaks are easily observable in the electron yield but not in the fluorescence yield, this clearly illustrates surface enrichment of the PFOS. We are currently developing a technique to extract the surface composition quantitatively from the electron yield spectra, by using a linear combination of the pure component spectra.

A potential concern about the NEXAFS experiments is the sensitivity of the model photo-resist to soft X-ray radiation. To test this we executed multiple

NEXAFS scans in the same spot. No chemical changes were observed with NEXAFS due to the multiple measurements, illustrating that the surface chemistry effects (observed in Figure 3) were not associated with X-ray induced degradation of the films.

A possible explanation for the differences in the surface and bulk reaction rates is the UV dose profile as a function of film thickness. This could potentially lead to a gradient in acid concentration through the film thickness, with higher acid content at the air interface due to higher doses near the resist surface. However, this is an unlikely explanation. Typical exposure doses for PBOCSt based resist are in the range of 20 to 50 mJ/cm^2. We used a dose of 500 mJ/cm^2. In addition, the film thickness here was ~ 200 nm, much thinner than the typical PBOCSt based resist thickness of roughly 1 μm. Considering the relatively high dose and thin film thickness, all the PFOS molecules in the film are likely converted to acid. Therefore a gradient in acid concentration due solely to a dose profile in the film does not explain the observed changes in surface deprotection rates. The change in surface acid content leading to faster surface deprotection must be due to a segregation of photo-acid generator at the air interface as Figure 3b illustrates.

In Figure 3, no delay times were incorporated between successive processing steps: spin coat, PAB, UV exposure, PEB. However, time delays between the various steps can have a significant impact on the resultant lithographic patterns. Figure 4 shows the carbon K-edge fluorescence (Figure 4a) and electron yield (Figure 4b) spectra for a PBOCSt / PFOS film after the PAB only, and after a PAB + UV + PEB sequence. The NEXAFS spectra in Figure 4a,b are from the same sample. However, for this sample a 10 min delay time was incorporated between the UV exposure and PEB. This is called a post exposure delay (PED). In the carbon K-edge fluorescence yield spectra, a carbonyl peak at 290.3 eV is observed in the PAB film. After UV exposure, 10 min PED, and a 2 min PEB at 100 °C, the peak area has dramatically decreased, indicating deprotection in the bulk of the resist film (Figure 4a). In the carbon edge electron yield spectra, the peak decreases only slightly after UV exposure, PED, and PEB, indicating incomplete deprotection at the film surface (Figure 4b).

The mechanisms leading to the incomplete surface deprotection reaction due to PED have not been determined and are a focus of our current research. One possible explanation is acid neutralization in the resist film due to atmospheric contaminants. Nalamasu et al. showed that the post exposure delay time was critical to the performance of chemically amplified resists (7). A PED of several minutes lead to an aqueous-base insoluble residue at the resist / air interface, while longer PEDs prevent the lithographic image from being developed (7). It was shown that resist performance deteriorated dramatically in basic environments, but could be improved by controlling the processing atmosphere or coating the resist with a base-neutralizing (weakly acidic) polymer layer (7). Incomplete deprotection near the resist / air interface was suggested as the cause of the insoluble residual layer (7). MacDonald et al. also showed that airborne amine contaminants degrade resist performance by leading to the formulation of a thin insoluble skin at the resist / air interface (8). Hinsberg et al. illustrated that the extent of base contamination in a resist film depends on the polymer solubility parameter, and the temperature difference between the post apply bake and the polymer glass transition (9). So the extent of resist contamination will depend on the polymer-contaminant interaction as well as the physical and thermal properties of the resist films.

PED is considered to be a critical factor in t-topping (7,10). These experiments illustrate that a PED can lead to incomplete surface deprotection. The PAB temperature of 100 °C for these PBOCSt / PFOS films was below the glass transition of bulk PBOCSt. By comparison to the work of Hinsberg et al. (9) this would lead to an uptake of atmospheric contaminants by the resist film, since more contaminant absorbs in resists with PAB temperatures well below the bulk polymer T_g. It is interesting that the incomplete surface deprotection was observed in these PBOCSt / PFOS films, where despite having significant excess PFOS at the film surface, atmospheric contamination can still neutralize the excess surface acidity.

Surface Depth Profiling with NEXAFS

The electron yield signal is surface sensitive. By adjusting a negative voltage bias on the electron yield detector, different effective surface sampling depths can be probed. Figure 5 shows a schematic of the process. When the polymer film is excited by the incident X-ray radiation, the entire region of the film that absorbs photons also emits electrons. The electrons emitted deep within the film cannot escape. Only the electrons emitted near the top (1 to 10) nm from the film surface have enough energy to escape the surface potential. The electron yield detector has a grid where a negative voltage bias can be placed across the grid. The electrons that escape the surface of the film, but were emitted from furthest within the film, will be low in energy due to inelastic interactions with other atoms. These low energy electrons will not have enough kinetic energy to pass the negative detector bias and will not be sensed. If the negative detector bias is gradually increased, progressively higher energy electrons are detected, and the effective electron yield sampling depth gets closer to the film surface.

We take advantage of this surface depth profiling capability in order to study the chemical composition profile of a model developed line edge region. For these experiments, bilayer samples of PBOCSt and PHS were spun cast onto

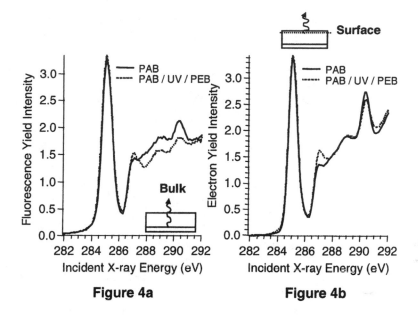

Figure 4. The carbon K-edge fluorescence yield spectra (Figure 4a), and carbon K-edge electron yield spectra (Figure 4b) are shown for the PBOCSt / PFOS film after PAB (solid line) and after UV exposure and a 2 min post exposure bake at 100 °C (dotted line). A post exposure delay of 10 min was incorporated in the processing.

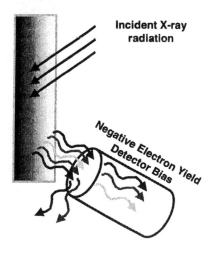

Figure 5. A schematic is shown of NEXAFS surface depth profiling.

silicon wafers as discussed in detail elsewhere (11,12). A schematic of these bilayer samples is shown in Figure 6. The wafers were cleaned by immersion in sulfuric acid and hydrogen peroxide solution followed by a rinse in deionized water. A hydrophobic surface was generated by treating the wafers with hexamethyldisilazane (HMDS) vapor in a vacuum oven. The bottom PBOCSt layer was spun cast from solution with PGMEA and soft baked for 60 s at 130 °C. A top layer of PHS (with a 5% mass loading of PFOS) was spun cast on PBOCSt from a solution of n-butanol. The samples were exposed to UV radiation, generating acid in the PHS feeder layer, and post exposure baked for various times at 90 °C. During PEB, the acid diffuses into the PBOCSt under-layer, and initiates a diffusion / deprotection front that propagates into the under-layer. After PEB, the soluble top portion of the bilayer film was removed (developed) by immersion in aqueous 0.26 N tetra-methyl-ammonium-hydroxide (TMAH) solution, leaving a residual deprotection profile in the final developed under-layer. NEXAFS measurements were then conducted on the developed bilayer (step number 5 in Figure 6) samples as a function of the electron yield detector bias. We suspect that the breadth of the diffusion / reaction front in the line edge region will impact the development process and corresponding line edge roughness. Therefore it is important to develop techniques to measure the composition profile of the line edge region. We start by utilizing NEXAFS surface depth profiling on the model bilayer interfacial regions.

Figure 7 shows NEXAFS pre- and post-edge jump normalized spectra in the carbonyl absorption region, between (288 and 292) eV, for the developed bilayer samples after various PEB times at 90 °C and development in TMAH solution. For these NEXAFS spectra, the detector bias was fixed at −200 eV (sampling depth of roughly 3 nm, or three monomeric layers). Since the electron density of both PHS and PBOCSt are similar, fixing the detector bias at −200 eV fixes the surface sampling volume to roughly 3 nm. The top spectra is for the PBOCSt / PHS bilayer with no PEB and after development in TMAH. Since no PEB was done the photo-generated acid in the PHS feeder layer does not diffuse into the PBOCSt under-layer. During development in TMAH, the top PHS layer is rinsed off, leaving the pure PBOCSt under-layer. In that case no deprotection occurs in the PBOCSt under-layer, and the carbonyl absorption is large. However, after incorporating a PEB, the acid in the PHS feeder layer can diffuse into the PBOCSt under-layer initiating the propagation of a deprotection – acid diffusion front into the under-layer. Upon development in TMAH, the under-layer will have a residual deprotection profile on the film surface due to this acid diffusion – deprotection across the PBOCSt – PHS interface. The carbonyl absorption clearly decreases in the developed bilayers with increasing PEB times, indicating that the average extent of deprotection in the surface sampling volume is increasing with the PEB time. Clearly the surface composition of the developed bilayer is changing with PEB times. Two scenarios can explain this

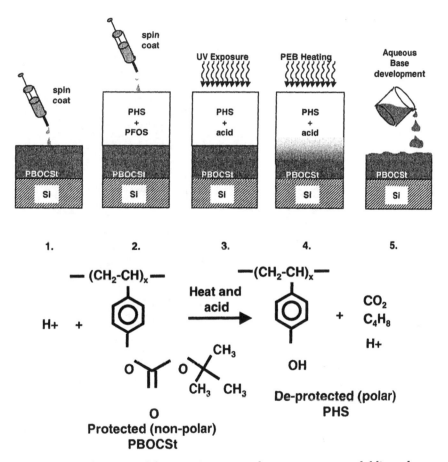

Figure 6. A schematic of the experiments used to generate a model line edge region. NEXAFS depth profiling was conducted on the developed bilayer sample, stage 5 above.

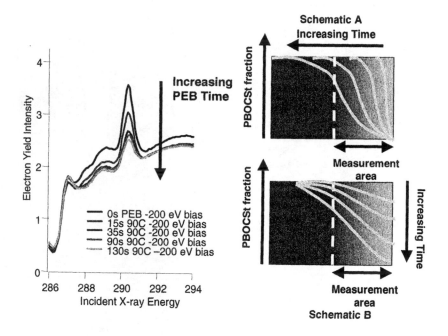

Figure 7. NEXAFS carbon K-edge electron yield spectra at a constant detector bias of –200 eV are shown for the developed bilayer samples with different PEB times at 90 °C.

change in average surface deprotection. One is that the decrease in the surface PBOCSt fraction (carbonyl) is due to broadening of the diffusion / reaction profile at short times leading to a broader surface composition profile in the developed bilayer, while the PBOCSt fraction at the resist – air interface remains constant. Schematic A in Figure 7 illustrates qualitatively this broadening profile through the effective measurement area as a function of bake time. Another possible explanation is that the actual PBOCSt fraction at the film - air surface is decreasing with increasing bake times, illustrated by Schematic B in Figure 7.

To distinguish between schematic A and B in Figure 7, bias dependent NEXAFS spectra were measured for developed bilayers with short and long PEB times. Figure 8 shows the NEXAFS spectra as a function of detector bias for a bilayer sample that was subjected to a short 15 s PEB at 90 °C. The spectra are both pre- and post-edge jump normalized so the carbonyl peak area represents the PBOCSt fraction in the sampling volume. As the negative detector bias increases, the effective electron yield sampling depth is progressively closer to the film surface (as shown by the dotted lines in the schematic in Figure 8). The carbonyl peak area for this short PEB sample does not change dramatically with detector bias. This clearly illustrates that the composition does not change with the changing surface sampling volume, and indicates a diffuse surface PBOCSt composition profile over the total sampling volumes scanned with the various detector bias settings (see schematic in Figure 8).

Figure 9 shows the NEXAFS spectra as a function of detector bias for a bilayer sample that was subjected to a 60 s PEB at 90 °C. Again the spectra are both pre- and post-edge jump normalized. However, with the longer PEB time, the carbonyl peak area clearly decreases with increasing detector bias. Since the sampling area in the electron yield is progressively closer to the film surface with increasing detector bias, a decrease in the carbonyl peak area with increasing bias indicates a decrease in the PBOCSt fraction as the effective sampling depth becomes closer to the surface. A comparison of the bias dependence of the carbonyl peak areas at 15 s (Figure 8) and 60 s (Figure 9) PEB times, illustrates a dramatic change in the surface composition profile after development with increasing bake times. Qualitative comparison of the carbonyl peak area (288 to 292) eV shows that for the short PEB (15 s at 90 °C) the peak areas are larger than for the longer PEB time (60 s at 90 °C), at a particular detector bias. This illustrates that the average surface PBOCSt fraction is higher for the short bake times and indicates that Schematic B in Figure 7 is the more appropriate representation of the surface composition profile changes with increasing bake times. While the surface composition profile changes with time, it is unclear how the breadth of the buried reaction / diffusion profile (before development) influences the corresponding dissolution process and the resulting surface composition profile and line edge roughness. While these areas are currently

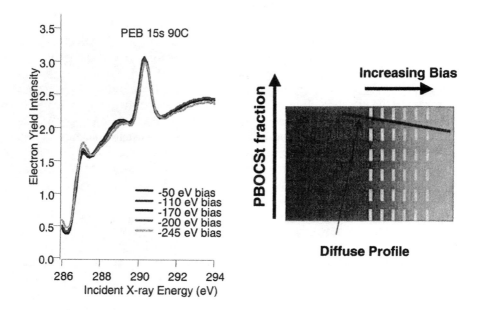

Figure 8. NEXAFS carbon K-edge electron yield spectra are shown as a function of detector bias for a bilayer with a 15 s PEB at 90 °C. The area of the carbonyl absorption does not change significantly with detector bias.

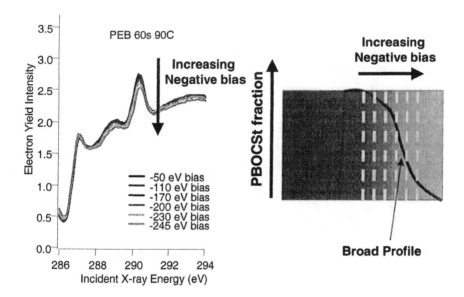

Figure 9. NEXAFS carbon K-edge electron yield spectra are shown as a function of detector bias for a bilayer with a 60 s PEB at 90 oC. The area of the carbonyl absorption decreases with increasing negative bias.

under investigation by a number of research groups, the development of new and unique measurement capabilities allows the potential to make these connections. We are currently developing a theoretical formalism to quantitatively extract the surface composition profile from the NEXAFS bias dependant spectra.

Conclusions

With the advent of sub-100 nm lithography interfacial phenomena will present significant problems for the resist performance. In order to understand and control the resist performance, it is necessary to have sensitive measurement capabilities that can elucidate the mechanisms causing common interfacial problems like t-topping, footing, undercutting, and line edge roughness. In this report, near edge X-ray absorption fine structure (NEXAFS) was utilized to probe interfacial chemistry in resist films. In particular, NEXAFS proved useful for qualitatively detecting segregation of a photo-acid generator to the resist / air interface and differences in the deprotection reaction rate at that interface. In addition, by adjusting the bias on the electron yield detector, surface depth profiling capability of NEXAFS was demonstrated, qualitatively illustrating changes of the developed deprotection profile with bake times in a model line edge region. The depth profiling technique offers potential to provide insight into the mechanisms of the dissolution process and the relationship between the line edge chemistry and line edge roughness.

Acknowledgement

This work was supported by Defense Advanced Research Projects Agency under grant N66001-00-C-8803. Sandia is a multiprogram laboratory operated by Sandia Corporation, a Lockheed Martin Company, for the United States Department of Energy under contract DE-AC04-94AL85000.

References

1. *2001 National Technology Roadmap for Semiconductors*, The Semiconductor Industry Association: San Jose, CA, (2001).
2. H. Ito, Jpn. J. Appl. Phys, **31**, 4273 (1992).
3. S. E. Hong, M. H. Jung, J. C. Jung, G. Lee, J. S. Kim, C. W. Koh, K. H. Baik, *Advances in Resist Technology and Processing XVII*, Proceedings of the SPIE, **3999**, 966 (2000).

4. For detailed information about the NIST/Dow Soft X-ray Materials Characterization Facility at NSLS BNL, see: http://nslsweb.nsls.bnl.gov/nsls/pubs/newsletters/96-nov.pdf

5. D. A. Fischer, J. Colbert, J. L. Gland, Rev. Sci. Instrum. **60(7)**, 1596 (1989).

6. J. Stöhr, *NEXAFS Spectroscopy*, Springer Series in Surface Science (Springer, Heidelberg, 1992), Vol. 25.

7. O.Nalamasu, E. Reichmanis, J. E. Hanson, R. S. Kanga, L. A. Heimbrook, A. B. Emerson, F. A. Baiocchi, S. Vaidya, Poly. Eng. Sci. **32**, 1565 (1992).

8. S. A. MacDonald, W. D. Hinsberg, R. Wendt, N. J. Dlecak, G. C. Willson, C. D. Snyder, Chem. Mater. **5**, 348 (1993).

9. W. D. Hinsberg, S. A. MacDonald, N. J. Dlecak, C. D. Snyder, Chem. Mater. **6**, 481 (1994).

10. H. Yoshino, T. Itani, S. Hashimoto, M. Yamana, N. Samoto, K. Kasama, A. G. Timko, O. Nalamasu. Microelect. Eng. **35**, 153 (1997).

11. D. L. Goldfarb, M. Angelopoulos, E. K. Lin, R. L. Jones, C. L. Soles, J. L. Lenhart, W.-L. Wu, J. Vac Sci. Technol. B. **19(6)**, 2699 (2001).

12. E. K. Lin, C. L. Soles, D. L. Goldfarb, B. C. Trinque, S. D Burns, R. L. Jones, J. L. Lenhart, M. Angelopoulos, C. G. Willson, S. K. Satija, W.-L. Wu, Science, **297**, 372 (2002).

Chapter 9

Photolabile Ultrathin Polymer Films for Spatially Defined Attachment of Nano Elements

B. Voit[1], F. Braun[1], Ch. Loppacher[2], S. Trogisch[2],
L. M. Eng[2], R. Seidel[3], A. Gorbunoff[3], W. Pompe[3], and M. Mertig[3]

[1]Institute of Polymer Research at Dresden, Hohe Strasse 6,
01069 Dresden, Germany
Institutes of [2]Applied Photophysics and [3]Material Science, University
of Technology at Dresden, 01062 Dresden, Germany

Photolabile terpolymers have been designed and used to prepare thin organic films covalently attached to glass and silicon substrates. For this, terpolymers containing the photolabile and charged diazosulfonate units, and on the other hand photolabile protected amino functions, as well as anchoring groups have been synthesized via free radical polymerization. Thin films from these terpolymers were structured imagewise by laser light irradiation in order to create defined functional areas at the template surface ready for further modification and attachment of nanostructures. The selective decomposition and deprotection, respectively, was verified by UV-imaging and selective modification reactions using gold colloids and fluorescence labeling.

Introduction

The present high-tech era requires faster and smaller elements and systems, e.g. in microelectronics, sensors, or medicine. Common imaging technologies for patterning have now reached their limits, therefore novel technologies are necessary to obtain pattern dimensions and functional elements on the nano and even the molecular scale. Today selective metallization e.g. by the use of template polymers and surface finishing, are important fields of application in our macroscopic world as well as in nanotechnology. Another point of interest is the spatially defined provision of functional groups as docking stations for defined attachment of functional elements like molecular switches or carbon nanotubes to build – in the future – nanomachines or at least microreactors containing nanostructures (1-8).

In classical lithographic methods, photoresists are applied which undergo reactions upon irradiation leading to enhanced dissolution behavior (positive) or crosslinking leading to non-dissolution (negative) (9). Commonly, a development step is then added which selectively removes the photoresist film from the substrate at the imaged or non-imaged areas.

The concept of our work differs from these approaches in that way that no selective removal of a polymer film from the substrate after irradiation was intended but rather the selective provision of specific functional groups at the substrate surface. This is comparable to soft lithographic methods (5) where functional molecules are literally printed onto substrates then being ready for further supramolecular assembling as well as reactions. However, this techniques requires the preparation of "stamps", whereas a lithographic approach allows a more flexible patterning to be realized on the substrate.

Experimental Part

Materials:
The synthesis of the diazosulfonate terpolymer via free radical terpolymerization is described in ref. 10. Amino terpolymers were synthesized by free radical polymerization of the three monomers MMA, 3-(triethoxysilyl)propylmethacrylate and the N-NVOC-aminopropyl-methacryl-amide (equimolar amounts) in dioxane at 75°C for 70 h as described in ref. 11. The protection of the amino monomer has been carried out according to the literature (12).

Characterization:
All polymers were analyzed by NMR, FT-IR, and UV-VIS spectroscopy as well as by thermal analysis (TGA) (10,11). The composition of the used co- and terpolymers was determined by ^{1}H-NMR and it was varied between 20-50 % photo active monomer, 0-50% anchor monomer and 50-80 % MMA depending on the final purpose. The imaged films shown in this manuscript had the

following composition (MMA/chromophor/anchor): diazosulfonate terpolymer (76/18/6), amino terpolymer (46/18/36). Molar masses of the used products were in general between 5000 and 15000 g/mol with Mw/Mn appr. 2.5. Glass transition temperatures could only be determined for amine copolymers (MMA/protected amine) with values in the range of 55-59°C depending on the composition. The diazosulfonate polymers do not show glass transitions temperatures before degradation of the diazosulfonate group (> 230 °C). Details on surface characterization of polymer films may be found in ref. 10, 11, and 13.

The silicon wafer and glass substrates used for polymer film deposition were cleaned using a standard procedure (10). Substrates were used immediately after the cleaning procedure.

Thin polymer films of the terpolymers were produced by spin-coating of the diluted (purified) reaction solution (1 wt.-% in DMSO for diazosulfonate terpolymer; 2 wt.-% in DMSO for amine terpolymers) at 3000 rpm for 30 s. Consecutively, the films were annealed for 30 min at 80°C in water saturated atmosphere and then for 2 hours at 120°C under ambient conditions. Non-attached polymer was removed by sonicating in Millipore-water for 5 to 10 min at room temperature.

UV decomposition and imaging experiments

UV spectra were recorded on a VARIAN Carry 100 UV – Visible Spectro-photometer (VARIAN Australia Pty. Ltd, Mulgrave VIC 3170 / Australia). Both the terpolymer solution and thin films of 20 - 30 nm thickness deposited onto glass slides were investigated. A 200 W HgXe arc lamp (Oriel Instruments, Stratford, CT / USA) was used for UV decomposition experiments. A water filter was used in the UV beam to avoid IR radiation at the sample. The distance between the light source and sample was kept at 50 cm. Samples were illuminated in a 5 second interval. The intensity of the UV light at the sample surface was determined to measure 20 - 100 mW/cm depending on the intensity filters in use (10,11). In some cases a TEM grid (1000 mesh) was used as a mask for UV structuring of the films.

For UV imaging, a HeCd laser (KIMMON IK5651R-G, cw, 24 mW at λ = 325 nm and / or 69 mW at λ = 441,6 nm) was focused onto the sample with a microscope objective (ZEISS Fluar 100x/1,30 Oil, UV transparent). In order to write structures of random shape, the sample was moved in the focal plane according to the desired pattern using a piezoelectric scanning stage. Created patterns were imaged with broad-band light from a Xe lamp (XBO 101) focused onto the sample. The transmitted light was collected by a multimode fiber coupled to the entrance slit of a grating monochromator (SPEX 1681B, 300 lines/mm), and the spectrum was monitored with a CCD detector (INSTASPEC4, ORIEL) (10,11,13).

Results and Discussion

In order to realize the proposed concept it is necessary to design polymers having suitable photolabile units and in addition anchoring groups. A covalent attachment of the final polymer film to the substrate is essential in order to allow further modification of the film surface even in water and other solvents. A polymer approach has the advantage that the different units can be incorporated by co- or terpolymerization and in addition, functional polymer films may be prepared by an easy spin-coating step. However, we do not expect a highly ordered film structure. Thus the functional units will be located not only at the film surface but also within the polymer bulk. Therefore, it is impossible to obtain a dense packing of functions at the surface. However, considering the resolution which can be achieved using laser irradiation even under best conditions, the structures intended here will have a size ranging between 100 nm and 1 μm, which is well above the molecular level. Thus even when using terpolymers containing only 10 - 30 mol% of the photolabile unit, a sufficiently high density of the functional groups within the spot of interest is maintained.

We choose two approaches to realize thin functional films, grouped into the two classes of polymers which are activated or deactivated by laser irradiation. In the first concept the photolabile diazosulfonate group is employed which decomposes upon UV irradiation. Thus, the polymer surface having polar and charged sulfonate groups converts (after imaging and loss of the diazosulfonate group) into a less polar, partially crosslinked and no longer charged film surface. Only the non-imaged areas can still be used for further modification e.g. by using electrostatic interactions (negative concept). In the second concept, pendent photolabile protected amino functions are incorporated into the polymer structure. By imaging the film, the protecting groups are removed and the free amine groups can be used to attach functional nanoelements like DNA strands (positive approach).

Diazosulfonate Polymers

Copolymers containing the photolabile diazosulfonate group were intensively studied already, e.g. as new water soluble photo resins for the preparation of offset printing plates (14,15). The diazosulfonate group is decomposed upon UV irradiation (@ $\lambda = \sim 330$ nm) under the loss of nitrogen and SO_2. For the purpose of spatially defined surface functionalization we synthesized a novel terpolymer containing two additional units besides the diazosulfonate group, the comonomer MMA which enhances film forming properties and mechanical stability of the film, and a siloxy anchoring group for the covalent attachment of the film to the glass or silicon substrate (10). Scheme 1 presents the structure of this terpolymer.

Usually about 20 mol% chromophore and 5 mol% anchor groups were incorporated into the polymers. However, the chromophore content may be increased up to 50 mol%.

It was possible to form thin films down to a 10 nm thickness from this polymer by optimized spin casting. The rms roughness of the polymer films measures less than 1 nm. The chemical composition was verified by XPS and UV spectroscopy. In order to demonstrate local UV patterning of thin polymer films, figure 1 illustrates the optical transmission intensity monitored over the wavelength interval ranging from 310 nm to 390 nm. Prior to UV-imaging, the letters "IAPP" and "IPF" denoting the two institute logos were inscribed into the film at a higher UV light dosage (@ λ = 325 nm). Using a confocal set-up, the line width achieved measures ~ 1.5 μm. Irradiated areas show a larger optical transmission within the denoted wavelength regime due to the loss of the absorbing azo functions. Note that topographic investigations over the patterned areas using contact AFM showed no measurable changes in sample topography (z-resolution < 50 pm) when films of only 10 nm thickness have been used. The observed contrast therefore stems from the locally changed optical properties only.

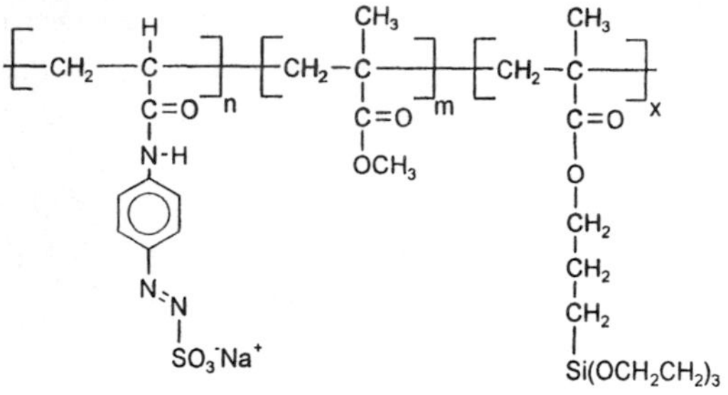

Terpolymer

Scheme 1

The structured diazosulfonate polymer films may further be used for functionalization. E.g. it is possible to exchange the sodium ions of the non-irradiated diazosulfonate groups by other metal ions, as for instance with silver ions, which may be chemically reduced to elemental silver clusters at the sample surface (13). However, silver cluster formation is not fully restricted to the non-irradiated film parts since Ag-clusters slightly tend to complex non-specifically with amide and ester groups even within the polymer film structure, as observed experimentally.

Fig. 1: Area selective UV imaging of a diazosulfonate terpolymer film;
presentation of the changes in optical density after UV laser light
patterning @ λ = 325 nm (line width of ~ 1.5 μm).

A higher selectivity is achieved when using electrostatic bonding between charged adsorbate and the non-imaged and hence negatively charged film areas. In previous work we proved that highly ordered films are prepared from diazosulfonate polymers (polyelectrolyte) – surfactant complexes (16). Here we selectively adsorb positively charged gold nanoparticles onto non-irradiated film areas. This set-up allows the growth of continuous metallic films (Scheme 2).

Figure 2 illustrates the formation of spatially defined gold metal layers after irradiation of the diazosulfonate polymer film with a UV lamp through a TEM grid, and the selective adsorption of gold clusters with subsequent development in a gold salt solution. Similarly UV laser irradiation may be used for writting patterns. As clearly shown only the non-imaged film areas (protected by the TEM grid structure) are free for gold adsorption (bright areas, shown in reflection) which proves the validity of our concept. Further details are published elsewhere (17).

Amine Polymers

Thin organic films containing labile protected amino groups were prepared similarly to the diazosulfonate films by spin coating. Photolabile amine protecting groups are widely used in peptide synthesis (18). We choose the nitroveratryloxycarbonyl (4,5-dimethoxy-2-nitrobenzyl-oxycarbonyl, NVOC) described previously by Fodor et al. (19) and Vossmeyer et. al. (20) for amino group containing self assembly monolayers (SAM) as the protecting group in our amine polymer films. By irradiation with UV light (400 nm > λ > 320 nm) the NVOC protecting group is decomposed and the free amine group is regenerated.

Again, together with the photolabile units, anchoring groups have been incorporated into the polymer structure by using a suitable triethoxysilane group containing methacrylate comonomer (11). Scheme 3 shows the resulting polymer structure.

124

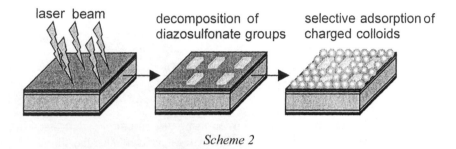

laser beam

decomposition of
diazosulfonate groups

selective adsorption of
charged colloids

Scheme 2

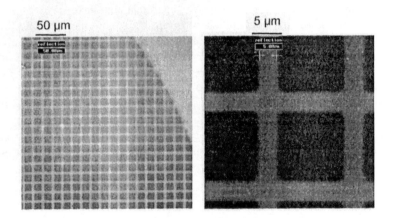

50 μm   5 μm

*Fig. 2: Laser scanning microscope pictures recorded in reflection from the
diazosulfonate terpolymer film after UV irradiation through a TEM grid
mask and subsequent adsorption of gold nanocolloids: brighter areas
indicate the gold colloid layer.*

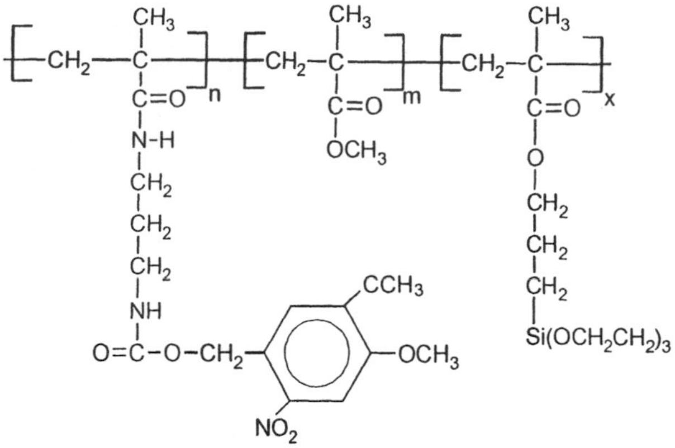

Scheme 3

A molar monomer feed ratio of 1:1:1 (protected amine, MMA, anchoring monomer) was used in the free radical polymerization for our polymers used in film preparation. Unfortunately, the use of the nitro group containing monomer results in a reduced polymerization rate. Therefore, the polymers had to be purified from unreacted monomers resulting in a reduced amine content within the polymer structure compared to the feed ratio (only ~ 18 mol% instead of 33 mol%).

Removal of the protecting group was achieved by UV irradiation both in solution and for the thin polymer films, as shown by the reduced UV absorption of the aromatic protecting group. Compared to the diazosulfonate groups the protecting group is less photosensitive: more than 180 sec irradiation time is necessary using a HgXe UV lamp (100 mW/cm^2 at the sample surface) in order to achieve full deprotection.

The protecting group can be removed also by UV laser irradiation both in the optical far field and the optical near-field. Thus, when using a SNOM tip emitting UV laser light we may achieve very tiny structures. One aim of this work is to provide spatially defined attachment points for DNA strands on the films (Scheme 4).

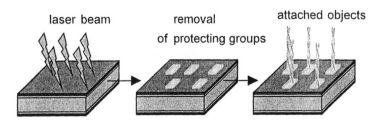

laser beam   removal   attached objects
             of protecting groups

Scheme 4

Fig. 3: a) (left) Area selective UV imaging of a protected amine terpolymer film; presentation of the changes in optical density after UV (325 nm) laser irradiation with a confocal set-up (line width of ~ 1.5 μm); b) (right) Confocal laser scanning fluorescence microscopy image of a selectively deprotected amino polymer film (20 – 30 nm) structured throught a SNOM tip and imaged after treatment with FITC (image-size: 62.5 μm x 62.5 μm)
(Reproduced with permission from reference 11. Copyright 2003 Wiley-VCH.)

The selective deprotection of the protected amino polymer film using a UV-laser beam for irradiation could be visualized as described already for the diazosulfonate films. Figure 3a) illustrates again the optical transmission intensity. Prior to UV-imaging in the spectral range from 340 nm to 390 nm "Olympic rings" have been written by a UV laser using the confocal set-up and an UV-transparent objectiv (line width ~ 1.5 μm, λ = 325 nm). Areas of increased transmittance after loss of the protecting groups are clearly discernable.

When the polymers were irradiated in solution, the deprotection and the generation of the amino groups could be verified by NMR spectroscopy. The regeneration of free amines in the polymer film was proven both by contact angle measurements and by the selective labeling of free amino groups within irradiated film parts using the fluorescence marker fluoresceine isothiocyanate (FITC, SCN-R', see Scheme 5).

$$R—NH_2 + S{=}C{=}N—R' \xrightarrow[\text{rt, 1 h}]{\text{DMF}} R—NH—\overset{\overset{\displaystyle S}{\|}}{C}—NH—R'$$

R' =

R = terpolymer

Scheme 5

The latter process was performed by immersing the patterned amino polymer films into a solution containing FITC, then the film was intensively washed, and the resulting fluorescent structure finally was visualized with a confocal laser scanning fluorescence microscopy. The result is displayed in figure 3b).

Conclusions

We demonstrated that the use of photolabile functional polymer films is a suitable approach to generate randomly shaped surface areas of functional units which may further be used for film modification, e.g. to introduce metallization, or to attach nanoelements. Using UV-lasers for imaging, the aimed functional structure can be written freely into the polymer film without using a mask or stamp as needed for soft lithography. Our polymer approach offers an extremely variable way of incorporating different functional comonomers during free radical polymerization as shown for both a negative and positive structuring process. Furthermore, we present for the first time that the surface functionality is maintained for polymer films of only 10 - 40 nm prepared on a solid substrate.

Acknowledgement

We would like to thank Dr. Komber for the NMR analysis, Dr. Grafström for XPS measurements, as well as Dr. K.-J. Eichhorn and Dr. K. Grundke for help with the surface characterization techniques. Financial support by the *DFG (Forschergruppe "Nanostructured functional elements in macroscopic systems")* is gratefully acknowledged.

References

1. Shipway, A.N.; Katz, E; Willner, I. *Chem.Phys.Chem.* **2000**, *1*, 18.
2. Chechik, V.; Crooks, R.M.; Stirling, C.J.M. *Adv. Mater.* **2000**, *12*, No. 16, 1161.
3. Ulman, A. *"An Introduction to Ultrathin Organic Films"*, Academic Press, Boston 1991.
4. Willner, I.; Katz, E. *Angew. Chem.* **2000**, *112*, 1230.
5. Xia, Y.; Whitesides, G.M. *Angew. Chem.* **1998**, *110*, 568; *Angew. Chem. Int. Ed.* **1998**, *37*, 550.
6. Antonietti, M.; Göltner, C. *Angew. Chem. Int. Ed.* **1997**, *109*, 944.
7. Park, M.; Harrison, C. K.; Chaikin, P. M.; Register, R. A.; Adamson, D. H. *Science* **1997**, *276*, 1401.

128

8. Spatz, J. P.; Mössmer, S.; Hartmann, C.; Möller, M.; Herzog, T.; Krieger, M.; Boyen, H.-G.; Ziemann, P.; Kabius, B. *Langmuir* **2000**, *16*, 407.

9. Niu, Q.J.; Frechet, J.M.J. *Angew. Chem.* **1998**, *110*, 685; *Angew. Chem. Int. Ed.* **1998**, *37*, 667.

10. Braun, F.; Eng, L.; Loppacher, Ch.; Trogisch, S.; Voit, B. *Macromol. Chem. Phys.* **2002**, *203*, 1781.

11. Braun, F.; Eng, L.; Trogisch, S.; Voit, B. *Macromol. Chem. Phys.*, submitted (2002).

12. Amit, B.; Zehavi, U.; Patchornik, A. *J. Org. Chem.* **1974**, *39*, 192.

13. Loppacher, Ch.; Trogisch, S.; Braun, F.; Zherebov, A.; Grafström, S.; Eng, L.M.; Voit, B. *Macromolecules* **2002**, *35*, 1936.

14. Nuyken, O.; Voit, B. *Macromol. Chem. Phys.* **1997**, *198*, 2337 and references therein.

15. van Aert, H.; van Damme, M.; Nuyken, O.; Schnöller, U.; Eichhorn, K.-J.; Grundke, K.; Voit, B. *Macromol. Mater. Eng.* **2001**, *286*, No.8, 488.

16. Antonietti, M.; Kublickas, R.; Nuyken, O.; Voit, B. *Macromol. Rapid Commun.* **1997**, *18*, 287; Thünemann, A.F.; Schnöller, U.; Nuyken, O.; Voit, B. *Macromolecules* **1999**, *32*, 7414.

17. Pompe, W.; Mertig, M.; Gorbunoff, A.; Opitz, J.; Seidel, R., Braun, F., Voit, B. *in preparation*.

18. a) Pillai, V.N.R. *Org. Photochem.* **1987**, *9*, 225; b) Pillai, V.N.R. *Synthesis* **1980**, 1.

19. Fodor, S.P.A.; Read, J.L.; Pirrung, M.C.; Stryer, L.; Tsai Lu, A.; Solas, D. *Science* **1991**, *251*, 767.

20. Vossmeyer, T.; Delonno, E. ; Heath, J. R. *Angew. Chem. Int. Ed. Engl.* **1997**, *36*, 1080; *Angew, Chem.* **1997**, *109*, 1123

Chapter 10

Soft Lithography on Block Copolymer Films: Generating Functionalized Patterns on Block Copolymer Films as a Basis to Further Surface Modification

Martin Brehmer[1], Lars Conrad[1], Lutz Funk[1], Dirk Allard[1], Patrick Théato[1], and Anke Helfer[2]

[1]Institute of Organic Chemistry,Johannes Gutenberg University, D–55099 Mainz, Germany (email: mbrehmer@mail.uni-mainz.de)
[2]Department of Chemistry, Bergische Universität Wuppertal, D–42097 Wuppertal, Germany

Functionalized patterns on the surfaces of amphiphilic diblock copolymer films were generated using polar/apolar interactions applied by soft lithographic techniques. Further modification of the patterned surfaces included e.g. the deposition of conducting and semiconducting material, which offers the opportunity to build sensor structures ranging from the micron to the submicron size.

Introduction

Block copolymers consist of two or more homopolymer chains that are linked by covalent bonds. One characteristic feature of these polymers is the phenomenon called micro-phase separation, which is due to the immiscibility of the different homopolymers on the one hand and their covalent connection on the other. The immiscibility arises from the reduced contribution of the polymer chains to the entropic term in the free energy term of the system. As a rule, the enthalpic term of the equation has a positive value, as long as there are not any attractive interactions between the different polymer chains. If the entropic term cannot compensate the value of the enthalpic term, which means that the free energy becomes negative, a miscibility gap will occur. The connectivity of the immiscible polymer phases leads to microscopic structures in the polymer sample with morphologies determined by the volume fraction of the phases[1]. The most well known morphologies are spheres or cylindrical morphologies, gyroid structures, perforated layers and lamellae. If one of the two phases is removed the remaining polymer framework can be used as an etch mask for the preparation of semiconducting capacitors[2].

The morphologies described are characteristic for the internal structure of a block copolymer sample so far. However things are different at the surface. Here the adjacent medium leads to an enrichment of the polymer phase that has the lowest surface energy towards this medium[3]. Consequently the surface of a block copolymer sample consist only of one sort of segments if enough material is available. As the enrichment is a dynamic process, it is necessary to heat the sample above the glass transition temperature of the polymer[4]. We use this property of enrichment of one polymer phase at the surface to generate functionalized patterns on thin spin-casted films of amphiphilic block copolymers. By using amphiphilic block copolymers we are able to switch from polar segments at the surface to nonpolar just by replacing the adjacent apolar by a polar medium or vice versa[5]. This process was studied by contact angle and AFM measurements. To get a patterned surface, we applied polar and apolar surface interactions with the help of soft lithographic techniques that were introduced by G. M. Whitesides et al.[6]. We think that a combination of these concepts of surface reconstruction on amphiphilic block copolymers and soft lithography is of special interest.

By doing so we should also be able to functionalize the pattern generated on the spin-casted block copolymer film. Therefore we synthesized diblock copolymers via nitroxide mediated controlled radical polymerization containing phenolic **(1)** or hydroxy **(2)** groups in the hydrophilic segments, whereas the hydrophobic segments remained unfunctionalized.

These functional groups show up where the hydrophilic segments come to the surface allowing various surface modifications. A few of these modifications like "grafting from" polymerization, electroless deposition of copper, polyelectrolyte multilayers or deposition of mesoporous transition metal oxides are shown later in this article. The deposition of conducting and semiconducting material on the patterned surface, with pattern size ranging in the micron and submicron scale, opens up prospects to construct various sensor structures.

Experimental Section

Reagents

4-Acetoxystyrene (**3**) was bought from Aldrich. 4-Octylstyrene (**4**) was synthesized according to **ref. (7.)**, the initiator (**5**) for the nitroxide mediated free radical polymerization according to **ref. (8.)**. The monomer acetic acid 2-[2-(4-vinyl-phenoxy)-ethoxy]-ethyl ester (**6**) was prepared as described below and outlined in **Figure 1**. All solvents were freshly distilled and all reactions were carried out under nitrogen. Diglyme was freshly distilled from potassium hydroxide to remove peroxides. Potassium carbonate was ground and dried overnight in vacuo at 150°C.

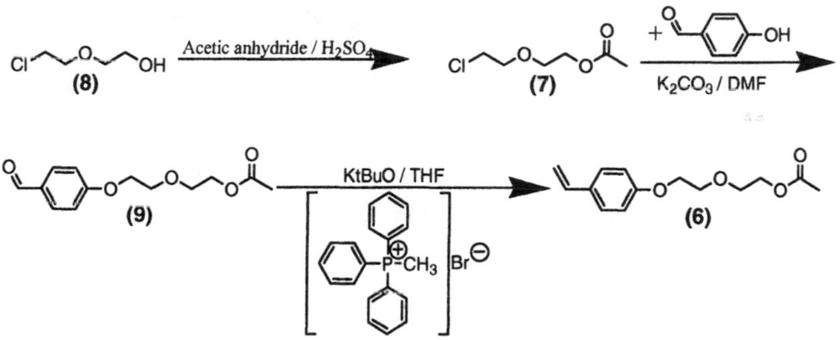

Figure 1: Synthesis of Monomer (6)

Synthesis of Monomer (6)

Synthesis of acetic acid 2-(2-chloro-ethoxy)-ethyl ester (7)

To a mixture of 114.5 g 2-chloro-ethoxy-ethanol (**8**) (0.92 mol) and 93.9 g acetic anhydride (0.92 mole) was added a few drops of sulfuric acid under stirring whereupon an exothermic reaction started. When the heat emission decreased, the mixture was stirred at 100°C for two more hours and poured on ice water afterwards. The aqueous phase was extracted three times with ethyl acetate and the combined organic phases were neutralized with sodium carbonate. The ethyl acetate was evaporated and the residue distilled in vacuo (112°C at $2*10^{-2}$ bar). The product, a colorless liquid, was given in 80% yield.

Synthesis of acetic acid 2-[2-(4-formyl-phenoxy)-ethoxy]-ethyl ester (9)

15.12 g 4-Hydroxybenzaldehyde (0.12 mol), 41.25 g acetic acid 2-(2-chloro-ethoxy)-ethyl ester (7) (0.25 mol) and 34.5 g potassium carbonate (0.25 mol) were suspended in 200 ml dimethylformamide and refluxed for 2 days. Afterwards the reaction mixture was poured in 200ml water and the aqueous phase was extracted three times with ethyl acetate. The combined organic phases were washed with 2 n potassium hydroxide in order to remove the remaining 4-hydroxybenzaldehyde. The organic solution was dried with potassium carbonate and the solvent was evaporated. The crude product was purified by column chromatography (petrolether:ethyl acetate 2:1). The product was obtained as white crystals. Yield 33%.

Synthesis of acetic acid 2-[2-(4-vinyl-phenoxy)-ethoxy]-ethyl ester (6)

22 g Methyltriphenylphosphoniumbromide (0.057 mol) were suspended in 100ml THF and 6.4 g KOtBu (0.057 mol) in 50 ml THF were added dropwise under stirring. The resulting yellow solution was stirred for another hour, before 9 g acetic acid 2-[2-(4-formyl-phenoxy)-ethoxy]-ethyl ester (9) in 100 ml THF was added dropwise keeping the temperature of the reaction mixture below 20°C. The solution was stirred at room temperature overnight. The solvent was then evaporated and ethyl acetate added, resulting the precipitation of triphenylphosphoniumoxide, which was filtered off. The solvent was again evaporated and the crude product purified by column chromatography (petrolether:ethyl acetate 2:1). The product was obtained as a white waxlike substance. Yield 46 %.

Synthesis of polymers (1) and (2)

All polymerizations were carried out under nitrogen, degassed by three freeze-pump-thaw cycles using Schlenk type flasks. For the homopolymerizations the monomer to initiator ratio was calculated for a Mw of 70000 (100% conversion). To get the amphiphilic block copolymers a two-fold molar surplus of the second monomer was added to the homopolymer.

4-Octylstyrene (4) and poly(4-octylstyrene) (10) together with 4-acetoxystyrene (3) were polymerized in substance for 20h to 30h at 123°C. The polymerization of poly(4-octylstyrene) (10) with the monomer acetic acid 2-[2-(4-vinyl-phenoxy)-ethoxy]-ethyl ester (6) was carried out in diglyme over five days at

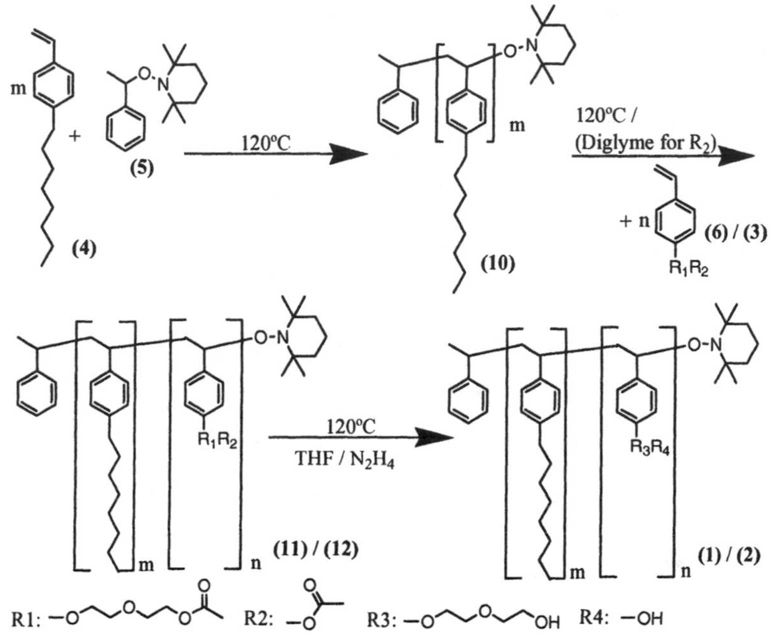

Figure 2: Synthesis route to block copolymers (1) and (2)

123°C. All polymers were dissolved in THF and precipitated in methanol after polymerization. This procedure was repeated three times in order to remove remaining monomer. The complete route of synthesis to block copolymer (1) and (2) is shown in **Figure 2**.

Hydrazinolysis of block copolymer (11) and (12)

Polymer (11) / (12) was dissolved in THF and the fivefold amount of hydrazine hydrate in respect to the polymer was added. The clear reaction mixture was then refluxed for 4h and stirred at room temperature overnight resulting in a blurry solution. The block copolymer was precipitated by pouring the THF solution into methanol, filtered off and dried in vacuo.

Characterization of block copolymers and block copolymer films

Glass transition temperatures

As the generation of a patterned surface goes along with a reorientation of chain segments, their mobility must be ensured. Therefore the restructuring of block copolymer **(2)** is carried out above the glass transition temperature. In the case of block copolymer **(1)** the high T_G of the hydrophilic segments is lowered below room temperature through swelling these segments with water.

The glass transition temperatures of all polymers described in this article were determined by DSC measurements and are listed in **Table 1**.

Table 1: Glass transition temperatures of block copolymers

Polymer	T_G hydrophobic phase	T_G hydrophilic phase
Poly(4-octylstyrene) **(4)**	-35°C	-
Poly(4-octylstyrene)-*block*-(4-acetoxystyrene) **(12)**	-33°C	90°C
Poly(4-octylstyrene)-*block*-(4-hydroxystyrene) **(1)**	-33°C	135°C
Poly(4-octylstyrene)-*block*-(acetic acid 2-[2-(4-vinyl-phenoxy)-ethoxy]-ethyl ester) **(11)**	-35°C	-8°C
Poly(4-octylstyrene)-*block*-(2-[2-(4-vinyl-phenoxy)-ethoxy]-ethanol) **(2)**	-35°C	12°C

Contact angle measurements

By measuring the contact angle of a drop of water on a surface one receives information about the hydrophilicity of this surface. Surfaces with a contact angle above 100° are called hydrophobic, otherwise they are called hydrophilic. By measuring the contact angle of a surface of a block copolymer film and then comparing it to the contact angle of the respective homopolymers, one gets information about the ratio of the polar and apolar segments at this surface.

The contact angle of the hydrophobic phases was measured after annealing the film overnight under air and above glass transition temperature. The contact angle of the hydrophilic segments was determined after exposing the film surface to water for 20h and then removing the water that sticks to the surface through spinning the sample in a spin coater. In case of polymer **(2)** no reliable value for the contact angle of the hydrophilic phase could be measured. The values varied, but were always below 50°.

Table 2: Contact angles of different polymers

Polymer	PDI	Mw	Contact angle hydrophobic phase	Contact angle hydrophilic phase
Poly(4-octylstyrene) **(4)**	1,23	42000	114°	-
Poly(4-octylstyrene)-*block*-(4-hydroxystyrene) **(1)**	1,48	63000	109°	96°
Poly-(acetic acid 2-[2-(4-vinyl-phenoxy)-ethoxy]-ethyl ester)	1,36	63000	-	76°
Poly(4-octylstyrene)-*block*-(2-[2-(4-vinyl-phenoxy)-ethoxy]-ethanol) **(2)**	1,33	58000	109°	<50°

Swelling of hydrophilic segments

As polymer **(1)** with its high T_G of the hydrophilic block is the one to be treated with water for pattern generation, we did some measurements to get an idea of the swelling process of the 4-hydroxystyrene segments. Therefore we spin-casted a thin film of polymer **(1)** on a gold sputtered glass substrate and measured the thickness of the film by Surface Plasmon Resonance Spectroscopy (SPR) in dependence on the time the film was exposed to water[9]. The results are outlined in **Figure 3**. It shows that the process of swelling is finished after 3.5h and results in a thickness of the swollen film that is about 75% bigger compared to the unswollen film.

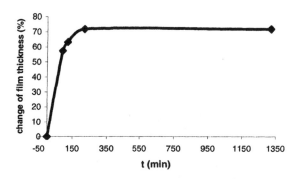

Figure 3: time dependence of swelling

Hydrophilic-hydrophobic pattern generation

Masking techniques should allow a reconstruction of the surface in selected areas only. This leads to a pattern at the surface consistent of differences in the polymer functionality, which appear at the surface. These functionalities can then be used for further chemical modifications. After the first modification a second reconstruction can take place. This gives us the possibility of complex pattern generation.

For the pattern creation we are using diblock copolymers consistent of a hydrophobic polymer, this is in most cases poly(4-octylstyrene) **(1)**, and a hydrophilic polymer, which bears the functional group we use for the surface modification. To achieve this pattern we have to spin-cast a thin film of the block copolymer on top of a flat substrate. As substrate we have used glass silicon or gold but it should also be possible to use a lot of other kind of materials like e.g. polymers. The solvent we use for the spin-cast process limits the materials, which can be used only. After the spin-casting process the surface is covered by the block, which has the lower surface energy in air. These are normally the hydrophobic alkyl groups of the poly(4-octylstyrene) **(1)** chain. As mentioned above the polymer on the surface changes if you change the surface from air (hydrophobic) to e.g. water (hydrophilic). We use this effect for the pattern generation by partially exchanging the interface. This is done by using a poly-dimethyl-siloxane-stamp (PDMS stamp), which bears the pattern as a height profile **(Figure 4)**. For the production of the PDMS stamps we are using Sylgard 184 from Dow Corning, which is poured on top of a silicon-master. After curing the PDMS over night at 80°C the stamp can be lifted of the silicon-master. Because of different polymer properties we are using two different types of modified soft lithography techniques.

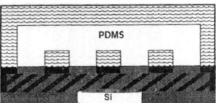

Figure 4: scheme of surface patterning

If the hydrophilic block of the polymer has a T_G above room temperature (poly(4-hydroxystyrene)) we put the PDMS stamp with its structures on top of the polymer film and put both into water for 24h at room temperature. The water will be drawn into the channels between the polymer and the stamp by capillary forces. This creates a pattern of hydrophilic, water covered, and hydrophobic, PDMS stamp covered, interfaces and a reconstruction of the polymer blocks in the water-covered areas is following. After removal of the water and lifting of the stamp the surface shows the negative pattern of the PDMS stamp. As mentioned above the water has two purposes in this process. On one hand it changes the interface so that the hydrophilic polymer has the lower surface-energy and on the other hand it swells the hydrophilic polymer and reduces its T_G below room temperature. This is necessary to achieve the mobility of this block for the organization process. Because of the swelling process the surface shows after the reconstruction a height profile formed by the swollen hydrophilic parts and the hydrophobic surface (**Figure 5**). After removal of the water the polymer deswells and the T_G of the polymer raises again above room temperature and freezes the pattern. Now the pattern is very stable and the substrate can be stored for weeks in air. One disadvantage of this method is that you have to have channels for the water. This means it is not possible to create single structures like e.g. spheres or triangles. Out of this we developed a second process.

 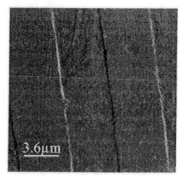

Figure 5: AFM-images of hydrophilic 5μm lines, left: height profile (100nm line height), right: vertical deflection

For the creation of closed structures we are using block copolymer **2** where both blocks have a T_G below room temperature. This makes the use of water obsolete and by this it is possible to create single closed structures because we don't need channels where water could be drawn into. For the patterning process it is still necessary to create structure difference on the surface with hydrophilic and hydrophobic interfaces on top of the polymer film to start the reorientation process. This makes a second modification in our procedure necessary. We are using now a PDMS stamp for the structuring process, which has been modified in an oxygen plasma to hydrophilize its surface. When we expose the surface of the substrate to such a stamp over night, we create hydrophobic (air)/ hydrophilic (PDMS) interfaces. This leads to a reconstruction like before but now we don't need water. The advantages are, as mentioned before, the possibility to create closed single structures and no additional reagents are needed. But there is although a problem with this kind of structuring. Because of the mobiltiy of the polymer chains at room temperature the pattern vanishes slowly in air. This means that you have to perform any further modifications immediately after the patterning process.

Surface modification

The surface of the patterned substrate bears the functional groups of the hydrophilic block. This means that they are accessible for further chemical modifications of the polymer.

When we use a hydrophilic polymer like poly(4-hydroxystyrene) or poly(2-(2-(4-vinyl-phenoxy)-ethoxy)-ethanol), which will show hydroxy groups at the surface after reconstruction, we are able to deposit selectively inorganic materials like titanium dioxide[10] or copper[11] on these parts.

For the deposition of titanium dioxide the substrate is exposed to a 5% (v/v) solution of titan-tetrabutoxide in isopropanol. First the titanium forms phenolates on the hydrophilic areas. When this solution is exposed to air the titanium alkoxide starts to hydrolyze and titanium dioxide is deposited at the hydrophilic parts of the substrate. If this process happens to fast because of high humidity or good air exchange above the reaction mixture the complete surface will be covered with titanium dioxide. But because of the bad interaction between the hydrophobic parts and titanium dioxide it is possible to remove the excess titanium dioxide by sonification. The titanium dioxide at the hydrophobic parts of the surface is much more resistant against this substrate treatment (**Figure 6**).

Figure 6: SEM Image of 5μm titanium dioxide lines

Another interesting possibility is the selective deposition of copper on the structures. We are using an electro less deposition process for this purpose. First of all we have to define the parts of the copper deposition by activating them with palladium, which acts as a catalyst. Because palladium is not able to coordinate directly to hydroxy groups it is necessary to treat the surface with a $SnCl_2$ solution. In a first step Sn^{2+} will coordinate to the hydroxy groups and then it will reduce the $PdCl_2$-solution in a second step. After the selective deposition of the catalyst the substrate is exposed to a mixture of a solution consistent of potasium-sodium tartrat copper(II)sulfate and potassium hydroxide in deionized water and a solution of formaldehyde in water for the reduction of the copper solution. The deposition of the copper is very fast and so the substrate has to be removed from the solution after 30 seconds (**Figure 7**).

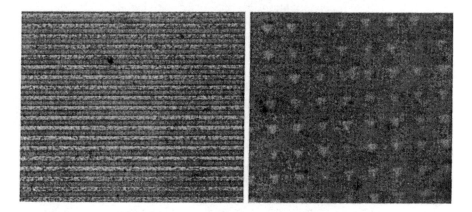

Figure 7: 5μm copper lines reflection and 10μm copper free triangles

In the case of hydroxy bearing hydrophilic polymers one other possible reaction is a grafting from via Atom-Transfer-Radical-Polymerization (ATRP). The ATRP polymerization of acrylates and acrylamides is initiated with a 2-bromopropionic ester[12]. The ester is prepared by esterification of the surface OH groups of hydrophilic areas with 2-bromopropionic acid. There for we are using an aqueous solution of 10 mmol 2-bromopropionic acid and 2.5mmol N´-(3-dimethylaminopropyl)-N-ethylcarbodiimide hydrochloride (EDC) as an activator in 10 ml water. A patterned substrate with polymer (**1**) is kept in this solution for 1h. The reaction time is much shorter than the reconstruction time of 24h, which is used for the patterning process. This is necessary because of the possible reconstruction process, which might take place with longer reaction times. After this reaction the former hydroxy group bearing parts of the surface are now able to initiate an ATRP-polymerization[13] (**Figure 8**).

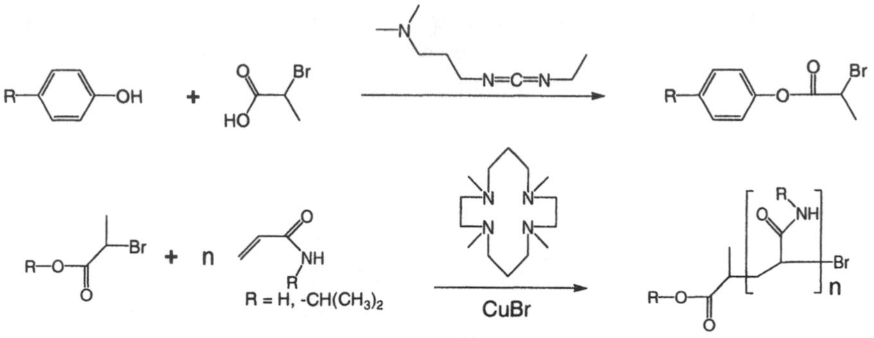

Figure 8: scheme of esterification and ATRP Polymerization of acrylamides

For the ATRP-polymerization we are using acrylamide and isopropyl acrylamide. These two monomers and its polymers are soluble in water and don't need other solvents, which could cause problems with the block copolymer on the substrate. The polymerization is carried out in a sealed and nitrogen flushed flask so that the reaction can take place in the absence of oxygen. The polymerization solutions consist of a solution of (28mmol) acrylamide or isopropyl acrylamide in 10ml water and a solution of (0.4mmol) 1,4,8,11-tetraaza-1,4,8,11-tetramethylcyclotetradecane and (0.4mmol) copper(I)bromide in 10ml water. Both solutions were degassed by bubbling nitrogen through them for at least 30min. After combining the solutions they were given to a sealed flask, which contains the substrate.

Poly(N-isopropylacrylamide) has an interesting property we can observe on the surface with SPR. This polymer is soluble in water at room temperature and becomes insoluble above 32°C. This effect is called the lower critical solution temperature (LCST)[14]. **Figure 9** shows the angle of the measured minimum of the SPR-curve. At 32°C we see a shift in the minimum position. This corresponds to a collapse of the grafted chains, which leads to an increase in the optical thickness at the interface.

In the case of patterned polymer (1) the phenolic hydroxy groups at the surface can easily be deprotonated. Due to the negatively charged surface partially charged polyelectrolytes can absorb to the surface[15]. The deprotonation can be done by exposing the substrate in a 0.1g/ml aqueous solution of sodium hydroxide for 5 seconds. For the deposition we were using cationically charged polythiophene as a 10^{-2} mol/l solution in water with 10% ethanol (**Figure 10**). The most interesting property of this kind of polymer is the conductivity of the thiophenes. This is another possibility to generate a conductive structure on top of a non-conducting polymer.

determination of the LCST of poly(N-isopropylacrylamide)

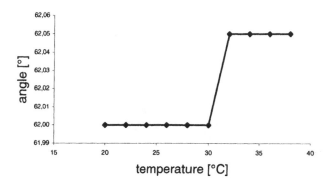

Figure 9: collapse of Poly(N- isopropylacrylamide) at 32°Cangle of minimum against temperature

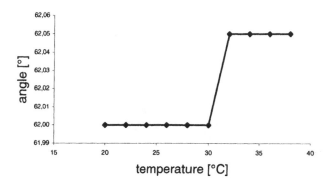

Figure 10: 5μm lines of deposited polythiophene

Polyelectrolytes can be used to easily create more complex structures. The polyelectrolyte can be applied to the surface by a stamp, whose channels are perpendicular to the anionic lines at the surface. After pouring some polyelectrolyte solution at the edge of the stamp the solution is drawn into the lines by capillary forces. This leads to a pattern shown in **Figure 11**. Because only parts of the hydrophilic lines are covered with polyelectrolyte it should be possible to deposit another kind of polyelectrolyte on the still deprotonated parts of the line.

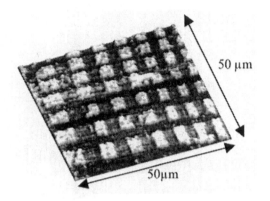

Figure 11: squared deposition of polyelectrolyte on 5μm lines

Conclusion

In this chapter we demonstrate the use of amphiphilic diblock copolymers in combination with soft lithography for pattern generation. This is based on surface reconstruction if the block copolymer film is brought into contact with alternating hydrophilic hydrophobic interfaces. This can be achieved by either using a hydrophobic PDMS stamp together with water, or a hydrophilic stamp together with air. For both processes diblock copolymers with different glass transition temperatures were prepared by nitroxide mediated radical polymerization. The created hydrophilic areas can be functionalized selectively. For the functionalization different reactions can be used. The materials applied contain metals, metalloxides, grafted polymers and polyelectrolytes.

References

1. Zhao, J.; Majumdar, B.; Schulz, M.F.; Bates, F.S. *Macromolecules* **1996**, 29, 1204.
2. Black, C. T.; Guarini, K. W.; Milkove, K. R.; Baker, S. M.; Russell, T. P.; Tuominen, M. T. *Applied Physics Letters* **2001**, 79, 409.
3. Krausch, G. *Mater. Sci. Eng. R-Rep.* **1995**, 14, 1.
4. Fryer, D. S.; Peters, R. D.; Kim, E. J.; Tomaszewski, J. E.; de Pablo, J. J.; Nealey, P. F.; White, C. C.; Wu, W. L. *Macromolecules* **2001**, 34, 5627.
5. Mori, H.; Hirao, A.; Nakahama, S. *Macromolecules* **1994**, 27, 4093.
6. Xia, Y.; Whithsides, G.M. *Angew.Chem.* **1998**, 110, 586.
7. Bartlett, P.D.; Benzing, E.P.; Pincock, R.E. *J. Am. Chem. Soc.* **1960**, 82, 1762.
8. Mori et al.; *Macromol. Chem. Phys.*, **1994**, 195, 3213.
9. Knoll, W. *MRS Bull.* **1991**, 16, 29.
10. Koumoto, K.; Seo, S.; Sugiyama, T.; Seo, W. S. *Chem. Mater.*, **1990**, 11, 2305.
11. Charbonnier, M.; Romand, M.; Harry, E.; Alami, M. *J. Appl. Electrochemistry*, **2001**, 31, 57.
12. Théato, P.; Preis, E.; Brehmer, M.; Zentel R. *Macromol. Symp.* **2001**, 164, 257.
13. Husemann, M.; Mecerreyes, D.; Hawker, C. J.; Hedrick, J. L.; Shah, R.; Abbott, N. L. *Angewandte Chemie International Edition,* **1999**, 38, (5), 647.
14. Wischerhoff, W.; Zacher, T.; Laschewsky, A. *Angew. Chem.* **2000**, 112, 24, 4771.
15. Decher, G.; Hong, J. D. *Berichte Der Bunsen-Gesellschaft-Physical Chemistry Chemical Physics*, 1991, 95, (11), 1430.

Chapter 11

Nanoporous, Low-Dielectric Constant Organosilicate Materials Derived from Inorganic Polymer Blends

R. D. Miller[1], W. Volksen[1], V. Y. Lee[1], E. Connor[1], T. Magbitang[1], R. Zafran[1], L. Sundberg[1], C. J. Hawker[1], J. L. Hedrick[1], E. Huang[2], M. Toney[1], Q. R. Huang[3], C. W. Frank[4], and H. C. Kim[1]

[1]IBM Almaden Research Center, 650 Harry Road, San Jose, CA 95120–6099
[2]IBM T. J. Watson Research Labs, Yorktown Heights, NY 10598
[3]Department of Food Science, Rutgers, The State University of New Jersey, 65 Dudley Road, New Brunswick, NJ 08901
[4]Department of Chemical Engineering, Stanford University, Stanford, CA 94305

Porous materials will be needed to reach the ultralow-k dielectric on-chip insulator objectives for advanced semiconductor chips. Nanoporous organosilicates can be prepared using thermally labile macromolecular porogens and silsesquioxane resins. This process readily achieves dielectric constants of 1.5. There is a significant difference in porous morphologies depending on whether the porogens function by nucleation and growth or an actual templating mechanism.

Introduction

The size of active transistors on semiconductor chips has become progressively smaller with time scaling according to Moore's Law, which states that the number of transistors/unit area on a chip will roughly double every 18 months. Devices with 130 nm minimum dimensions are already in production while others with critical dimensions below 100 nm (e.g., 90 nm) have been produced in a manufacturing environment. Figure 1 shows a visual comparison of these dimensions with those of some common familiar objects.

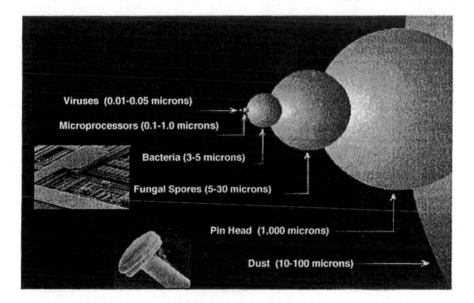

Figure 1: Relative size representations of common objects

As device dimensions decrease and densities increase, chip performance degrades due to crosstalk and capacitive coupling between the metal lines connecting the active devices with the outside world.[1] This metal wiring and the on-chip insulators, hard masks, caps, etc., constitute the back-end-of-the-line (BEOL) in semiconductor terminology. The copper interconnect wiring minus the dielectric insulator for an advanced IBM CMOS 7S logic chip is shown in Figure 2a, while a typical back end wiring schematic together with insulators and etch stops is depicted in Figure 2b. The intralayer metal wiring is joined by vias that connect one wiring level to another.

146

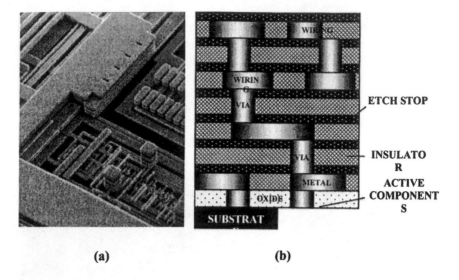

(a)   (b)

Figure 2: (a) a copper CMOS logic chip from which the SiO$_2$ insulator has been selectively removed (b) A schematic depiction of the back-end-of-the line wiring

Performance issues are summarized schematically in Figure 3 which describes the interconnect decay (RC delay) in terms of the metal resistivity, the dielectric constant of the insulator (until recently SiO$_2$) and the wiring dimensions. Also shown is the relationship describing power consumption. It is significant that both delay and power consumption depends on the capacitance, which scales directly with the dielectric constant of the on-chip insulator. It is the eroded performance predicted for future chips which is driving the somewhat frantic search for new low dielectric constant insulators (low k) to replace silicon dioxide (k = 3.9–4.2).[3] The first device generation utilizing a truly low-k material will be the 90 nm technology node.[4] The plethora of low-k candidates may be divided into those deposited by chemical vapor deposition (CVD) and those applied by spin-on techniques.[3] The former are mostly inorganic materials (e.g., organosilicates or carbon-doped oxides while the latter class is comprised of both inorganic and organic polymer candidates. Although the semiconductor manufacturing world is currently divided between CVD and spin-on candidates,[5] the unexpected difficulties experienced in integrating replacement candidates for SiO$_2$ has put a premium on dielectric extendibility (i.e., the use of the same elemental composition for multiple technology nodes).

In this regard, there is no true dielectric extendibility without the incorporation of porosity to progressively lower the insulator dielectric constant. Dielectric extendibility issues are currently fueling the interest in nanoporous, thin film dielectrics.

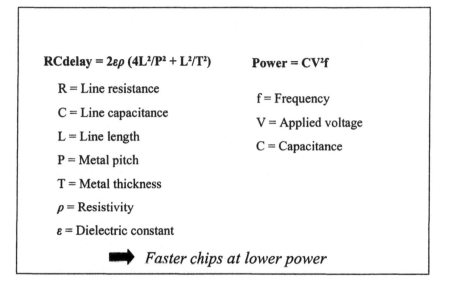

Figure 3: Performance issues for advanced chips

For some time now, we have been studying the introduction of nanoporosity in thin films using a sacrificial macromolecular pore generator (porogen) approach.[6] These systems function by a nucleation and growth mechanism[7] (NG) where an initially miscible porogen dissolved in a thermosetting prepolymer undergoes phase separation during curing and is subsequently removed during heating to generate porosity. Upon curing the matrix, the porogen phase separates into nano-domains whose growth ideally is limited by the viscosity of the thermally curing matrix (Figure 4). The ultimate generation of porosity comes from the thermolysis of the porogen domains and diffusion of the fragments. For organosilicates, the porous morphology is determined by that of the inorganic-organic nanohybrid

Since initial polymer miscibility between the porogen and the matrix is required for nucleation and growth, considerable care is exercised in the selection of both components. For the resin, we have generally utilized low molecule weight organosilsesquioxanes (SSQ, $(RSiO_{1.5})_n$) with an abundance of chain ends available for interaction with the porogen. When the organic porogens do not contain mainchain functionality, which is strongly interacting with the resin, low molecular weight materials with multiple arms and chain ends[8] are utilized. Linear polymers can also be employed but only if they strongly interact with the resin prepolymer to promote miscibility.[9]

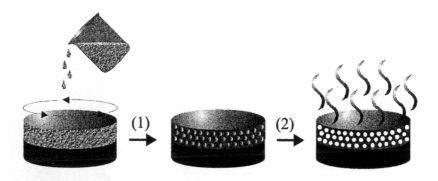

Phase 1: Initial Cure & Consolidation (200 – 250 °C)

Phase 2: Porogen Volatilization & Final Cure (300 – 450 °C)

Figure 4. Formation of nanoporosity using blending and sacrificial porogens

Results and Discussion

The thermosetting organosilicate matrix resin is a critical component of the dielectric blend. For this purpose, we have chosen low molecular weight silsesquioxanes $(RSiO_{1.5})_n$ prepared by the acid-catalyzed hydrolysis of labile organosilicates $(RSi(X)_3)$; the hydrolytically labile substituents (X) may be halo, alkoxy or acetoxy depending on the desired properties of the resin.[10] The SSQ derivatives so derived have a complex structure comprised of various functional units as depicted schematically in Figure 5.

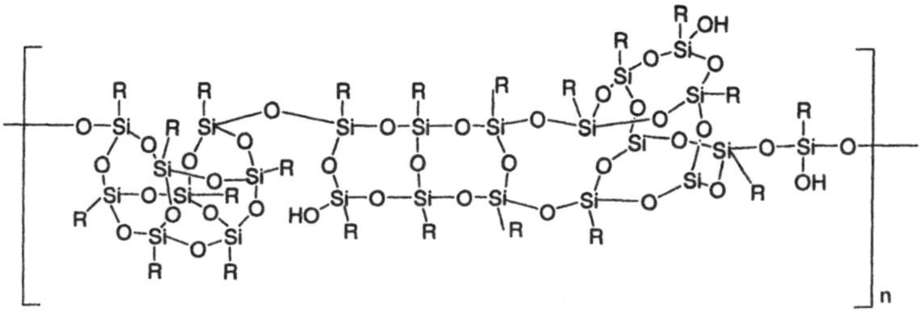

Figure 5: A representative silsesquioxane structure

Low molecular weight versions usually have abundant chain end functionality, although reaction procedures have been developed to promote intramolecular condensation leading to soluble low molecular weight SSQ derivatives with relatively little SiOH functionality.[11] Here we have studied three different low molecular weight resins, two were pure methyl silsesquioxane (MSSQ) while the third was a copolymer containing 25–35% of SiO_2 linkages which improve the mechanical properties of the resin.

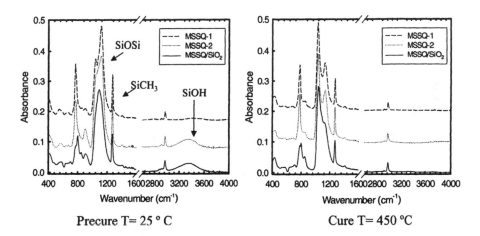

Precure T= 25 ° C   Cure T= 450 °C

Figure 6: FT-IR comparison of various MSSQ homo- and copolymers; (---) MSSQ-1, MSSQ-LO, (···) MSSQ-2, MSSQ-HI, (—) MSSQ-SiO$_2$ copolymer

The homopolymers are distinguished by the relative amounts of SiOH ends groups as determined by FT-IR analysis and are designated as MSSQ-HI and MSSQ-LO on this basis. The copolymer also contains substantial quantities of SiOH by IR analysis. The IR spectra of the respective resins normalized for thickness are shown in Figure 6. In the figure, the designations MSSQ-1 and MSSQ-2 refer to the low and high SiOH contents respectively as determined by the characteristic IR bands around 3360 and 920 cm^{-1}. The copolymer is also characterized as a high SiOH resin. Both MSSQ-HI and MSSQ-LO cure to the same structure as also shown in Figure 6. The spectrum of the cured copolymer is similar, but reflects the presence of additional SiO_2 linkages.

GPC's of the three resins are shown in Figure 7 and are calibrated relative to linear polystyrene standards. All of the resin samples are low molecular weight with relatively broad polydispersities. Given the inappropriate nature of the linear polystyrene GPC standards for SSQ derivatives, the data in Figure 7 are useful for relative comparison only.

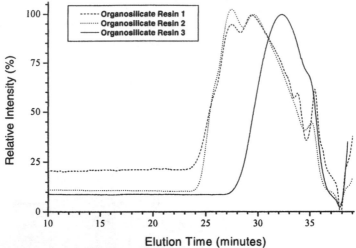

Figure 7: Size Exclusion Chromatography traces of various MSSQ homo and copolymers in tetrahydrofuran at ambient temperature. Resin 1 – low SiOH MSSQ (M_n = 1816, PDI = 3.5), resin 2 – high SiOH MSSQ (M_n = 1625, PDI = 3.5, and resin 3 – MSSQ/SiO$_2$ copolymer (M_n = 1056, PDI = 2.1)

A wide variety of porogens ranging from linear materials[9] to multiarm stars, dendrimers and hyperbranched systems,[8] and crosslinked nanoparticles[12] have been studied in these resins. Most linear polymers are incompatible with the matrix resin in the absence of mainchain functionality, which interacts strongly with the resin. Examples of weakly interacting porogens would be linear methyl methacrylate and poly-caprolactone and lactide. For these systems, multiarm stars are necessary for miscibility. Linear polymers which interact strongly with the resin, however, can form miscible blends as exemplified by copolymers of methyl methacrylate and dimethylaminoethyl methacrylate (DMAEMA-MMA) and various alkylene oxide homo and copolymers.[9] Figure 8 shows the thermal and mechanical properties of a blend of MSSQ-HI with a low molecular weight (M_w < 6000g/mol) poly(alkylene oxide) porogen. Superimposed on the thermal curves is the dynamic mechanical drive signal for the pure resin as a function of curing temperature. From the data, the resin loses ~ 10% of its weight upon curing beginning slightly above 100°C. When the porogen (20 wt.%) is incorporated, the total weight loss, as expected, is the sum of the porogen loss and resin condensation. An interesting feature of the figure is that the resin stiffness increases below the porogen decomposition temperature, a necessary condition to maintain the porous structure. Film refractive index changes

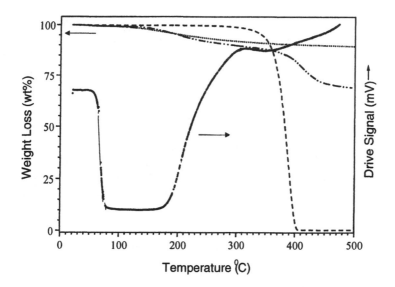

Figure 8: Thermal and mechanical properties of a MSSQ-HI resin and a poly(alkylene oxide) porogen; (···) MSSQ-HI resin alone; (---) porogen; (-··) 80/20 hybrid

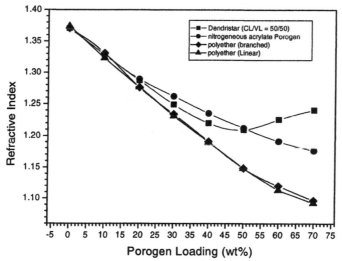

Figure 9: Comparison of porogen foaming efficiencies

may be used to assess the relative pore forming efficiencies of various porogens. Data on the change in refractive index as a function of porogen loading is shown in Figure 9 for a variety of porogens. The porogens include a 4-arm star copolymer of caprolactone-co(50)-valerolactone (dendristar), the nitrogenous copolymer DMAEMA-co(75)-MMA, and two poly(alkylene oxides (branched and linear). While the foaming efficiencies, as estimated by refractive index, are similar below 30 wt.% loading, they begin to diverge strongly at higher loading levels. From the figure, it is obvious that the alkylene oxides are more efficient pore generators at loading levels above 45 wt.%. The relationship between refractive and dielectric constant (100 KHz) is shown in Figure 10 for one of the alkylene oxides from Figure 9. From the data, it is obvious that a very low dielectric constant of < 1.5 can be reached at a loading level of ~ 50 wt.%.

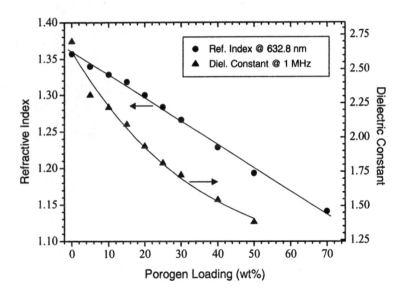

Figure 10: The relationship between refractive index and dielectric constant for porous MSSQ produced using a poly(alkylene oxide) porogen

Figure 11 shows a comparison of TEM and FE-SEM micrographs of porous MSSQ at the 20% porosity level. The details of the porous morphology are much more evident in the TEM sample, although interpretation is complicated by the 3D representation depicted by the micrograph.

Figure 11: TEM (left) and FESEM micrographs of porous MSSQ (20%) prepared from a poly(alkylene oxide) porogen

Small angle x-ray scattering (SAXS) has become the method of choice in studying the morphology in porous organosilicates.[13] Scattering data for the MSSQ-LO resin with DMAEMA-MMA porogen as a function of loading level is shown in Figure 12. From the data, we see that the average pore diameter increases with loading level. This would be expected for a classic nucleation and growth mechanism where earlier phase separation would be expected to lead to larger polymer domains and hence bigger pores.

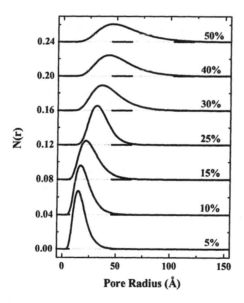

Figure 12: SAXS size distributions from model fits for porous MSSQ-LO as a function of porogen loading level

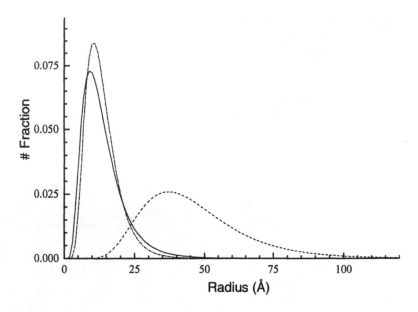

Figure 13: SAXS porous model fits for a variety of porous (35%) resin samples; (—) MSSQ-SiO$_2$, (⋯) MSSQ-HI, (---) MSSQ-LO

The pore size also depends on the structure of the resin as shown in Figure 13. Although the average calculated pore size for the MSSQ-HI and copolymer resins at 35 wt.% porogen loading are similar, those in the MSSQ-LO sample are substantially larger (3-4X), a feature consistent with earlier phase separation in this resin, which lacks polar functional chain ends. All of the SAXS data is consistent with a classical nucleation and growth (NG) mechanism for the phase separation. In such a process, the polymer domain and pore size will be a function of porogen structure, loading level, molecular weight and polydispersity as well as resin structure and processing conditions. The end result is a process which can be difficult to control because of the number of dependent variables.

An alternative to nucleation and growth processes is one where the porogen actually templates the vitrification of the matrix. Templated processes have been described for the generation of mesoporous silica where surfactant structures control the formation of silica from small molecule precursors.[14] Highly regular structures can result, although film morphologies are dependent on dynamic self organization. In principle, highly crosslinked nanoparticles could control matrix polymerization provided particle aggregation can be avoided.[15] In such a process, the particles are never really miscible in the polymerizing medium, but rather simply "go along for the ride." Previously, we have described use of such particles prepared by intramolecular crosslinking.[12]

Scheme 1

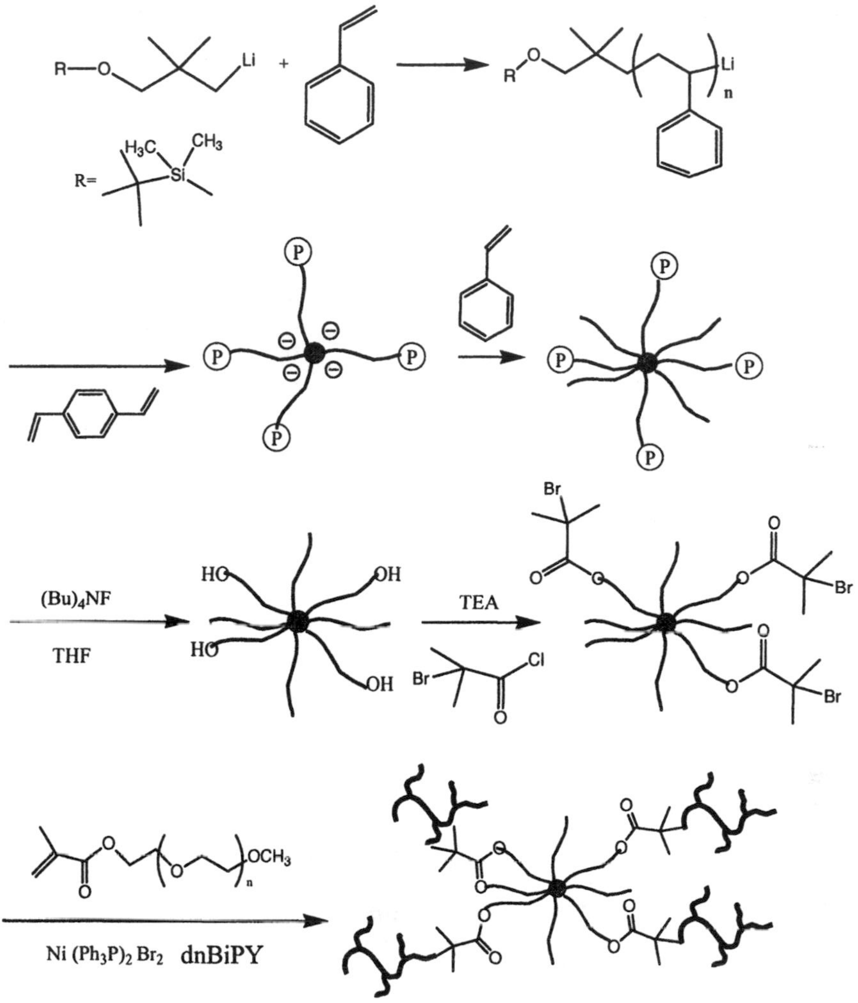

Extensive particle crosslinking would, in principle, not be necessary for templating provided the polymer is very insoluble in the polymerizing media and the chains collapse upon mixing. Chain collapse without compensation would, however, lead to rapid aggregation, a situation which is unacceptable for the formation of controlled nanoporosity. For such systems, functionality, which is compatible with the polymerizing media, needs to be present on the surface of the collapsed particles to stabilize the dispersion and prevent aggregation. For admixture with polar silsesquioxanes, a micellar structure with a hydrophobic core and hydrophilic corona is ideal. In order to avoid the complex dynamic equilibra associated with small molecule and linear polymeric amphiphiles,[16] we sought synthetic routes to unimolecular polymeric micelles,[17] i.e., polymer molecules with many arms, each with amphiphilic character. In such a molecule, collapse of the inner core would be offset by the presence of a compatibilizing corona to prevent precipitation and aggregation. Any synthetic route to such materials would ideally also provide control of the polymer molecular weight and polydispersity.

The combination of anionic polymerization[18] and atom transfer polymerization[19] provides a versatile route to such materials. It is known that living polystyrene anionic chains produce star polymers in the presence of a crosslinking reagent such as divinyl benzene.[20] Initial reaction produces a polyanionic star with linear polymer arms emanating from a small polyanionic core. Subsequent addition of more polymerizable monomer leads to the growth of additional arms from the core. Ideally, the number of arms added at each stage would be equal and the polymer is basically a star where the small crosslinked core serves only as a scaffold for the polymer arms. If the initiating, living, linear macromonomer has additional functionality, this is incorporated into 1/2 of the total polymer arms and can be used for post polymerization elaboration. Such a route is depicted in Scheme 1. Here the initiator was 3-t-butyldimethylsiloxy-2,2-dimethyl propyl lithium available from FMC Corporation. The functionalized star polymer decorated with t-butyldimethyl siloxy functionality is produced in a controlled fashion. GPC data shown in Table 1 suggests a M_n=80K g/mol and a polydispersity index = 1.05. The GPC molecular weight is inaccurate, however, since it is based on linear polystyrene standards and static light scattering yielded a M_w = 269K g/mol. In this case, the calculated number of functional arms was 15 and the hydrodynamic radius (THF) measured by dynamic light scattering (DLS) was 7.8 nm. The t-butyldimethylsilyl protecting groups were removed with tetrabutyl ammonium fluoride (1M in THF) and the pendant hydroxy functionality esterified with □-bromo isobutyryl chloride in methylene chloride in the presence of 4-N,N-dimethylaminopyridine (DMAP) and triethylamine (TEA).

The material so functionalized constitutes a multiarm macroinitiator for ATRP processes. From this core was grown a polar corona by the polymerization of polyethylene glycol methacrylate (PGG-DP = 7) using bis-triphenylphosphine nickel II bromide as the catalyst in toluene (95°C, 40% solids, initiator/catalyst = 0.23 , nickel/monomer = 0.019). The amphiphilic product was isolated by precipitation and the polymer data listed in Table 1. Here GPC analysis is particularly troublesome, since the materials stick to the column resulting in very low estimates for the molecular weight. From ^{1}H-NMR, the degree of polymerization/arm was estimated at ~14 and the measured R_h (THF) = 10.1 nm (~ 20% size increase relative to the polystyrene core).

	GPC Molecular Weight (Mn)[c]	PDI (GPC)	Light Scattering (Static) Molecular Weight (Mw)[c]	PDI	Dynamic Light Scattering (DLS)[c] (diameter nm)
Protected PS-Star	99K	1.11	268K	1.06	15.5
Hydroxy PS-Star	80K	1.05	265K	1.06	15.8
PS-Star Initiator	80K	1.05	270K	1.07	15.6
PS-PEG Core-Shell[a,b]	20K	1.35	355K	1.09	20.2

Table 1: Functionalized stars prepared by anionic/ATRP polymerization (Scheme 1). (a) Total number of arms = 44; (b) number of functionalized arms = 22 (c) measured in THF.

These materials were mixed with the MSSQ-SiO$_2$ copolymer resin (solids content = 30 wt.%) at various loading levels based on dry weight and spun as thin films. After heating at 80°C/1h to remove the solvent, the wafer were heated to 450°C (5 °/min) to produce the porous films. Figure 14 shows a plot of refractive index vs. porogen loading for the amphiphilic stars versus a standard poly(alkylene oxide)(NG) porogen. Where directly compared, the refractive index numbers are practically identical.

Comparison of film morphologies by FE-SEM as a function of loading level (Figure 15) suggests that the polymeric unimolecular micelles are functioning via a templating mechanism. In this regard, Figure 15 shows FE-SEM cross-sectional micrographs of porous MSSQ-SiO$_2$ films demonstrating that the morphologies are largely compositionally invariant, i.e., higher loading levels just produce more holes of the same size in the film. Finally, measurement of the average hole size in the film corresponds reasonably well to that of the initial particle determined by DLS, although the holes are slightly smaller. The latter is not surprising since sizes measured in solution by DLS represent those for solvent-swollen systems.

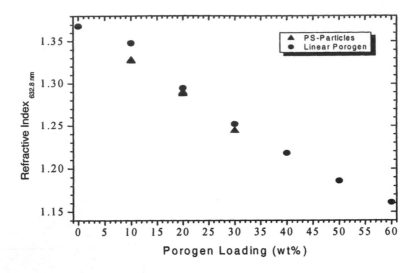

Figure 14: Refractive index vs. porogen plots in MSSQ-SiO$_2$ copolymer. Porogens were either the amphiphilic particles (▲) or a poly(alkylene oxide) (●)

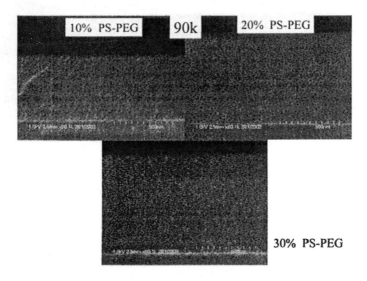

Figure 15: TEM micrographs of porous MSSQ-SiO$_2$ copolymer generated using amphiphilic particles at various loading levels

In summary, we have discussed the efficacy of producing porous organosilicates via the sacrificial macromolecular porogen route. By this technique, we have achieved dielectric constants of < 1.5 and pore sizes < 6 nm. Depending on the nature of the interactions between the porogens and resins, nanoporosity has been demonstrated from linear, branched and star polymers via a nucleation and growth mechanism. On the other hand, polymeric unimolecular micellar materials prepared by tandom-anionic procedures appear to function by a templating mechanism. A manifestation of this is the observation of a composition independent morphology and the observation that one polymeric molecule produces one hole of comparable size. Polymer templating eliminates many of the process variables endemic to the more complex nucleation and growth procedures.

ACKNOWLEDGMENTS: The authors gratefully acknowledge partial funding for this work from NIST-ATP contract No. 70NANB8H 4013.

REFERENCES

1. Wilson, S. R.; Tracy, C. J.; Freeman, Jr., J. L.; In *Handbook of Multilevel Metallization of Integrated Circuits*; Wilson, S. R.; Tracy, C. J.; Freeman, Jr., J. L., Eds.; Noyes Publications: Park Ridge, NF, 1993, Chap. 1.
2. Reference 1, Chap. 12.
3. Miller, R. D. *Science* **1999**, *286*, 421 and supplementary material.
4. McCoy, M. *C&E News* August 12, 2002, p. 17.
5. Corbett, M. A.; Davis, J. C. *Solid State Technology* **2001** (October).
6. Volksen, W.; Hawker, C. J.; Hedrick, J. L.; Lee, V. Y.; Magbitang, T.; Toney, M.; Miller, R. D.; Huang, E.; Lui, J.; Lynn, K. G.; Petkov, M.; Rodbell, K.; Weber, M. H. In *Low Dielectric Constant Materials for IC Applications*; Ho, P. S.; Lui, J.; Lee, W. W., Eds.; Springer Series in Advanced Microelectronics; Springer: Berlin, 2002, Chap. 6.
7. Kiefer, J.; Hedrick, J. L.; Hilborn, J. G. *Adv. Polym. Sci.* **1999**, *147*, 161.
8. Hedrick, J. L.; Magbitang, T.; Connor, E. F.; Glauser, T.; Volksen, W.; Hawker, C. J.; Lee, V. Y.; Miller, R. D. *Chem. Eur.* **2002**, *15(8)*, 3309 and references cited therein.
9. Huang, Q. R.; Volksen, W.; Huang, E.; Toney, M.; Frank, C.W.; Miller, R. D. *Chem. Mater.* **2002**, *14*, 3676.
10. Baney, R. H.; Itoh, M.; Sakakibara, A.; Suzuki, T. *Chem. Rev.* **1995**, *95*, 1409.
11. a) Frye, C.L.; Collins, W.T. J. Am. Chem. Soc. **1970**, 92, 5586 b) Weiss, K.D.; Frye, C.L. US Patent # 4,999,397 (**1991**).
12. Mecerreyes, D.; Hawker, C. J.; Hedrick, J. L.; Miller, R. D. *Adv. Mater.* **2001**, *13(3)*, 204.

160

13. Huang, E.; Toney, M. F.; Volksen, W.; Mecerreyes, D.; Brock, P.; Kim, H.-C.; Hawker, C. J.; Hedrick, J. L.; Lee, V. Y.; Magbitang, T.; Miller, R. D.; Lurio, L. B. *Appl. Phys. Lett.* **2002**, *81(12)*, 2232.

14. a) Brinker, C. J.; Lu, Y.; Sellinger, A.; Fan, H. *Adv. Mater.* **1999**, *11(7)*, 579. b) Raman, N. K.; Anderson, M. T.; Brinker, C. J. *Chem. Mater.* **1996**, *8*, 1682. c) Bruinsma, P. J.; Hese, N. J.; Bontha, J. R.; Lui, J.; Baskaran, S. *Proc. Mater. Res. Soc.* **1997**, *443*, 105. d) Liu, J.; Kim, A. Y.; Wang, L. Q.; Palmer, B. J.; Chen, Y. L.; Bruinsma, P.; Bunker, B. C.; Evankos, G. J.; Rieke, P. C.; Frysell, G. E.; Virden, J. W.; Laraservich, B. J.; Chick, L. A. *Adv. Colloid, Interf. Sci.* **1996**, *69*, 131. e) Lu, Y.; Ganguli, R.; Drewien, C.A.; Anderson, M. T.; Brinker, C. J.; Gong, W.; Guo, Y.; Soyeh, H.; Dunn, B.; Huang, M. H.; Zink, J. I. *Nature* **1997**, 389, *364.* f) Zhao, D.; Huo, Q.; Leng, J.; Chmelka, B. F.; Stucky, G. D. *Science* **1998**, *279*, 548. g) Zhao, D.; Yang, P.; Melosh, N.; Feng, J.; Chmelka, B. F.; Stucky, G. D. *Adv. Mater.* **1998**, *10(16)*, 1380.

15. Gallagher, M.; Adams, T.; Allen, C.; Annan, N.; Blankenship, R.; Calvert, J.; Fillmore, W.; Gore, R.; Gronbeck, D.; Ibbitson, S.; Jehoul, C.; Lamola, A.; Prokopowicz, G.; Pugliano, N.; Sullivan, C.; Talley, M.; You, Y. Proc. Polym. Mater. Sci. Eng, **2002**, 87,442.

16. Yang, S.; Mirau, P.A.; Pai, C.-S.; Nalamasu, O.; Reichmanis, E.; Pai, J.C.; Obeng, Y.S.; Seputro, J.; Lin, E.K.; Lee, H.-J.; Sun, J.; Gidley, D.W. Chem. Mater. **2002**, 14, 369.

17. a) Heise, A.; Hedrick, J. L.; Frank, C. W.; Miller, R. D. *J. Am. Chem. Soc.* **1999**, *121*, 8647. b) Kikuchi, A.; Nose, T. *Macromolecules* **1996**, *29*, 6770. c) Hawker, C. J.; Wooley, K. L.; Fréchet, J. M. J. *J. Chem. Soc., Perkin Trans. 1* **1993**, 1287. d) Newkome, G. R.; Morrefield, C. N.; Baker, G. R.; Saunders, M. J.; Grossman, S. H. *Angew. Chem. Int. Ed.* **1991**, *30*, 1178. e) Tomalia, D. A.; Berry, V.; Hall, M.; Hedstrand, D. M. *Macromolecules* **1987**, *20*, 1164. f) Stevelmans, S.; van Hest, J. C. M.; Jansen, J. F. G. A.; van Boxtel, D. A. F. J.; de Brabander-vanden Berg, E. M. M.; Meijer, E. W. J. *J. Am. Chem. Soc.* **1996**, *118*, 7398. g) Gitsov, I.; Fréchet, J. M. J. *J. Am. Chem. Soc.* **1996**, *118*, 3785.

18. Morton, M. *Anionic Polymerization: Principles and Practice*; Academic Press: New York, NY, 1983.

19. a) Patten, T. E.; Matyjaszewski, K., K. *Adv. Mater.* **1998**, *10*, 961. b) Matyjaszewski, K. In *Controlled Radical Polymerization*; Matyjaszewski, K., Ed.; ACS Sym. Ser. No. 685; American Chemical Society: Washington, DC, 1998.

20. a) Lutz, P.; Rempp, P. *Makromol. Chem.* **1988**, *189*, 1051. b) Gnanou, Y.; Lutz, P.; Rempp, P. *Makromol. Chem.* **1988**, *189*, 2885. c) Hadjichristides, N. *J. Polym. Sci.: Part A: Polym. Chem.* **1999**, *37*, 857.

Chapter 12

Porous Low-k Dielectrics: Material Properties

C. Tyberg[1], E. Huang[1], J. Hedrick[1], E. Simonyi[1], S. Gates[1],
S. Cohen[1], K. Malone[2], H. Wickland[2], M. Sankarapandian[2],
M. Toney[3], H.-C. Kim[3], R. Miller[3], W. Volksen[3], P. Rice[3],
and L. Lurio[4]

[1]IBM Thomas J. Watson Research Center, Yorktown Heights, NY 10598
[2]IBM Microelectronics, Hopewell Junction, NY 12533
[3]IBM Almaden Research Center, San Jose, CA 95120
[4]Northern Illinois University, Dekalb, IL 60115

Abstract

Improvements in back end of the line (BEOL) interconnect
performance require the reduction of resistance and
capacitance. The semiconductor industry, led by IBM, has
migrated from aluminum to copper wiring in a scaled manner
to lower resistance and capacitance in appropriate wiring
levels and enhance performance. More recently, industry
focus has centered on decreasing capacitance by reducing the
dielectric constant of the insulator. Numerous low k dielectric
candidates exist ranging from PECVD materials to organic
thermosets. Unfortunately, no potential candidate possesses
properties comparable to silicon dioxide, which has been the
primary insulator in semiconductor chips for over 30 years. In
this paper, properties and challenges of integrating low k
dielectrics will be described. In addition, extendibility of low
k dielectrics by incorporating porosity is discussed.

Introduction

For over 30 years silicon dioxide (SiO_2) has been the dielectric insulator of choice for the semiconductor industry. Silicon dioxide possesses excellent dielectric breakdown strength, a high modulus, good thermal conductivity and excellent adhesion to metallic liners, PECVD barrier cap layers, etc. However, with ground rule reductions and the need for improved interconnect performance, SiO_2 is being replaced with materials possessing lower permittivity to achieve reduced capacitance. Fluorosilicate glass (FSG), for instance, has replaced SiO_2 in high performance logic and SRAM technologies. IBM integrated FSG with copper and implemented the technology at the 0.18 μm technology node (1). For the 130 nm technology generation IBM selected the SiLK™ Semiconductor Dielectric integrated with copper for advanced BEOL interconnects (2). The combination of SiLK™ and copper reduces the normalized resistance/capacitance (RC) delay by 37% compared to silicon dioxide and aluminum structures.

At the 90 nm technology generation, the target effective dielectric constant (k_{eff}) according to the 2001 International Technology Roadmap for Semiconductors (3) is 2.6-3.1. The k_{eff} is a composite value comprised of the dielectric, hardmask (if present), barrier cap layer, and etch stop layer (if present) as shown in Figure I. To achieve a k_{eff} of ~3.0 the dielectric constant of the intermetal dielectric (IMD) must be ~2.7. Therefore, the introduction of ultra low k dielectrics is not expected until the 65 nm technology generation, where the target effective dielectric constant is 2.3-2.6 (3). To achieve this, ultra low dielectric constant, materials with k < 2.2 will be required. In order to attain sufficiently low dielectric constants, porosity must be incorporated.

Extendibility of unit processes and tooling in the fabrication of interconnects is dependent on the extendibility of the dielectric material and integration scheme. Therefore, the dielectric choice for the 90 nm generation ideally should provide a pathway toward ultra low k porous dielectrics for the 65 nm generation and beyond.

Low k Dielectrics

The integration of low-k dielectrics into BEOL interconnects is not trivial. In fact, the immensity of the task parallels the transition from aluminum to copper wiring. The challenges can be attributed to the absence of low-k dielectrics available having electrical, thermal, mechanical, or thermal conductivity properties comparable to silicon dioxide. Low-k materials, in general, are less dense and typically possess a lower modulus and hardness as well as decreased thermal conductivity.

Table I shows a comparison of the properties of an organic low-k dielectric (SiLK™), SiCOH (a class of materials containing Si, C, O , and H), and silicon

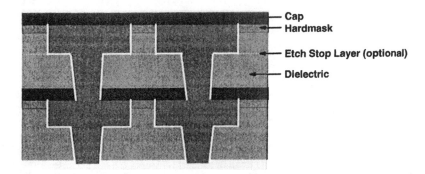

Figure I. Back end of the line (BEOL) wiring structure.

Table I. Comparison of Low-k Dielectric Material properties to Silicon Dioxide

Property	Organic Dielectric (SiLK™)	$Si_wC_xO_yH_z$	Silicon Dioxide
Dielectric Constant	2.62	2.7-3.0	3.9-4.5
Leakage Current at 1 MV(A/cm²) at 150°C	3.3×10^{-10}	$\sim 1 \times 10^{-10}$	$\sim 2 \times 10^{-11}$
Breakdown Field (MV/cm²)	~4	6-10	>8
Modulus (GPa)	2.7	9-15	72
Hardness (GPa)	0.25	1.3-2.4	8.7
Toughness MPa-m$^{1/2}$	0.62	~ 0.28	0.8
Thermal Conductivity (W/mK)	0.19	0.40	1.07

dioxide. Comparison of the organic low-k dielectric and SiCOH demonstrates that SiCOH possesses a significantly higher modulus and hardness, but the organic low-k dielectric (SiLK^{TM}) possesses superior toughness. However, in terms of material properties, SiO_2 is far superior to both low k dielectrics. Silicon dioxide possesses: a modulus that is 5 to 25 times higher; thermal conductivity that is 2.5 to 5 times higher; and a hardness value that is 4 to 30 times higher than SiCOH and SiLK^{TM}.

Extendibility of Low-k technologies with Porous Dielectrics

With the prospect of changing dielectrics, liners, hardmasks and barrier cap layers in future technologies to enable performance enhancements and maintain reliable interconnects with ground rule reductions, it is essential to evaluate the extendibility of current technologies. Extendibility is important to reduce development costs and capital expenditures, thus it is critical to maximize the lifetime of development efforts and avoid re-tooling of the semiconductor infrastructure with each technology generation. Thus, 90 nm generation unit processes and structures ideally should enable evolution to the 65 nm technology node and beyond.

With k_{eff} targets of ~2.3-2.7 for the 65 nm generation, the incorporation of porosity into dielectrics to enable reduced permittivity is inevitable. The utilization of porous dielectrics in BEOL interconnects will pose additional challenges beyond conventional low-k technology. A thorough characterization of the porous structure along with the accurate measurement of electrical and mechanical properties will be critical in the selection of ultra low-k porous dielectrics.

Integration of ultra low dielectric constant porous dielectrics is extremely challenging due to the reduction of mechanical properties in comparison to dense low-k dielectrics. Mechanical properties such as modulus, hardness, fracture toughness, adhesion, and coefficient of thermal expansion are critical parameters in screening materials to achieve successful integration. Porous materials with a dielectric constant of 2.2 may have a ~30% reduction in modulus and hardness due to the incorporation of porosity. Similarly, the fracture toughness and adhesion may also be lowered by the incorporation of porosity. This reduction in the mechanical properties can result in delaminations during chemical mechanical polishing or failure during reliability testing or packaging.

Another property of importance is the coefficient of thermal expansion. Any mismatch in the coefficient of thermal expansion of the dielectric insulator and the metal interconnects can create stress on liners and metal lines during processing and reliability testing. This is especially critical for organic dielectrics that have significantly higher thermal expansion than the metal wiring.

Table II shows a comparison of the properties of an organic spin-on ultra low dielectric constant material (porous SiLK^{TM}), to silicon based ultra low-k dielectrics by both spin-on (primarily silsesquioxane materials) and CVD processes (4). Porous organic dielectrics have slightly lower modulus and hardness, lower breakdown voltages, and higher coefficients of thermal expansion than the SiCOH type dielectrics of the same k value. In addition, the pore size of porous organic dielectrics is often larger than that of the porous SiCOH based dielectrics of an equivalent dielectric constant. However, porous organic dielectrics, such as porous SiLK^{TM}, are significantly tougher than porous SiCOH dielectrics and are not susceptible to cracking; whereas, most silsesquioxane (spin-on porous SiCOH) based porous dielectrics have a crack threshold of less than 2 microns, and are susceptible to stress corrosion cracking. In addition, the porous organic dielectric evaluated exhibits superior adhesion behavior during blanket metalized chemical mechanical polishing, as delaminations are not observed at down forces up to 9 psi. Porous silsesquioxane dielectrics tend to show delaminations at down forces as low as 1-2 psi.

Therefore, as with the non-porous low-k dielectric candidates, the competing ultra low-k porous dielectrics exhibit similar trade-offs in properties as organic dielectrics have lower modulus, higher CTE, and larger pore size, but significantly improved toughness and adhesion in comparison to silicon based porous dielectrics. These trade-offs make the choice of the best material for integration difficult. However, due to the need for an extendable integration approach, the choice of porous dielectric will likely depend on the dielectric choice and integration success of the 90 nm technology generation.

Characterization of Porous Dielectrics

Mechanical and thermal characterization of ultra low k dielectrics is very similar to the characterization of dense low k dielectrics; however the introduction of porosity requires the development of new characterization techniques in order to understand the pore structure. The dielectric constant, dielectric breakdown and coefficient of thermal expansion can be measured using the same techniques used for dense low k dielectrics. Modulus and hardness can also be measured by the same techniques, however, if using nano-indentation, measurements from porous dielectrics may have larger substrate contributions at equivalent film thicknesses. Therefore, the modulus values

Table II. Mechanical Properties of Porous Dielectric

Property	Porous Organic (Porous SiLK^{TM})	Porous $Si_wC_xO_yH_z$ (Spin-on)	Porous $Si_wC_xO_yH_z$ (CVD)
Dielectric Constant	2.15	1.4-2.3	2.05-2.5
Modulus (GPa)	2.1-2.5	1.5 – 3	3 - 8
Hardness (GPa)	0.10-0.15	0.10 – 0.35	0.2-1.2
CTE (ppm/°C)	~ 65-70	~14	~12
Crack Threshold (μm)	> 25	< 2	< 2
Blanket metalized CMP evaluation	No delaminations tested up to 9 psi.	Delaminations observed at 1-2 psi.	No delaminations at 6 psi.
Avg. Pore Size (nm)	~8	< 5	< 3

reported for porous dielectrics are relative values that are dependent on the film thicknesses. The modulus and hardness values reported in this paper were obtained from one-micron thick films to be consistent with the current standard.

Pore Characterization

Perhaps the most important criteria for nanoporous dielectrics is that the pore size must be substantially smaller than device structures. Pores comparable or larger than the line width of the dielectric layers may lead to numerous issues including: localized deformation or structural collapse due to insufficient support, formation of shorts, and degraded performance associated with fringing electrical fields resulting from the inhomogeneous structure between lines. Generally, closed celled structures are also required to minimize the introduction of contaminants into the dielectric during wet cleans, chemical mechanical polishing steps, and other processes. Therefore, an understanding of the pore size, size distribution, pore morphology, and their formation is necessary to successfully utilize nanoporous dielectrics.

Numerous methodologies to characterize the pore structure in ultra low-k dielectrics have been employed on a variety of systems. Each of the methods has their inherent strengths and drawbacks; consequently, it is often desirable to utilize multiple techniques to obtain complementary information. The applicability of the various techniques will be dependent on numerous factors including the nature of the porous dielectric, the information desired, whether a non-destructive technique is required, and accessibility to instrumentation. Generally, these characterization approaches can be delineated into two classes according to the information obtained; visualization methods and methods that provide averaged information about the pore structure in the film.

First, visualization approaches including the various microscopy techniques: scanning electron microscopy (SEM), transmission electron microscopy (TEM), and atomic force microscopy (AFM) have commonly been utilized. These techniques can potentially provide: pore size, information regarding how pores are distributed through the film, porosity levels, and insight into the morphology, e.g., open celled structures. Figure II a and b show cross sectional scanning electron micrographs of a porous methyl silsesquioxane (MSSQ), i.e., one type of porous SiCOH, and a porous polyarylene polymer, respectively. The porous MSSQ sample, having a porosity of ~20%, was simply cleaved prior to imaging. The porous organic film was cleaved and polished prior to imaging. In both images, the pores are clearly evident from the dark regions in the micrograph and a pore size can be estimated to be ~ 7 to 10 nm and 10 to 12 nm for the porous MSSQ and porous organic film, respectively.

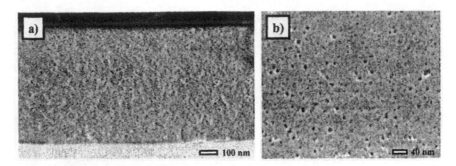

Figure II. Cross sectional scanning electron micrographs of a) a 20% porosity MSSQ film after film cleavage b) porous organic dielectric after film cleavage and polish

Figure III a and b show cross sectional transmission electron micrographs of the identical systems. The film specimens were defined by focus ion beams (FIB) to generate cross sectional thicknesses of (50-100 nm). Areas containing pores appear lighter in color due to lower attenuation of the electron beam in those regions. The pores in these images are consistent in size to the values observed by SEM.

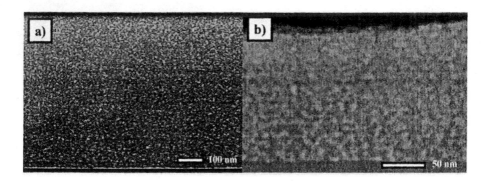

Figure III. Transmission electron micrographs of a) a 20% porosity MSSQ film b) porous organic dielectric

Although, these images provide excellent insight into the pore structure and size, careful interpretation of the images is often warranted. Artifacts introduced during sample preparation may preclude accurate assessment of the pore structure. For example, with SEM, artifacts may be generated during film cleavage or polishing steps. For TEM, FIB preparation must be performed carefully to avoid damage of the dielectric or redeposition of sputtered materials into the porous structure. Furthermore, quantitative analysis of the pore structure to obtain a pore size distribution can be difficult. Limitations in resolution or contrast tend to overemphasize the presence of larger pores. In addition, pores intersecting the cleavage or FIB interface may appear smaller in cases where this interface does not coincide with the center of the pore. For films studied by TEM where the pores are substantially smaller than the specimen thickness, superimposition of pores in the direction of the transmitted electron beam creates ambiguity in defining the pore structure.

The second type of characterization approach includes methods that provide averaged information of the pore structure. Unlike the visualization techniques described above, these techniques can sample substantially larger numbers of pores as they are not limited to the local field of view and consequently can provide a better statistical representation of the structure. These include: small angle x-ray scattering (SAXS) (5), small angle neutron scattering (SANS) (6,7), ellipsometric porosimetry (EP) (8), and positron annihilation lifetime spectroscopy (PALS) (9,10). These approaches can also be faster and less tedious to perform, making them better suited for detailed studies involving numerous samples. Furthermore, they may be applied to in-situ studies in order to probe the evolution of the pore structure. The primary drawback of these methods, however, is that data analysis is often nontrivial and assumptions and models regarding the pore structure, e.g., pore shape, often must be made.

Although EP and PALS have been demonstrated as very useful techniques for porous silicate based systems, these techniques are less effective for the porous SiLK systems. For EP, these limitations may possibly be attributed to swelling effects from the solvent probe that may complicate analysis. For PALS, it is observed that the ratio of positron annihilation by 3γ emission to 2γ processes, which is proportional to the porosity, is markedly lower than expected values based on known porosity levels. In contrast, SANS, which is sensitive to the scattering length density differences between the matrix material and pore, i.e., air, is effective with porous organic materials but is not effective in evaluating many silicate systems. This limitation is attributed to the low scattering length densities of many of the silicate systems having a high hydrogen content resulting in a lack of contrast between the pore and matrix material.

Thus, in this treatment, small angle x-ray scattering (SAXS) is used to obtain quantitative values for the pore size and size distributions. SAXS, which is sensitive to the electron density difference between the matrix and pores, is effective for both silicate and organic systems. Figure IV (left) shows the scattering intensities (symbols) for a porous silsesquioxane and SiLK system, as a function of the scattering vector, $q=(4\pi/\lambda)\sin(\theta)$, where 2θ is the angle between the scattered photon and transmitted beam and λ is the wavelength of the x-ray. Model fits (solid lines), using treatments described by Pedersen (11), whereby pores are assumed to be spherical and are locally monodisperse in size, provide pore size distributions of the films as shown in Figure IV (right) and are in good agreement with the electron microscopy images. A detailed description of SAXS on porous dielectrics has been presented elsewhere (5).

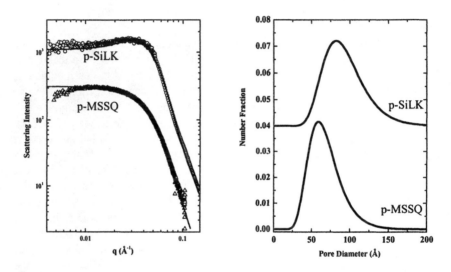

Figure IV. (Left) SAXS intensities (symbols) plotted as a function of the scattering vector with best model fits (lines). (Right) Pore size distributions obtained from model fits.

It should be noted that the data above compares only a single version of the porous organic dielectric (porous SiLK[TM]) and porous SiCOH dielectric and is not intended to generalize these classes of systems. Depending on system design and processing, the pore size can vary dramatically. In many porous SiCOH systems, for example, substantially smaller pores may be attained for comparable porosities.

Conclusions

The significant reduction in capacitance required for enhanced interconnect performance has resulted in an extensive effort to understand low k dielectrics and ultra low k porous dielectrics. Although the transition from silicon dioxide to dielectrics with k < 3 was extremely challenging, the next step to porous dielectrics is expected to be even more difficult. In order to facilitate the conversion to porous dielectrics, it is critical to consider extendibility when selecting the dense low k dielectric for the 90 nm technology generation. The low k dielectric proven to be the most successful for the 90 nm generation, whether it's spin-on or CVD, organic or SiCOH based, will likely dictate the first generation of ultra low k porous dielectrics.

Acknowledgements

The authors would like to acknowledge the support of the National Institute of Standards and Technology (NIST) for their financial support on the development of porous dielectrics under the Advanced Technology Program (ATP) in collaboration with The Dow Chemical Company. The SAXS experiments were performed on the IMM-CAT beamline 8-ID at the Advanced Photon Source and was supported by the U. S. Department of Energy, Office of Science, Office of Basic Energy Sciences, under Contract No. W-31-109-ENG-38.

References:

1. E. P. Barth, T. H. Ivers, P. S. McLaughlin, A. McDonald, E. N. Levine, S. E. Greco, J. Fitzsimmons, I. Melville, T. Spooner, C. DeWan, X. Chen, D. Manger, H. Nye, V. McGahay, G. A. Biery, R. D. Goldblatt and T. C. Chen, *Proceedings of the International Interconnect Conference*, **2000**, 219-221.
2. R. D. Goldblatt, B. Agarwala, M. B. Anand, E. B. Barth, G. A. Biery, Z. G. Chen, S. Cohen, J. B. Connolly, A. Cowley, T. Dalton, S. K. Das, C. R. Davis, A. Deutsch, C. DeWan, D. C. Edelstein, P. A. Emmi, C. G. Faltermeir, J. A. Jitzgimmons, J. C. Hedrick, J. E. Heidenreich, C. K. Hu, J. P. Hummel, P. Jones, E. Kaltalioglu, B. E. Kastenmeier, M. Krishnan, W. F. Landers, E. Linger, J. Liu, N. E. Lustig, S. Malhotra, D. K. Manger, V. McGahay, R. Mih, H. A. Nye, S. Purushothaman, H. A. Rathore, S. C. Seo, T. M. Shaw, A. H. Simon, T. A. Spooner, M. Stetter, R. A. Wachnik and J.

G. Ryan, *Proceedings of the International Interconnect Conference*, **2000**, 261-263.

3. The National Technology Roadmap for Semiconductors, Semiconductor Industry Association: San Jose, CA, 2001.

4. A. Grill, V. Patel, K.P. Rodbell, E. Huang, S. Christiansen, *Mat. Res. Soc. Symp. Proc.*, **2002**, 716, B12.3.

5. E. Huang, M. F. Toney, W. Volksen, D. Mecerreyes, P. Brock, H.-C. Kim, C. J. Hawker, J. L. Hedrick, V. Y. Lee, T. Magbitang, R. D. Miller. *Applied Physics,* **2002**, 81, 2232.

6. W.-l. Wu, W. E. Wallace, E. K. Lin, G. W. Lynn, C. J. Glinka, E. T. Ryan, and H.-M. Ho, *J. Appl. Phys.*, **2000**, 87 (3), 1193.

7. G. Yang, R. M. Briber, E. Huang, H.-C. Kim, P. Rice, R. D. Miller, and W. Volksen, Appl. Phys. Lett. (In press).

8. M. R. Baklanov, K. P. Mogilnikov, V. G. Polovinkin, and F. N. Dultsev, *J. Vac. Sci. Technol. B* , **2000**, 18, 1385.

9. D. W. Gidley, W. E. Frieze, T. L. Dullo, A. F. Yee, E. T. Ryan, and H.-M. Ho, *Phys. Rev. B* , **1999**, 60 (8), R5157.

10. M. P. Petkov, M. H. Weber, K. G. Lynn, K. P. Rodbell, and S. A. Cohen, *J. Appl. Phys.,* **1999**, 86 (6), 3104.

11. J. S. Pedersen, *J. Appl. Crystallogr.* , **1994**, 27, 595.

Chapter 13

Ultra Low-k Dielectric Films with Ultra Small Pores Using Poragens Chemically Bonded to Siloxane Resin

Bianxiao Zhong and Eric S. Moyer

Dow Corning Corporation, Midland, MI 48686–0994

Burning Out Sacrificial Spacers or BOSS technology has been developed to fabricate porous ultra low-k (ULK) dielectrics having ultra small pores. The technology utilizes BOSS resin, which has a curable resin backbone and poragen groups chemically linked to the resin backbone. The chemical bond between the resin and poragen prevents the phase separation during curing, allowing the molecular level control of the pore size and distribution. Porous ultra low-k thin films were fabricated from the resin using conventional spin coating and furnace cure. A proto-type BOSS material, a SiH siloxane resin having isophytol poragen groups, led to films having a dielctric constant of 1.8, breakdown voltage of 4 MV/cm, and an average pore size of 2.2 nm by PALS. The film mechanical strength was identified as the key challenge for this technology, or any porous ULK technology.

ULK dielectrics are needed by IC industry to reduce the IC interconnect delay. One of the approaches toward these materials is incorporating porosity into spin-on thin films. An ideal ULK material should redeem itself to porous during conventional thermal cure. For easy integration, the pore size of the films should be no more than a few nanometers, or substantially smaller than the IC feature size. Due to the large internal surface area of a porous material, physical adsorption of moisture onto the surface may cause moisture sensitivity of dielectric constants. Therefore, the internal surface of a porous material needs to be hydrophobic to reduce moisture adsorption.

Most of the known porous low-k technologies utilized the blending approach that was pioneered by IBM researchers [1, 2]. In this approach, nonvolatile organic molecules were blended into a curable siloxane resin to serve as "molecular templates", or poragens. When the blend was cured thermally, the resin matrix became rigid and the poragen molecules were decomposed to leave pores behind. In this way, the dielectric constants of the matrix materials were significantly reduced.

A modified blending approach, Dow Corning XLK^{TM} technology [3], used a volatile solvent having a high boiling point as the poragen. Ammonia treatment was utilized to facilitate the resin network formation at a temperature below the boiling temperature of the poragen. The porosity and dielectric constants of XLK films were controlled by varying the loading of the poragen. The average pore diameter as measured by PALS [4] also varied with the poragen loading. For a film with a dielectric constant of 2.0, the average pore diameter was determined to be 4 nm.

In this paper, we will discuss a technology for fabrication of ULK films having ultra small pore sizes, and the work that led to the technology.

Blending Approach

It is known that the small pore size and narrow pore-size distribution is one of critical requirements for ULK materials. In order to develop an approach to porous ULK materials with ultra-small pore sizes, the blending approach was first examined using hydrogen silsesquioxane (HSQ) resin as the model matrix. HSQ resin has been used as a spin-on dielectric material with a dielectric constant of 2.9 (Dow Corning FOx^{TM}). A number of organic molecules were evaluated as poragens. Selection of poragens was based on their compatibility with HSQ resin and their thermal stability. An incompatible poragen may phase-separate from the resin during coating or curing, causing non-uniform films or very large pore sizes. Poragens that volatilize or decompose before the resin is cured are not capable of inducing significant porosity in the films.

Samples of HSQ/poragen blends were cured in crucibles at 470 °C for 1 hour in nitrogen, and the resulting solids were analyzed by nitrogen adsorption at

77 °K. All of the samples displayed type IV isotherms, indicating their mesoporous characteristics [5] as shown in Figure 1. Average pore diameters of these porous solids ranged from 2.6 to 5.3 nm based on BJH analysis [6] as shown in Table I. Among the poragens studied, laurone yielded the smallest pore size, 2.6 nm, and the most narrow pore size distribution. Despite a great effort, porous solids with pore size smaller than 2.6 nm were unable to be prepared using the blending approach. Even the smallest pore size, 2.6 nm, was much larger than that of individual poragen molecule, suggesting that the aggregation of the poragens was not avoidable. The sizes of the poragen aggregates should have been impacted by the compatibility between the resin and the poragen.

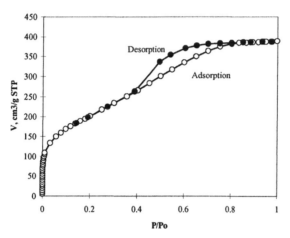

Figure 1. N$_2$ adsorption isotherm a cured HSQ/triaconene blend containing 40 wt.% triaconene.

One may hypothesize that a perfect compatibility would result in the smallest pore size for a given poragen and a given poragen loading. The best way to achieve compatibility between a resin and a poragen is to link the two together chemically. This hypothesis was supported by a simple experiment, a direct comparison between two cured HSQ/triaconcene samples: one was a simple blend of HSQ and triaconcene in 1:1 weight ratio, and the other was the reacted blend via hydrosilylation catalyzed by a Pt catalyst, where the poragen groups were chemically attached to the resin. As shown in Figure 2, the cured sample of the blend displayed a Type IV nitrogen adsorption, indicating a mesoporous solid, and an average pore diameter of 5 nm was obtained from the isotherm by using BJH analysis. On the other hand, the otherwise identical system having poragen linked to the resin backbone displayed a type I isotherm, indicating a microporous solid [7], and an average pore diameter of 1 nm was obtained from the isotherm using H-K analysis [8].

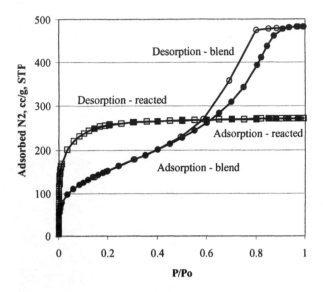

Figure 2. Nitrogen adsorption isotherms at 77K for cured HSQ/triaconcene (in 1:1 weight ratio) samples: blend and reacted by hydrosilylation in presence of a Pt catalyst.

Table I. Nitrogen adsorption data on cured HSQ/poragen blends.

Poragen	Poragen wt %	BET S, m2/g	Total porosity, %	BJH pore diameter, A
Laurone	40	679	32.5	26.0
Hexadecyl hexadecanoate	50	625	41.3	34.4
Squalane	50	421	27.5	32.1
Didecylph-thalate	50	500	36.9	36.9
Triaconene	40	738	47.3	34.8
Triaconene	50	568	52.8	52.5

BOSS Approach

The hypothesis discussed in last section led to the development of a generic approach for fabrication of ULK materials with ultra-small pore sizes, Dow

Corning BOSS technology [9]. BOSS resins have poragen molecules chemically linked to the resin backbone. Therefore, BOSS resins have a "perfect" compatibility between the resin backbone and the poragens, and even volatile molecular precursors can be used as poragens. A resin backbone in a BOSS resin can be a siloxane resin of different types such as HSQ and methyl silsesquioxane (MSQ) resins. A poragen can be different in type, size and loading. The basic requirements for a BOSS resin are: (1) the resin backbone is thermally curable; (2) the resin backbone becomes rigid during cure before the poragen is decomposed; (3) the resin is coatable; (4) the removal of the poragen does not yield polar groups after cure; and (5) the cured resin is hydrophobic. In this section, synthesis, thin film evaluation and integration of selected BOSS resins will be discussed.

Synthesis of BOSS Resins

There are two synthetic methods to produce a BOSS resin. Method I is to graft a siloxane resin with a porgen. Method II is to co-hydrolyze silane mixtures, one of which has an attached poragen group. An example for method I is to graft HSQ resin with poragen molecules having vinyl groups by hydrosilylation in presence of a Pt catalyst [10, 11]. Olefins having terminal vinyl groups are more suitable for this reaction. Examples are linear, branched or substituted 1-alkenes. For most olefins except very small ones, about 30% SiH groups of HSQ resin are accessible.

Examples for method II [12-14] included the hydrolysis of a mixture of $HSi(OEt)_3$ or $MeSi(OMe)_3$, $Si(OEt)_4$, and $RSi(OEt)_3$, where R was an alkyl poragen group. $RSi(OEt)_3$ can be prepared by hydrosilylation of $HSi(OEt)_3$ with a vinyl-containing poragen.

Evaluation of BOSS Powder Samples

It was found that BOSS resins became microporous solids upon thermal cure in an inert atmosphere. For example, a sample of $[HSiO_{3/2}]_n[(n\text{-}C_{20}H_{21})SiO_{3/2}]_m$ resin having 46 wt% of $n\text{-}C_{20}H_{21}$ prepared using method I, discussed in last section, was cured at 470 °C in nitrogen for 1 hour. The resulting solid was analyzed for nitrogen adsorption at 77 °K using a Micromeritics ASAP2000 Porosimetry System. A Type I isotherm indicating a micropous solid was obtained as shown in Figure 3. The sample displayed a BET surface area of 620 m^2/g, a porosity of 31%, a median pore diameter of 0.6 nm and a narrow H-K pore size distribution as shown in Figure 4.

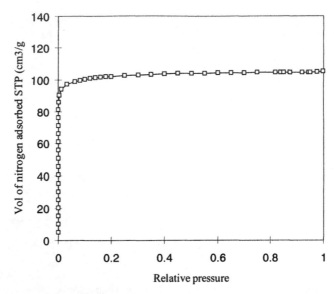

Figure 3. N_2 adsorption isotherm for cured $[HSiO_{3/2}]_n[(n\text{-}C_{20}H_{21})SiO_{3/2}]_m$ resin sample.

S

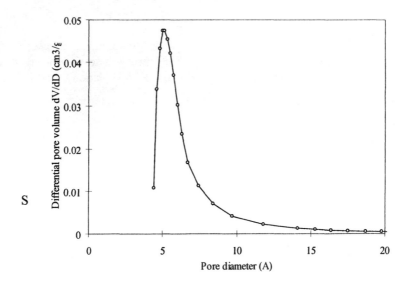

Figure 4. H-K pore size distribution for cured $[HSiO_{3/2}]_n[(n\text{-}C_{20}H_{21})SiO_{3/2}]_m$ resin sample.

Volatiles evolved from the BOSS sample during cure were collected using a liquid-nitrogen cold trap, and analyzed by GC in order to understand the decomposition chemistry of the poragen group. The volatiles consisted of all alkenes having 19 and fewer carbon atoms, suggesting a random radical decomposition. The absence of C20 hydrocarbon suggested that few Si-C bonds cleaved and $(n\text{-}C_{20}H_{21})SiO_{3/2}$ units became $MeSiO_{3/2}$ units during cure, predicating hydrophobic surfaces for the cured material, which were later confirmed for the cured films.

BOSS resins having 46 wt% of poragen of varying sizes, $(CH_2)_nCH_3$ with n = 11, 13, 15, 19, 23 and 29, were studied. The pore sizes of cured samples determined by nitrogen adsorption and H-K analysis varied only in small degree, 0.6 – 0.8 nm for the smallest poragen $(CH_2)_{11}CH_3$ to the largest poragen $(CH_2)_{29}CH_3$. This 33% increase in pore size was comparable to the increase of the poragen diameter; 36%, as estimated from their formula weights.

Evaluation of BOSS Thin Films

It appeared that the size of poragen group did not dramatically impact porosity and dielectric constants of the thin films made from $[HSiO_{3/2}]_n[(n\text{-}C_pH_{2p+1})SiO_{3/2}]_m$ with p = 12 – 24. When the films were cured in nitrogen at 470 °C for one hour, they all displayed low refractive indexes in the range 1.18 – 1.22, indicating the highly porous films. The FTIR spectra of the cured films indicated they all had SiH and SiMe functionality. The dielectric constants of these films were all determined to be low, between 1.8 and 2.2.

On the other hand, a poragen smaller than a certain size did dramatically impact the film porosity and dielectric constants. As shown in Table II, resins having small groups such as isobutyl or trifluoropropyl led to thin films having dielectric constant as high as that of dense HSQ films, indicating little porosity in these films. For linear alkyls, the threshold size appeared to be C_8H_{17} for producing films with a low k of 2.2 or smaller. For a given R, the k value of the

Table II. Dielectric constants of cured films from $[HSiO_{3/2}]_n[RSiO_{3/2}]_m$ resins. All films were cured at 470 °C for one hour in nitrogen.

R	R mole%	k
$n\text{-}C_{18}H_{37}$	24	1.97
$n\text{-}C_{12}H_{25}$	24	2.00
$i\text{-}C_4H_9$	24	3.05
$i\text{-}C_4H_9$	60	2.82
$_n\text{-}C_3H_4F_3$	60	3.02

cured BOSS films was decreased as the R loading was increased until a plateau was reached (1.8 – 2.2 depending on the types of resin and poragen).

The cure temperature of BOSS resin is primarily determined by the decomposition temperature of its poragen. Resins having linear alkyl poragen groups have to be cured at a very high temperature, 470 °C, for one hour to remove the poragen groups. This temperature is too high for IC manufacturing. A lower cure temperature can be achieved by using a poragen group having some weak bonds in the chain. For example, when isophytol, a branched C20 alkyl having an OH group on the third carbon, was used as the poragen group in a HSQ-based BOSS resin, the cure temperature could be as low as 425 °C for an 1-hour cure (Type 2 in Figure 5). Further reduction of the cure temperature to 400 °C (1 hour cure) was achieved by utilizing a poragen group containing oxygen atoms in the main chain such as polyethers (Type 3 in Fig. 5).

It was found necessary to cure the BOSS films in inert atmospheres such nitrogen. Curing in air would yield a significant amount of silanol in the film, causing a high moisture sensitivity of the dielectric. BOSS resins with poragen groups linked to Si atoms through Si-C bonds yielded hydrophobic surfaces after curing in nitrogen as indicated by their high water wetting angles. This was due to the Si-Me functional groups formed from the poragen groups as discussed in last section. The existence of Si-Me groups in cured films is confirmed by the CH absorbance in FTIR spectra.

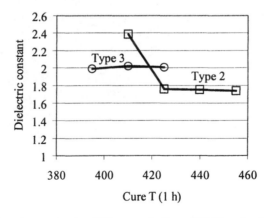

Figure 5. Dielectric constants of films made from BOSS resins having Type 2 or Type 3 poragen group and cured at different temperatures in nitrogen for 1 hour.

Studies on A Proto-Type BOSS Resin, $[HSiO_{3/2}]_n[(C_{20}H_{41}O)SiO_{3/2}]_m$

A proto-type BOSS resin, $[HSiO_{3/2}]_n[(C_{20}H_{41}O)SiO_{3/2}]_m$, made using method II was studied for thin film properties. The film was composed of 60% Si-Me, 30% SiO2 and 10% SiH as revealed by Rutherford Backscattering Spectrometry (RBS), Hydrogen Forward Scattering Spectroscopy (HFS) and FTIR analysis. Effectively, the methyl groups made the films hydrophobic. A summary of the thin film properties was shown in Table III. The films displayed a high cracking threshold, > 2 microns, an ultra-low dielectric constant of 1.8, a high breakdown voltage of 4 MV/cm, a high adhesion of > 60 MPa on Si or SiN, and a low film stress of 20 MPa. The high breakdown voltage for this highly porous material having an ultra-low k must be attributed to its ultra-small pore size. The downside of the performance was the low film modulus, 1.2 – 1.5 GPa.

Table III. Thin film properties for $[HSiO_{3/2}]_n[(C_{20}H_{41}O)SiO_{3/2}]_m$ resin. Films were cured at 450 °C for one hour in nitrogen.

Property	Performance
Maximum crack-free thickness	> 2.0 microns
Dielectric constant	1.8
Breakdown voltage	4 MV/cm
Modulus by nanoindentation	1.2 – 1.5 GPa (0.8 μm film)
Cohesive strength by Stud pull	> 60 MPa
Adhesion on SiN by Stud pull	> 60 MPa
Pore feature by PALS	Connected
Pore size by PALS	2.2nm
Initial stress	20 MPa

The pore characteristics of the film were studied using Positron Annihilation Lifetime Spectroscopy (PALS) [4] by Prof. Gidley at University of Michigan. It was found that all pores were interconnected, and the average pore diameter was 2.2 nm. Porosity was measured as 53% using Ellipsometric Porosimetry (EP) [15, 16] at IMEC.

Integration studies on this material were carried out, focusing on the critical process steps where a direct contact existed between the process and the porous low-k material [9]. The test structures consisted of a stack of 50 nm a-SiC:H, 390 nm BOSS film and dual hard mask of 50 nm a-SiC:H and 150 nm SiO2. The BOSS films were cured at 425 or 450 °C. An oxygen-free etch chemistry, Ar/CF4/CHF3, was shown to yield a straight vertical profile and a flat front with sufficient selectivity towards the photoresist. Ashing was successfully conducted by using highly anisotropic N2/O2 plasma at low pressure and low temperature. A suitable polymer stripper that was compatible to the BOSS material was identified. A unique property of this BOSS film was its superior compatibility with the metal barrier. Sheet resistance and Ellipsometric Spectrometry measurements were used for blank films to study the sealing performance of thin Ta(N) layers, and Ellipsometric Porosimetry and aqueous hydrofluoric acid dip tests were used to study patterned structures [17]. It was found that a Ta(N) layer as thin as 10 nm was capable of sealing the porous film efficiently. The ultra-small pore size of this material is likely one of the reasons for the unique property. The weakness of this BOSS film was its less than desired mechanical strength. The low modulus of the films and a relatively weak adhesion between the BOSS film and some of the hard masks might have caused some CMP damage.

It is well known that the mechanical strength is greatly reduced from that of the dense matrix when a high porosity is incorporated into a dielectric film. The most effective way to improve the mechanical strength of a porous material is to increase the mechanical strength of the matrix. One method for doing this is to increase the crosslink density of the matrix material by using a cure catalyst. As shown in Figure 6, when a potassium hydroxide catalyst was added to the proto-type BOSS material, the cured films displayed much higher modulus, 2.5 – 7 GPa for K of 1.8 – 2.2. The drawback for using such a cure catalyst is its negative effects to resin shelf life and potential contamination issues. Another potential approach under investigation is to use a BOSS resin based on a resin backbone capable of forming highly cross-linked matrix upon cure.

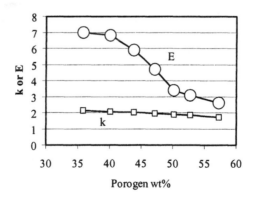

Figure 6. Dielectric constants and modulus measured by nanoindentation for the films made from catalyzed $[HSiO_{3/2}]_n[(C_{20}H_{41}O)SiO_{3/2}]_m$ resins having varying R level.

Theoretical Considerations On Open Pores *vs.* Closed Pores

A topic of much discussion concerning porous materials has been the ability to make low dielectric constant films with ultra small but closed pores. Closed pore materials are preferred by IC industry for perceived ease of integration. The issue of closed versus open pores is addressed here due to the improbability of making a "closed" pore system as the pore sizes approach 5 nm or less. A simple but effective way of looking at the pores and sidewalls of the pores is by using a "packing" model where the pores are assumed to be idealized spheres packed in a polymer matrix. Other assumptions used in this study were all the pores have identical pore diameter, D in angstroms, all pores are packed in a regular pattern, and the distance (T, angstroms) between the two neighboring pores (thickness of the polymer matrix between pores) remains constant in the entire film (no aggregation of pores). The mathematical analysis is done by using two packing models, cubic and hexagonal, which are most frequently seen for materials such as metals and some ceramics, which consist of spherical building blocks. However, since the hexagonal model is the closest packing possible it has been chosen here to help illustrate the effect that the pore size has on the thicknesses between pores for a given void volume fraction. For a given pore size and porosity, the thickness of a matrix between pores calculated for hexagonal model is the greatest possible thickness for packing considerations.

Note, as shown in Figure 7, that in the hexagonal arrangement, each sphere in the next layer sits on top of the center point of the three neighboring spheres as shown by the solid sphere. Four spheres form a tetrahedron, and each sphere is shared by 24 tetrahedrons. So each tetrahedron contains 1/6 sphere.

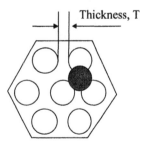

Figure 7: Pores in regular hexagonal arrangement.

The volume of a tetrahedron Vt:

$$Vt = 1/3[(3^{1/2})(D + T)^2/4][(2/3)^{1/2}(D + T)]$$

$$= (2^{1/2}/12)\,(D + T)^3 \quad \text{units } (A^3)$$

Each of the voids has a volume Vv:

$$Vv = \pi D^3/6 \quad \text{units } (A^3)$$

Porosity P:

$$P = (Vv/6)/Vt$$

$$= (\pi D^3/36\,)/\,[(2^{1/2}/12)\,(D + T)^3]$$

Therefore,

$$T = \{[\pi/(3*2^{1/2}\,)/P]^{1/3} - 1\}D$$

$$= [0.9047/P^{1/3} - 1]D \quad \text{units } (A)$$

Where D is the diameter of pores in angstroms; P is the porosity of the material; and T is the thickness of a matrix between pores in the porous material.

The average thickness of a matrix between pores for materials with various pore diameters as a function of porosity (pore volume fraction) is shown in Figure 8. For a material having 35% porosity (minimum for 2.1 k material) and 2 nm of pore diameter, the hexagonal model predicates a thickness of 5.7 angstroms between pores. Therefore, in an ideal case where the pores are

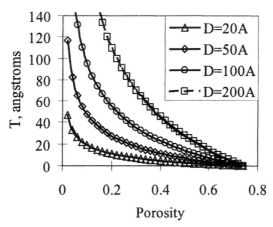

Figure 8. Thickness of the polymer matrix between two pores for a given pore diameter as a function of porosity.

perfect spheres there is an extremely low probability for closed pore structure (5 angstroms is the minimum molecular dimension for a siloxane resin system). Since it is in the molecular dimension the pores are being separated by a matrix that is one molecule thick.   In these dimensions it becomes a density consideration where one has a resin molecule separated by a "space" which is being called a pore.   For a material with 35% porosity and 5 nm of pore diameter, the cubic model predicates a thickness of 7.2 angstroms, and thus a low probability for making closed pore material; however, the hexagonal model predicates a thickness of 14 angstroms, and thus perhaps an increased probability for making a closed pore structure. For a material with 35% porosity and 10 nm of pore diameter or larger, the cubic model predicates a thickness of 14 angstroms, the hexagonal model predicates a thickness of 28 angstroms, and thus a high probability for making a closed pore structure.

Conclusions

In conclusion, BOSS technology is shown to be a generic approach toward ULK dielectrics having dielectric constants as low as 1.8 and pores as small as 2 nm using conventional spin coating and furnace cure processes. The demonstrated advantages of the small pore size included a high dielectric strength.

Acknowledgements

We are grateful to P. Schalk, C. Bargeron, M. Spaulding, J. Abaugh, S. Grigoras, K. Weidner, H. Meynen and SFM Application Center at Dow Corning, to F. Iocopi, S. Mailhouitre, M. Van Hove, K. Maex at IMEC, and to D. Gidley at University of Michigan for their help and discussions.

References

1. Hedrick, J. L.; Miller, R. D.; Hawker, C. J.; Carter, K. R.; Volksen, W.; Yoon, D. Y.; Trollsas, M., *Adv. Mater.* **1998**,*10*, 1049.
2. Nguyen, C. V.; Carter, K. R.; Hawker, C. J.; Hedrick, J. K.; Jaffe, R. L.; Miller, R. D.; Remenar, J. F.; Rhee, J. F.; Rice, P. M.; Toney, M. F.; Trollsas, M. F.; Yoon, D. Y., *Chem. Mater.* **1999**, *11*, 3080.

3. Chung, K.; Moyer, E. S.; Spaulding, M., "A Method of Forming Coatings". US patent 6,231,989, **2001**.
4., Gidley, D. W.; Frieze, W. E.; Dull, T. L.; Sun, J.; Yee, A. F.; Ngyen, C. V.; Yoon, D. Y., *Appl. Phys. Lett.* **2000**, 76, 1282.
5. *Adsorption, surface Area and Porosity* by S. J. Gregs and K. S. W. Sing, **1982**, Academic Press, Inc, 111.
6. Barrett, E. P.; Joyner, L. S.; Halenda, P. P., *J. Am. Chem. Soc.,* **1951**, *73*, 373.
7. *Adsorption, surface Area and Porosity* by S. J. Gregs and K. S. W. Sing, **1982**, Academic Press, Inc, 195.
8. Horvath, G.; Kawazoe, K., *J. Chem. Eng. Jpn.*, **1983**, *16*, 470.
9. Zhong, B.; Meynen, H.; Iocopi, F.; Weidner, K.; Mailhouitre, S.; Moyer, E.; Bargeron, C.; Schalk, P.; Peck, A.; Van Hove, M.; Maex, K., "A New Ultra-Low K ILD Material Based On Organic-Inorganic Hybrid Resins", *MRS Meetings*, April 5, **2002.**
10. Zhong, B., "Method For Making Nanoporous Silicone Resins From Alkylhydridosiloxane Resins", U.S. Patent 6,1484,260, **2001**.
11. Zhong, B., "Method For Making Microporous Silicone Resins With Narrow Pore-Size Distributions", U.S. Patent 6,197,913 **2001**.
12. Zhong, B.; King, R. K.; Chung, K.; Zhang, S., "Soluble Silicone Resin Compositions Having Good Solution Stability", U.S. Patent 6,232,424, **2001.**
13. Zhong, B.; King, R. K.; Chung, K.; Zhang, S., "Nanoporous Silicone Resins Having Low Dielectric Constants and Method for Preparation", U.S. Patent 6,313,045, **2001**.
14. Zhong, B.; King, R. K.; Chung, K.; Zhang, S., "Silicone Resin Compositions Having Good Solubility and Stability", U.S. Patent 6,359,096, **2002.**
15 Dultsev, F. N.; Baklanov, M. R., *Electrochem. Sol. St. Lett.*, **1999**, *2* 192.
16. Baklanov, M. R.; Mogilnikov, K. P.; Polovinkin, V. G; Dultsev, F. N., *J. Vac. Sci. Technol.*, **2000**, *B18(3)*, 1385.
17. Iacopi, F.; Tőkei, Zs; Le, Q. T.; Shamiryan, D.; Conard, T.; Brijs, B.; Kreissig, U.; Van Hove, M.; and Maex, K. submitted to *J.Appl.Phys.* Nov.2001.

Chapter 14

Design of Nanoporous Polyarylene Polymers for Use as Low-k Dielectrics in Microelectronic Devices

H. Craig Silvis[1], Kevin J. Bouck[1], James P. Godschalx[1],
Q. Jason Niu[2], Michael J. Radler[2], Ted M. Stokich[2], John W. Lyons[1],
Brandon J. Kern[2], Joan G. Marshall[1], Karin Syverud[1],
and Mary Leff[2]

[1]Corporate Research and Development and [2]Advanced Electronic
Materials, The Dow Chemical Company, Midland, MI 48674

Future progress within the microelectronics industry will
require the successful integration of insulating materials
possessing dielectric constants below that of silicon dioxide to
reduce interconnect capacitance, cross-talk, and power
dissipation. This will result in enhanced performance due to
an increase in the density of transistors and other devices. One
such class of low dielectric materials are spin-on organic
polyarylene polymers, which possess the inherent thermal and
chemical stability to perform in this demanding application.
Once cured, these materials form a polyaromatic network
structure with glass transition temperatures in excess of 450 °C
and have an inherent dielectric constant of 2.65. We have now
explored methods for preparing nanoporous polyarylene
polymers through the use of a sacrificial porogen polymer.

Introduction

The microelectronics industry has continued to regularly improve both the speed and performance of advanced microprocessors over the last two decades. Further progress within this industry will require the successful integration of higher conductivity interconnect metals combined with insulating materials possessing dielectric constants below that of silicon dioxide in order to reduce interconnect capacitance, cross-talk, and power dissipation, and thus allow the density of transistors and other features within the device to increase (1). Interconnect conductivity has recently been increased through the replacement of aluminum with copper metallization, thus attention is now focused on the dielectric material. One such class of low dielectric materials are spin-on organic polyarylene polymers, tradenamed SiLK (2,3), which possess the inherent thermal and chemical stability to perform in this demanding application. These interlayer dielectric polymers are polyarylene thermosetting resins that have a dielectric constant of 2.65 versus ca. 4.0 for silicon dioxide (4). The methodology used to generate porosity involves the use of a sacrificial porogen polymer that forms a discrete nanophase morphology as the SiLK/porogen polymer blend is cured on a semiconductor wafer through a phase nucleation and growth mechanism. Subsequent thermal processing to temperatures above the decomposition point of the sacrificial porogen then gives rise to nanopores within the SiLK polyarylene matrix, thus lowering the overall effective dielectric constant. This paper will detail early stage methods used to generate nanoporous polyarylene polymers, as well as some of the relevant properties and integration characteristics obtained when they are employed as insulating dielectrics in microelectronic devices.

Experimental

In general, the nanoporous films of polyarylene polymers were prepared by first reacting polycyclopentadienone and polyphenylacetylene monomers, along with a functionalized porogen polymer, together in gamma-butyrolactone solvent for 24-72 hours at 200 °C. After reaction, the prepolymer solutions were diluted to ~15 wt% solids with cyclohexanone, filtered through a teflon filter membrane (0.5 μm), and then spun onto a silicon wafer in a clean room environment. The films were then baked on a hot plate under nitrogen at 150 °C for two minutes, and then placed in an oven with a nitrogen atmosphere and heated to 430 °C at a ramp rate of 7 °C/min to simultaneously cure the polyarylene polymer and effectively burnout the sacrificial porogen polymer in order to generate nanopores. Subsequent characterization of the films involved standard measurements of refractive index, light scattering index, and transmission electron microscopy to evaluate the nanopore morphology. The sacrificial porogen used in this work was a functionalized star polystyrene. The basic preparation of this particular polystyrene molecular architecture has been described elsewhere (5,6) and these published synthetic methodologies were employed to prepare the various porogen materials used in these studies with minor modification. The description given below is representative of their preparation.

Reagents

HPLC grade cyclohexane and anhydrous THF were passed through activated alumina. Styrene was purchased from either Ashland Chemical Co. or Aldrich. Ashland styrene was passed through activated alumina prior to use and transferred using stainless steel vessels. Styrene purchased from Aldrich, was passed through activated alumina and distilled from calcium hydride prior to use and transfered using airless syringe techniques. The initiator sec-butyllithium (sec-BuLi) was purchased as a 1.3M solution (Aldrich) and was used as received or diluted with cyclohexane. The initiator concentration was determined using the standard Gilman double titration method. Divinylbenzene (80%) (Aldrich) was passed through activated alumina and distilled from calcium hydride prior to use. Ethylene oxide (EO) (Aldrich) was used as received. Acetyl chloride (Aldrich) was distilled from phosphorous pentachloride. γ-butyrolactone and cyclohexanone were electronics grade and used without further purification. The endcapping reagent 4-(phenylethynyl)benzoyl chloride (PEBC) was prepared by coupling 4-bromo-methylbenzoate with phenylacetylene using palladium acetate /triphenylphosphine as catalyst in triethylamine. The resulting 4-(phenylethynyl)methyl benzoate was converted to the corresponding potassium salt by treatment with potassium hydroxide in isopropanol. The potassium salt was then readily converted to the acid chloride by treatment with an excess of thionyl chloride. The acid chloride could be purified by either recrystallization from dry hexane or by sublimation at reduced pressure prior to use.

Detailed Synthesis of Functionalized Star Polystyrene Polymer

A 2.5 liter glass polymerization reactor, which had been washed with hot cyclohexane and dried under vacuum, was charged with 2 liters of cyclohexane. The reactor was heated to 50 °C and 27.9 mL (12.0 mmoles) of 0.43 M sec-BuLi was added followed by 49.74 g of styrene and 75 mL of THF. The dark orange solution was stirred for 15 min. The polymerization was sampled and 6.42 g (39.45 mmoles of active DVB, 3.3 eq) of 80% divinylbenzene, diluted with cyclohexane, was added to give a very dark red solution. After 30 minutes, the 44.63 g of styrene was added to give a dark orange solution. After 15 minutes, the reactor was sampled and 4.3 g of EO was added to give a colorless, viscous solution. After 1 hour, 5.44 g (22.60 mmoles, 1.9 eq) of phenylethynylbenzoyl chloride contained in THF was added. After an additional hour, the reactor was cooled and the contents were removed. An aliquot of the final star was isolated by precipitation into MeOH. GPC analyses are listed in Table 2. The solution was then filtered through a 5 μm Teflon membrane housed in a Teflon coated filtration device. Next, the volume of the polymer solution was reduced to ~500 mL (~20% solids) and washed with equal volumes of 1 N HCl (3X) and deionized H_2O (6X). The polymer was precipitated by adding the solution, via an addition funnel, to 4 L of, vigorously stirred, reagent grade MeOH contained in a 5 liter Morton flask, equipped with an overhead stirrer, glass stirring shaft and Teflon paddle. The polymer was collected by filtration using a clean 5 μm Teflon membrane housed in a Teflon coated filtration device. The polymer was dried under vacuum at 80 °C overnight. UV analysis indicated that the star contained an average of 2.6 wt% diphenylacetylene (DPA) functionality, or 16-19 DPA units per star.

A number of other star polystyrene porogens were also synthesized in a similar manner. They were functionalized to a lesser degree with 4-phenylethynylbenzoyl chloride and then remaining hydroxyl functionality on the star molecule was capped with an excess of acetyl chloride prior to precipitation and final purification of the porogen material.

Porogen Molecular Weight Determination

Gel permeation chromatography (GPC) was conducted on one of two Hewlett-Packard 1090 LC systems. Each system was equipped with a combination autosampler-injector and a diode array detector. One of the systems was additionally equipped with a Waters 2410 refractive index detector, and, for absolute molecular weight determinations, a Wyatt Technology MiniDawn laser light scattering detector. The eluent (1.0 mL/min) was tetrahydrofuran (THF) and the operating temperature was 40 °C, for both systems. The systems were calibrated with EasiCal PS-2 (Polymer Labs) narrow molecular weight polystyrene standards. The system with the single detector utilized a PL-Gel Mixed D (5 μm) (Polymer Labs) column in series with a PL-Gel Mixed E (3 μm) column. The other systems had two PL-Gel Mixed C (5 μm) columns in series. Software consisted of HP Chemstation (Hewlett-Packard) and PL GPC/SEC software for HP Chemstation (Polymer Labs).

Results and Discussion

Matrix Curing Chemistry

SiLK polyarylene polymers crosslink thermally *via* a concerted 4+2 Diels-Alder cycloaddition reaction as illustrated in Figure 1. This reaction leads to a bicyclic intermediate with a bridging carbon monoxide group that can be eliminated, thus allowing the remaining ring system to aromatize. These favorable energetics are the driving force for the crosslinking reaction which is obviously nonreversable in nature. The glass transition temperature of the matrix after cure is in the range of 450-500°C, which affords a dielectric material with enhanced thermal stability. Chemical and thermal stability results from the lack of any aliphatic carbon-hydrogen bonds in the final structure, and dimensional stability arises from the high degree of crosslinking achieved. Some of the relevant physical properties of SiLK polyarylene dielctric are given

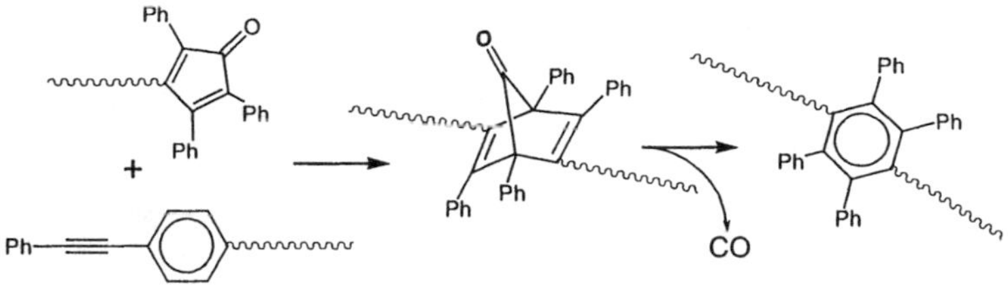

Figure 1. Schematic of polyarylene curing chemistry via a Diels-Alder cycloaddition reaction.

in Table 1. One can conclude from these data that SiLK polyarylene is indeed a robust material. In addition to the matrix curing chemistry that occurs at elevated temperatures, the reaction of the cyclopentadienone monomer or growing oligomer with diphenylacetylene groups on the termini of the star polystyrene arms allows the porogen to be covalently bonded to the polyarylene matrix. This, in turn, prevents the porogen from undergoing macro-phase separation during the curing process. In principle, the relative reactivity of both the monomer and porogen phenylethynyl groups should be nearly equivalent, which should allow for grafting of polyarylene chains to the porogen at a comparable rate to that of matrix polymerization.

Table 1. Selected Physical Properties of SiLK Polyarylene Dielectric (4)

Property	Value
Dielectric constant	2.65
Voltage Breakdown	4 MV/cm
Leakage current @ 1 MV/cm	0.33 nA/cm^3
Refractive index at 632.8 nm	1.63
Thermal stability	>425 °C
Weight loss @ 450 °C	0.7 wt%/hr
Glass Transition	>450 °C
Young's modulus	2.45 Gpa
Toughness	0.62 Mpa • m$^{1/2}$
CTE	66 ppm/°C

Porogen Synthesis

Styrene is polymerized anionically to give the star "arms." of relatively low average molecular weight. Divinylbenzene is then added to give an "in-star" with a living crosslinked anionic core. The polymerization can be quenched at this stage, or additional monomer can be added which then polymerizes off of the anionic core to give an "in/out-star." The molecular weight data for three representative star porogens is illustrated in Table 2. In this case, ethylene oxide is added to cap the living chain ends with aliphatic hydroxyl groups. The homopolymerization of ethylene is suppressed under these conditions due to the tightly bound lithium cation, thus only one monomeric unit of ethylene oxide is added per living chain end.

A general schematic of the porogen synthesis and functionalization is shown in Figure 2. in order to illustrate the stepwise nature of its preparation. Three versions of star polystyrene porogen with differing levels of DPA functionality were prepared in large enough quantities to allow subsequent incorporation into SiLK polyarylene. In each case the relative size of the porogen was held reasonably constant at 7-8 nm. The size was calculated based on the molecular weight of the product and assuming a roughly spherical shape with a density of ca. 1.0.

In principle, the primary variable is the degree of DPA functionality, which should strongly influence the level of grafting of polyarylene to the porogen molecules. It is anticipated that this will in turn affect the porogen-matrix phase separation process resulting in observable differences in final pore morphology.

Table 2. Molecular Weight Data on Selected Star Porogens

Star PS	Arm Mn (g/mol x 10⁻³)	Star Mn (g/mol x 10⁻³)	Calculated Size (nm)	DPA Groups per/Star[1]
Star 1	3.9	126	7.4	3-5
Star 2	4.2	141	7.6	7-10
Star 3	4.2	154	7.9	16-19

[1]determined by quantitative UV spectroscopy

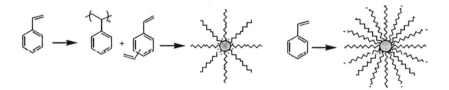

Figure 2. Schematic of star polystyrene synthesis

Pore Morphology

The functionalized star polystyrene porogens were then reacted together with the polyfunctional substituted cyclopentadienone and phenylethynyl monomers used to prepare the polyarylene polymers at 25 weight percent in γ-butyrolactone solvent at ca. 200 °C for 48 hours, spun onto a silicon wafer, and cured as described previously. Upon thermal removal of the porogen, the films were examined by transmission electron microscopy (TEM) to ascertain the pore morphology. Utilizing the star polystyrene porogens detailed in Table 2, the dependence of pore morphology versus the degree of DPA functionality was determined and the results are given in Figure 3. Star 3, possessing the highest

194

level of functionality, gave rise to virtually no porosity. In this case, it is hypothesized that a high level of functionality prevents a sufficient degree of porogen phase aggregation *via* nucleation and growth to generate a pore domain large enough for the polyarylene matrix to support throughout the entire curing process. Conversly, a similar treatment of star 1, possessing the lowest level of DPA functionality, affords a very course pore morphology. These results suggest that too little grafting is taking place in this case to prevent a high degree of phase agglomeration. More favorable results were obtained with porogen star 2, where the intermediate level of DPA functionality apears to strike a balance between the extremes of stars 1 and 3. Obviously, the degree of

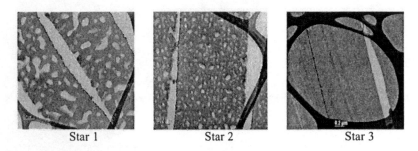

| Star 1 | Star 2 | Star 3 |

Figure 3. Effects of star porogen DPA functionality on polyarylene pore morphology as determined by TEM.

reactive functionality on the porogen molecule has a profound effect on the phase separation process; one that is clearly based on a nucleation and growth mechanism.

Not only can the extent of polyarylene grafting to the star porogen be manipulated by degree of reactive functionality, it can also be affected by the extent of reaction at a given functionality level. One would expect that longer reaction times would allow the molecular weight of the grafted chains to increase and thus have a direct impact on the phase separation process. Typically, the kinetics of the phase nucleation and growth process is controlled by both the density and size of grafted polymer chains at the interfacial boundary (7,8). In the present system, the molecular weight of the polyarylene chains grafted to the star polystyrene porogen can be controlled through reaction time at 200 °C. Specifically, the reaction time was varied from 24 to 72 hours using the low functionality star 1 porogen described above. This particular material gave a relatively course pore morphology when it was reacted with the polyarylene matrix monomers for 48 hours at 200 °C. The morphological results obtained by varying the reaction time are illustrated in Figure 4. It can be concluded from these data that reaction time has a profound effect on the phase separation process. The short reaction time of 24 hours produced a very course pore morphology with pores averaging ~200 nm in size. The 48 and 72 hour reaction times afforded progressively smaller pore sizes consistent with the expected higher degree of interfacial grafting between the functionalized polystyrene star and the polyarylene matrix. Longer polymerization times would undoubtedly afford even smaller pores, however this approach is impractical due to gelation of the reaction mixture if the level of conversion becomes too high.

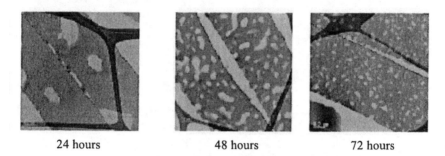

24 hours   48 hours   72 hours

Figure 4. Effect of reaction time between matrix and functionalized porogen on resulting pore morphology.

The integration of advanced materials into semiconductor devices is currently a very complex process involving hundreds of individual steps. In order to generate the interconnect architecture, it is often desirable to apply multiple layers of materials prior to the untimate high temperature treatment to cure the matrix. These intermediate steps to partially cure the matrix and allow deposition of additional layers are typically done on programmable hot plates at specific temperatures under an inert atmosphere. In order to test the effects of multiple hot plate treatments on the morphology of porous polyarylene, an experiment was conducted using polystyrene star 2 as the porogen at a 25 wt% loading in polyarylene. In the first case, the polymerized solution was given the standard 150 °C hot plate and then a 7 °C/minute ramped burnout to 430 °C. In the second case, the coated wafer was subjected to a sequence of hot plate bakes at 150, 320, and 400 °C prior to the ramped burnout described above. The second and third hot plate bakes serve to gel the matrix to the point that a second layer can be spun onto the polyarylene without danger of stripping it off. However, these intermediate temperature bakes can have the potential to significantly impact the pore morphology. By instantaneously exposing the polyarylene film to an elevated temperature on a hot plate that is significantly above its glass transition temperature, one creates conditions that could lead to phase agglomeration, since the matrix modulus drops significantly and porogen mobility is increased as a result. Such a phenomenon was indeed observed when polystyrene star 2 porogen was employed. As evidenced in Figure 5, a significant increase in the average pore size was observed when the multi-step sequential hot plate bake sequence was used.

 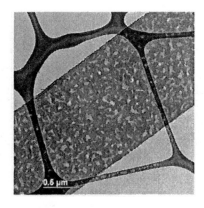

150 °C bake   150, 320, 400 °C bake

Figure 5.   Effect of hot plate bake sequence on final polyarylene pore morphology with polystyrene star 2 at 25 wt% loading.

Film Characterization

The obvious objective of this work was the preparation of nanoporous polyarylene polymers that could be utilized in advanced microprocessor assembly. The rationale for making the polymer porous was to reduce the effective dielectric constant (k) by introducing small voids with k $\approx$ 1.0, i.e. the dielectric constant of air. In addition, one can easily measure the refractive index of the porous film, as well as the light scattering index as a means of quickly determining whether the film is indeed porous and whether the pores are large or small, respectively. What is desired is a material which affords both a low refractive index, indicating a high degree of porosity, and a low light scattering index, which signifies small pores. As an example of the types of values obtained for polyarylene polymers with various pore morphologies, these parameters were obtained on the set of film samples used in Figure 3 and the results are listed in Table 3. As expected, the sample involving the high functionality star 3, which showed little evidense of porosity, gave a high refractive index and a low light scattering index. The sample employing the low functionality star 1, which possessed relatively large pores, gave a low refractive index, but a high light scattering index. Only the intermediate level functionality star 2 gave both a low refractive index *and* a low light scattering index, indicative of a significant level of small pores.

Table 3. Film Characterization Data for Polyarylene Containing 25 wt% Polystyrene Star Porogens

Porogen	Refractive Index	Light Scattering Index	Pore Size Range (nm)	Dielectric Constant
star 1	1.53	200	40-200	2.3
star 2	1.56	40	20-100	2.4
star 3	1.64	10	n/a	2.7

Conclusions

Nanoporous polyarylene thermoset polymers have been successfully prepared through the combination and reaction of a sacrificial functionalized porogen polymer with the multifunctional cyclopentadienone and phenylethynyl monomers used to form the polyarylene polymer. Upon heating, the polyarylene matrix is cured and the sacrificial porogen material is depolymerized to monomer, which migrates rapidly out of the film leaving nanopores behind. The process appears to involve a traditional nucleation and growth mechanism for phase separation. The relative pore size can be manipulated by several variables, including the level of porogen reactive functionality, the reaction time, and the thermal treatment sequence used in applying and curing the polyarylene film on wafer. Conditions have been found that are conducive to the formation of relatively small pores that lead to a material with a significantly reduced refractive index and dielectric constant compared to the dense polyarylene matrix. Future work in this area is focused on further reducing the average pore size obtainable with polyarylene matrix polymers.

Acknowledgements

The authors would like to acknowledge the support of the National Institute of Standards and Technology (NIST) for their financial support on the development of porous dielectrics under the Advanced Technology Program (ATP) in collaboration with IBM Corporation.

References

1. *The National Technology Roadmap for Semiconductors*, Semiconductor Industry Association, San Jose, CA **1994**.
2. Trademark of The Dow Chemical Company.
3. Godschalx, J. P. *et al.*, US Patent 5,965,679.
4. Martin, S.J.; Godschalx, J.P; Mills, M.E.; Shaffer, E.O.; Townsend, P.H. *Advanced Materials* **2000**, *12(23)*, 1769.
5. Tsitsilianis, C; Graff,S.; Rempp, P. *Eur. Polym. J.* **1991**, *27*, 243.
6. Okay, O.; Funke, W. *Macromolecules* **1990**, *23*, 2623.
7. O'Shaughnessy, B.; Vavylonis, D. *Macromolecules* **1999**, *32*, 1785.
8. Orr, C.A.; Cernohous, J.J.; Guegan, P.; Hirao, A.; Jeon, H.K.; Macosko, C.W. *Polymer* **2001**, *42*, 8171.

Chapter 15

Molecular Brushes as Templating Agents for Nanoporous SiLK* Dielectric Films

Q. Jason Niu, Steven J. Martin, James P. Godschalx, and Paul H. Townsend

The Dow Chemical Company, Midland, MI 48674

Starting with methacrylate or styrene-capped polystyrene macromonomers, we have successfully prepared a series of polystyrene molecular brushes via living atom-transfer radical polymerization (ATRP). The molecular weights of the resulting polystyrene brushes ranged from 80,000 to 200,000 g/mole with narrow molecular weight distributions. Porous SiLK dielectric films prepared from SiLK matrix and functional polystyrene brush poragen represent a dramatic improvement over the non-functional linear polystyrene system used previously. We have obtained spherical pores with no interconnectivity, even with high loading of such templating agents. Branched or dendritic-structured templates from molecular brush approach appear to have less agglomeration than linear templating molecules. These therefore yield closed spherical pores, rather than opened and interconnected pores formed in the linear systems. The mean pore size is around 40 nm. The refractive index of resulting films has dropped from 1.64 to 1.52 with the dielectric constant of 2.2-2.3 when 20% of molecular brush was loaded.

The demand of the microelectronic industry for polymeric materials as organic dielectrics has fueled much novel work in polyimide, polyphenylene and polyarylene ethers. Polyphenylene oligomers and polymers have an outstanding combination of properties, including thermal and chemical stability, excellent processability, good adhesion to a variety of surface and low dielectric constants (2.7).[1] SiLK semiconductor dielectric resins from such polymeric materials have been successfully applied to the fabrication of integrated circuit interconnects at various semiconductor companies. To develop the next generation materials with a dielectric constant less than 2.7, incorporation of porosity to current SiLK resin has been designed.

The general approach to this process involves the use of certain thermoplastic materials as the templates of the pore in SiLK resins. This second phase is removed by thermolysis to create the pores. While there exists significant literature[2] on such nanophase separated inorganic-organic hybrids derived from the oligomeric/polymeric organosilicates, relatively few examples are of nanoporous materials based on organic thermosetting resins.[3] For the organosilicate case, IBM has demonstrated that the degree of branching in the templates is crucial, as is polydispersity and molecular weight, to insure the nanophase separation.[4]

Molecular brushes are multibranched macromolecules characterized by extremely high branch densities along the backbone.[5] They can have unique molecular morphologies ranging from star-shaped spheres to rod-like cylinders, depending on the degree of polymerization (DP) of the backbone and branch chains. Recent developments in controlled/living radical polymerization, such as atom-transfer radical polymerization (ATRP), has provided possibilities for the synthesis of well-defined polymers with low polydispersities as well as polymers with novel and complex architectures.[6] Utilization of these methods in the synthesis of functional molecular brushes based on polystyrene macromonomers is the focus of this report.

Experimental Section

Synthesis of ATRP initiator with phenylethynyl functional group.

To a 500 ml three neck round flask was added 21.5 g (0.10 mole) of methyl 4-bromobenzoate, 12.2 g of phenylacetylene (0.12 mole), 25.0 g (0.25

mole) of triethylamine and 120 ml of N,N'-dimethylformide. The resulting mixture was purged with nitrogen for 15 minutes, and 1.57 g (0.006 mole) of triphenylphosphine was added, and after purging with nitrogen for another 5 minutes, 0.22 g (0.001 mole) was added to the flask. The reaction mixture was heated to 80°C for 24 hours under nitrogen, allowed to cool to room temperature, and poured into 200 ml of water. After extracting several times with toluene (total 300 ml), the organic layer was dried over sodium sulfate and passed through a silica gel column. Removal of the solvent affords crude product, which can be recrystallized in toluene/hexane to give the pure product (19.2g, 81%). NMR spectra confirmed the structure of the product.

A solution of 18.7g (0.079 mole) of methyl benzoate, prepared above, in 200 ml of anhydrous ether was added dropwise to a slurry of 3.8g (0.10 mole) of LiAlH$_4$ in 20 ml of ether. After addition was completed, the reaction mixture was continued to stir for half an hour before being quenched by aqueous NaOH solution. The solution was filtered and washed with saturated NaHCO$_3$ and water. After solvent is removed, the solid was recrystallized from hexane/ether mixture to give a white crystalline product (12.1g, 68%). NMR spectra confirmed the structure of product.

To a 250 ml three neck flask was added 2.24 g (0.01 mole) of benzyl alcohol prepared above, 50 ml of THF, and 4 ml of triethylamine. The mixture was allowed to cool at 0°C while 2.8 g (0.012 mole) of 2-bromosiobutyryl was added dropwise. After the addition was completed, the reaction mixture was warmed to 40°C for 1 hour and then quenched with water. The mixture was extracted with ethyl acetate and separated by silica gel column to give a colorless solid (3.36g, 94%). The structure of the product was confirmed by NMR spectra. This is the final step for the synthesis of mono-phenylethynyl ATRP initiator. ^{1}H NMR (CDCl$_3$), δ 7.55 (m, 4H), 7.37 (m, 5H), 5.22 (s, 2H), 1.98 (s, 6H). ^{13}C NMR (CDCl$_3$), δ 171.4, 135.4, 131.7, 131.6, 128.3, 127.8, 123.3, 123.0, 89.9, 88.9, 67.1, 55.5, 30.7.

Procedure of ATRP.

To a 1000 ml two neck round flask was added 250 grams of macromonomer (MA2, 0.104 mole) and 250 g of anhydrous anisole. The resulting solution was purged with argon for 15 minutes, while during the purging time, PMDETA (1.502 g, 0.00867 mole), Copper (I) bromide (1.243g, 0.00867 mole), and finally mono-phenylethynyl initiator were added. The mixture was immediately degassed 10 times by freeze-pump-thaw cycles. The flask was then stirred at 55°C under argon for 24 hours. After the reaction mixture was cooled to room temperature, 200 ml of THF was added to the solution. The polymer mixture was passed through silica gel filter and precipitated into 2 liters of methanol to give a white powder (232g, 93%). The polymer was further purified by column to remove trace amounts of copper salt. Absolute molecular weight measurement indicates a Mp of 154,000 g/mole with the polydispersity of 1.09. Based on SEC analysis, there remains 16% of unreacted macromonomer.

Synthesis of Functional Initiators

Development of ATRP offers the possibility of the synthesis of well-defined telechelic and end-functional macromolecules. Functional ATRP initiators were prepared in three steps as shown in **Scheme 1**. For example, 4-bromo methylbenzoate was converted to 4-phenylethynyl methylbenzoate via Pd coupling. This intermedate was reduced to the alcohol form by lithium aluminum hydride and, after further reacting with 2-bromoisobutyl bromide, functional initiator was obtained as white crystals. Depending on the starting materials, these initiators contain one or two phenyl acetylene groups that can graft to the SiLK matrix during subsequent oligomeration of matrix monomers (B-staging). Because such molecular brushes were prepared from phenylethynyl initiators, they always have reactive endgroups on each macromolecule. The overall yield is very good and the molecules behave like typical ATRP initiators.

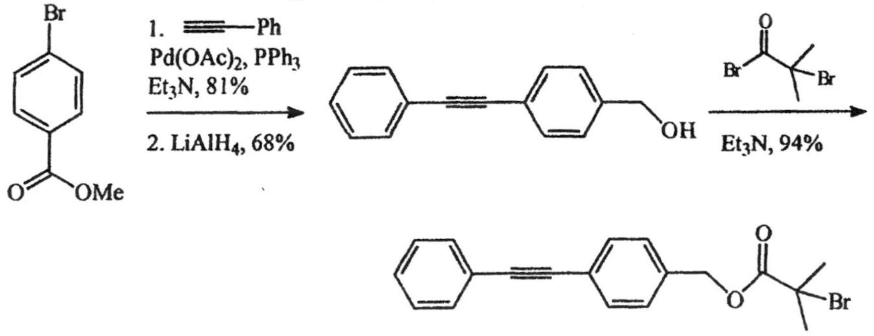

Scheme 1. Synthesis of Functional ATRP Initiator

Synthesis of Macromonomers

The oligostyrene-based macromonomer carrying a methacryloyl (MA) group or styrenic (VB) group was prepared by the end-capping of well-established living anionic polymers as shown in **Scheme 2**. These macromonomers were synthesized from polystyryl living anions by reaction (in situ) with ethylene oxide, followed by methacryloyl chloride or 4-vinyl benzoyl chloride.

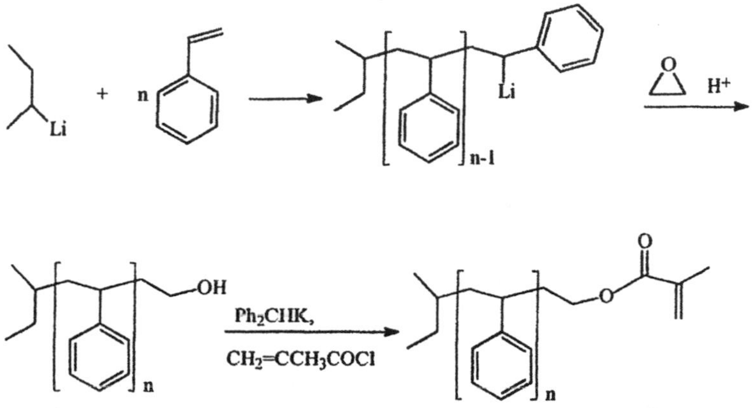

Scheme 2. Synthesis of Macromonomers

Table 1 summarizes the properties of macromonomers. Apparently, all of the macromonomers produced have a narrow molecular weight distribution (MWD) and controlled DP, with mole masses around 3000 g/mole.

Table 1. Characterization of Macromonomers

Code	Structure	Mw	Mw/Mn	Supplier
MA1	Methacryloyl	3,400	1.05	SP2
MA2	Methacryloyl	2,400	1.04	SP2
VB	Styrenic	2,100	1.11	Polymer Sources

ATRP of Macromonomers

As with ATRP for low molecular-weight monomers, the polymerization of MA-PSt or VB-PSt was conducted under homogeneous conditions using functional ATRP initiators and N,N,N',N',N'-pentamethyldiethylenetriamine (PMDETA) as the ligand; this coordinates copper (I) and soulblizes the resulting complexes in the polymerization medium (**Scheme 3**). The polymerization was carried out using functional 2-bromoisobutyrate as the initiator in conjunction with copper (I) bromide in anisole (50 wt%) at 50°C (in most cases) and the reaction was nearly completed after 24 hours.

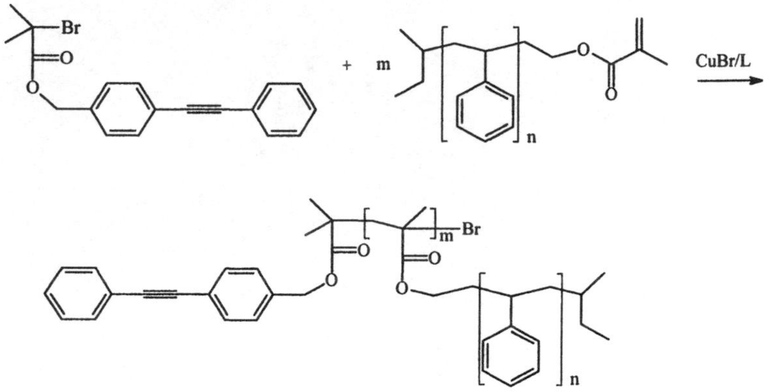

Scheme 3. ATRP of Macromonomers

Experiments can also be carried out with high [M]/[I] at different reaction times (**Table 2**). As seen from **Table 2**, the weight-average molecular weight of the obtained molecular brushes increased with increasing M/I ratio. These Mw and Mw/Mn values were estimated by polystyrene-calibrated Size-Exclusion Chromatography (SEC), hence, they are apparent values. Absolute molecular weight and absolute molecular weight distributions were also extimated for these polystyrene brush polymers with universal calibration. This technique is based on separation of polymers according to hydrodynamic volume using SEC followed by a concentration detector based on refractive index and an on-line viscometer detector. Their peak molecular weight (Mp) values are shown in **Table 2** and they agree reasonably well with the theoretical values at the low end of the M/I ratio. All these results support the "living" nature of the ATRP of MA-PSt or VB-PSt. Given the nature of low polydispersity for such polystyrene brushes, the small value of molecular weight estimated by SEC compared to absolute molecular weight obtained from universal calibration suggests a multibranched structure of the molecular brush that is more compact in hydrodynamic volume than a linear analogue with a similar molecular weight.

SEC trace of a typical experiment run on these molecular brushes has shown that there is 15-16% of unimer left after polymerization; this could be due to unfunctional oligostyrene in the macromonomer. Therefore the typical conversion is around 80% as there is no attempt to separate the unimer from polystyrene brush.

Table 2. Characterization of Polystyrene Brushes

Monomer	M/I/C[a]	Time(h)	Yield	Mw[d] g/mole	Mw/Mn	Mp(abs.) g/mole
MA1	30/1/5	50	84 %	43,000	1.27	96,000
	50/1/5	48	79 %	55,000	1.08	127,000
	70/1/5	80	80 %	63,000	1.05	185,000
	90/1/5	96	82 %	65,000	1.05	196,000
MA2	10/1/5	24	90 %	29,200	1.41	------
	20/1/5	24	93 %	37,000	1.41	84,000
	30/1/5	24	84 %	42,000	1.27	101,000
	60/1/5	24	83 %	49,000	1.09	154,000
VB	44/1/5	18[c]	80 %	46,000	1.17	101,000

Key: a. M, macromer; I, initiator; C, CuBr, [M] =50% for all cases.
b. The polymerization was run at 50°C.
c. The polymerization was run at 100°C.
d. Relative Mw vs polystyrene standard from SEC.

Evaluation of Molecular Brushes as Template

SiLK dielectric resins involve the synthesis of crosslinked polyphenylenes by the reaction of polyfunctional cyclopentadienone and polyfunctional acetylene-conatining materials.[7] Polyphenylenes are known to have excellent thermal stability. Typically, however, polyphenylenes need to be substantially substituted in order to achieve solubility and the processability. By preparing the polyphenylenes from cyclopentadienone and acetylene-containing monomers, the initial oligomers formed are soluble without undue substitution and can be processed. The cyclopentadienones react with the acetylenes in a 4+2 cycloaddition reaction followed by the expulsion of CO to form a new aromatic ring. The multifunctional nature of the monomers leads to crosslinked polyphenylene system after full cure on wafer (**Scheme 4**) .

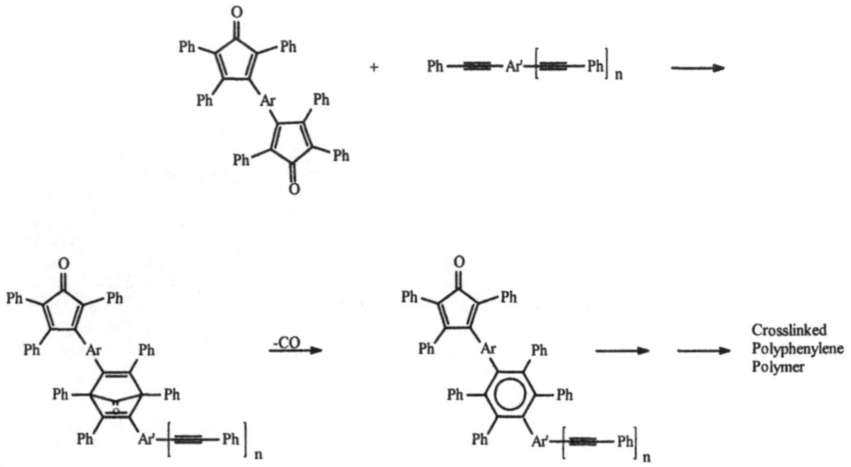

Scheme 4, Formation of Crosslinked Polyphenylene Matrix

The utility of functionalized polystyrene brushes prepared by ATRP was demonstrated by the synthesis of polystyrene brush/SiLK block copolymers. The phenylethynyl-terminated polystyrene brushes are ideally suited for the preparation of graft copolymers, and the reaction of polystyrene brush (20%) with CPD groups of SiLK resin in GBL (30%) was performed under b-staged conditions (200°C, 24-48 hours). This process not only produces SiLK oligomers, but also allows for the formation of polystyrene brush/SiLK graft copolymer (containing about 20% by weight polystyrene). Because SiLK oligomers tend to be low molecular weight and polystyrene brushes tend to be large, one can usually obtain at least some degree of separation between ungrafted SiLK chains and grafted or ungrafted polystyrene brushes in SEC separation mechanism. By using both refractive index and UV absorbance detectors after a SEC separation column, one can measure and verify the formation of grafted polymers. The results from such experiments have demonstrated the absence of polystyrene brush homopolymer, and revealed features consistent with graft copolymer formation and microphase separation.

The solution containing polystyrene brush/SiLK block copolymer was applied to a silicon wafer and cast by spin-coating to form a 1 micron-thick coating. The film was baked on a hotplate at 150°C for 2 minutes under nitrogen and then placed in an oven to ramp to 430°C and cure for 40 minutes. The latter heating step removed nearly 100 % of the polystyrene as measured by FTIR. The RI of resultant films has dropped from 1.64 to 1.52 with the dielectric constant of 2.2-2.3 when 20% of polystyrene brush was loaded. The estimated average pore size was about 40-50 nm with spherical morphology. TEM of porous films made with the above discussed chemistry are shown in **Figure 1**.

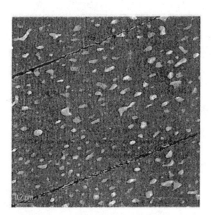 

Figure 1, Left, SiLK-*block*-polystyrene brush (127 k), B-staged for 30 hours and 20% loading of PS; Right, SiLK-*block*-polystyrene brush (127 k), B-staged for 30 hours and 17% loading of PS brush.

Conclusions

Starting with methacrylate-capped polystyrene macromonomers, we have successfully prepared a series of polystyrene molecular brushes via living atom-transfer radical polymerization. The molecular weights of the resulting polystyrene brushes ranged from 80,000 to 200,000 with narrow molecular weight distributions. Based on the porous film evaluations, 50:1 polystyrene brush (127k) was selected for use in porous SiLK resin, with 20% loading of polystyrene brush and 30 hours of b-staging. Porous SiLK resin prepared from functional polystyrene brush represents a dramatic improvement over the non-functional linear polystyrene system used previously. We have obtained spherical pores with no interconnectivity, even with high loading of poragen. Branched or dendritic-structured templates from molecular brush approach appear to have less agglomeration than linear templating molecules. These therefore yield closed spherical pores, rather than opened and interconnected pores formed in the linear systems. The mean pore size is around 40 nm. The RI of resulting films has dropped from 1.64 to 1.52 with the dielectric constant of 2.2-2.3 when 20% of polystyrene brush was loaded.

Acknowledgements

The authors thank Dave Meunier for the measurement of absolute molecular weight, Brandon Kern for the processing of these dielectric films, Joan Marshall for TEM, John Lyons and Kevin Bouck for the valuable discussions. The authors also thank NIST for their ATP grant to support the development of porous SiLK dielectric films and finally our ATP grant partner IBM Yorktown and Almaden Research Center.

*SiLK is a trademark of The Dow Chemical Company

References

1. Godschalx, J. P. *et al.* US Patent 5965679.
2. Chujo, Y.; Ihara, E. *Macromolecules*, **1993**, *26*, 5681.
3. Hedrick, J. L.; Hawker, C. J.; Miller, R. D.; Tweig, R.; Srinivasam, S. A.; Trollsas, M. *Macromolecules*, **1997**, *30*, 7607.
4. Remenar, J. F.; Hawker, C. J.; Hedrick, J. L.; Miller, R. D.; Yoon, D. Y.; Kim, S. M.; Trollsas, M. *Poly. Prep.*, **1998**, *39* (1), 631.
5. Wintermantel, M., Fisher, K., Gerle, M., Ries, R., Schmidt, M., Kajiwara, K., Urakawa, H., Wataoka, I. *Angew. Chem. Int. Ed. Engl.,* **1995**, *34*, 1472.
6. Yamada, K.; Miyazaki, M.; Ohno, K.; Fukuda, T.; Minoda, M. *Macromolecules*, **1999**, *32*, 290
7. Martin, S. J., Godschalx, J. P., Mills, M. E., Shaffer, E. O., Townsend, P. H. *Adv. Mater*, **2000**, *12*, 1769.

Chapter 16

X-ray Reflectivity as a Metrology to Characterize Pores in Low-k Dielectric Films

Christopher L. Soles, Hae-Jeong Lee, Ronald C. Hedden,
Da-wei Liu, Barry J. Bauer, and Wen-li Wu

Polymers Division, National Institute of Standards and Technology,
Gaithersburg, MD 20899–8541

A form of porosimetry is described whereby X-ray reflectivity is a highly sensitive metrology to quantify the capillary condensation of toluene vapor inside porous low-K dielectric films on a silicon substrate. As the partial pressure of the toluene environment over the film increases, capillary condensation occurs in progressively larger pores. This results in an appreciable increase in the electron density of the film. By monitoring the changes in the critical angle for total X-ray reflectance, one can directly calculate the average electron density, and therefore the toluene uptake. By invoking traditional porosimetry absorption/desorption procedures, characterstics such as porosity and the distribution of pore sizes can be extracted. To illustrate, the porosity in a number of low-K films are characterized using this methodology.

Introduction

Low-k dielectrics and the microelectronics industry

Increased miniaturization of the integrated chip has largely been responsible for the rapid advances in semiconductor device performance, driving the industry's growth over the past decade. Soon the minimum feature size in a typical integrated circuit device will be well below 100 nm. At these dimensions, interlayers with extremely low dielectric constants (k) are imperative to reduce cross-talk and increase device speed. State-of-the-art non-porous silicon based low-k dielectric materials have k values on the order of 2.7. However, k needs to be further reduced to keep pace with the demand for increased miniaturization. There are a number of potential material systems for these next generation low-k dielectrics, including organosilsesquioxane resins, sol-gel based silicate materials, CVD silica, and entirely polymeric resins. However, it is not yet evident as to which material(s) will ultimately prevail. Nevertheless, decreasing k beyond current values requires generating large-scale porosity; lower dielectric constant materials are not feasible with fully dense materials.

The demand of increased porosity in reduced dimensions creates many difficulties. To generate extensive porosity in sub-100 nm films and features one must have exacting control over the pore generation process. The first step towards achieving this control, even before addressing materials issues, is to be able to accurately characterize the physical attributes of the pore structure (porosity, pore size, pore distribution). While there are several mature porosimetry techniques (gas adsorption, mercury intrusion, quartz crystal microbalances, etc.) capable of characterizing pores significantly smaller than 100 nm, most methods do not have sufficient sensitivity for these porous low-k films. The sample mass in a 100 nm thick film is exceedingly small and the usual observables (i.e., pressure in a gas adsorption experiment or mass in a quartz crystal microbalance) exhibit extremely small changes as the pores are condensed or filled. Thin films require a porosimetry technique with extraordinary sensitivity.

Basics of Porosimetry

In the following we describe an X-ray reflectivity metrology to characterize pores (both porosity and size distribution) in sub-100 nm films. The experiments are similar to traditional gas adsorption techniques where the partial pressure P of a given vapor, in this case toluene, is increased until condensation occurs inside the smallest pores via capillary condensation. Further increasing P increases the critical radius for capillary condensation r_c, progressively filling the larger pores with liquid toluene. There are several possible relations to relate r_c to P, the simplest being the Kelvin equation:

$$r_c = \frac{2V_m\gamma}{-RT} \frac{1}{\ln(P/P_o)} \tag{1}$$

where γ is the liquid surface tension, V_m is the molar volume of the liquid, P is the partial pressure, and P_o is the equilibrium vapor pressure over a flat liquid surface at a temperature T. By knowing the amount of toluene condensed in the sample as a function of P/P_o, one can calculate a pore size distribution from Eq. (1). This is precisely the starting point for most forms of porosimetry, like the traditional gas adsorption techniques where the amount of condensed vapor is inferred from the pressure drop of a fixed volume of gas as the molecules leave the vapor phase and form a liquid. However, as mentioned above, the pressure drop upon adsorption is so small in these thin films that the traditional forms of porosimetry are no longer feasible.

It is well known that the Laplace pressure across a curved liquid-air interface (i.e., meniscus) shifts the equilibrium vapor pressure to lower values, causing sub-equilibrium vapors to condense inside small pores. Eq. (1) assumes that the pressure difference across the radius of curvature is the only driving force for condensation of toluene inside a pore. However this ignores the possibility of other effects such as preferential wetting, interactions between the toluene and the substrate, increases in the viscosity of the toluene absorbed into pores or channels that approach several macromolecular diameters [1], or even the possibility of spinodal evaporation/condensation [2]. It has been shown that ignoring these effects by using the Kelvin equation, especially in pores smaller than 20 nm, can lead to errors in the pore size of 100 % or more [2-4]. It is beyond the scope of this chapter to address which analysis is most appropriate for interpreting the adsorption/desorption isotherms; to a large extent this depends upon the details of the low-k material, adsorbate, and pore structure. Rather we intend to demonstrate that X-ray reflectivity can be used as a highly sensitive metrology to generate the adsorption/desorption isotherms that can then be analyzed by Eq. (1) or more sophisticated data interpretation schemes.

In Eq. (1) one of two thermodynamic parameters, either T or P, can be used to vary the critical radius for capillary condensation. Isothermally one can affect changes of r_c by adjusting the toluene concentration in the atmosphere.

Experimentally this is achieved by mixing feed streams of dry and toluene-saturated air in various ratios. Analytically this isothermal method is attractive since both γ and V_m are held constant. However, the experimental set-up for isothermal mixing can be somewhat elaborate since accurate mass flow control valves are required. It may be easier to flow a room temperature stream of toluene-saturated air over the sample and vary film temperature. When the film is at room temperature, the vapor will be condensing and toluene should fill all the voids. By then heating the sample the equilibrium vapor pressure P_o of the condensed toluene increases, causing P/P_o, and therefore r_c, to decrease. However, this non-isothermal approach is complicated by the fact that both γ and V_m are functions of temperature.

From a thermodynamic perspective, it is immaterial whether T or P is used to probe the range of critical pore sizes. The only relevant variables in the Kelvin equation, in addition to the radius of curvature, are γ and V_m and it should be possible to correct for the T variations of these parameters. In principal, the pore size distribution should not depend upon the combination of T and P used to probe a particular critical pore size. For example, suppose that

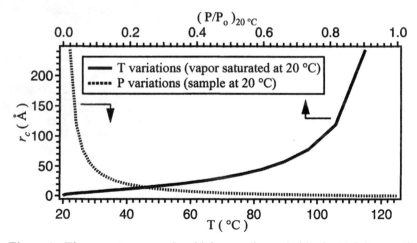

Figure 1. There are two ways in which r_c can be varied in the Kelvin equation. r_c values from approximately (1 to 250) Å can be obtained by mixing ratios of dry and toluene saturated air at 20 °C. Likewise, a comparable range of pore sizes can be achieved by flowing air saturated in toluene at 20 °C across the sample and heating the film between 20 °C and 125 °C.

two porosimetry experiments are performed: one utilizing traditional P variations where the toluene vapor concentration is varied while the film is isothermally held at 20 °C and another where toluene vapor saturated at 20 °C flows across the samples while the film is heated from 20 °C to elevated temperatures. In this scenario, for each value of r_c at a given P/P_o (under the isothermal conditions) there should be an equivalent r_c at a temperature above 20 °C. Graphically this is depicted in Figure 1. The vertical axis indicates the critical radius r_c while the upper and lower horizontal axes denote the P and T variations respectively. The red solid line shows how r_c varies isothermally with P/P_o (upper horizontal axis) while the blue dotted line depicts the non-isothermal variations (lower horizontal axis) for air saturated with toluene at 20 °C. In this simple model of the Kelvin equation it should be possible to compare the non-isothermal adsorption/desorption data with the isothermal data through the equality of r_c, which yields:

$$\left(\frac{P}{P_o}\right)_{20°C} = \left(\frac{P}{P_o}\right)_T^{\left(\frac{\gamma_{20°C}V_{m,20°C}T}{\gamma_T V_T 293}\right)} \tag{2}$$

In the following we present both T and P variation data, with the P variations presented as 20 °C isothermal data through the use of Eq. (2).

X-ray Porosimetry

X-ray reflectivity can be used to monitor the increase in the electron density that occurs when toluene condenses in the porous thin film. When a beam of X-rays is incident onto a smooth surface at a shallow or grazing angle, total reflection occurs and X-rays do not penetrate the film. As the angle of incidence increases, a critical angle is encountered at which point the X-rays begin to penetrate the film. This critical angle, θ_c, is related to the electron density ρ_e through:

$$\theta_c = \lambda \left(\rho_e r_e / \pi\right)^{0.5} \tag{3}$$

where λ is the X-ray wavelength and r_e is the classical electron radius. While the absolute pressure or mass changes as toluene condenses into a thin, low-k dielectric film may be very small, the change in the film's electron or mass density is significant. Therefore, θ_c is very a sensitive indicator of the amount of toluene condensed in the film.

214

Data collection

To demonstrate the sensitivity X-ray porosimetry, Figure 2 displays a series of reflectivity curves as a function of the toluene partial pressure for a typical hydridosilsesquioxane (HSQ) low-k dielectric film. Each reflectivity curve (as shown) requires approximately 15 min to collect; we allowed 30 min between partial pressure jumps for the film to equilibrate. We have verified that 30 min is more than sufficient (by a factor of about 3) for the film to come to equilibration for a largest possible pressure jump, i.e., from $P/P_o = 0$ to $P/P_o = 1.0$. Returning to Figure 2, at low angles all of the incident X-rays are reflected such that $R = I_{incident}/I_{reflected} = 1$, or $log (R) = 0$. Then at the critical wave vector ($Q_c = 4\pi sin (\theta_c) / \lambda$) the X-rays penetrate the film and the

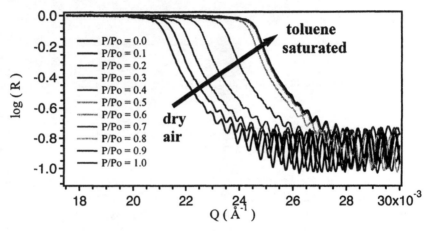

Figure 2. X-ray reflectivity curves as a function of the toluene partial pressure for a porous HSQ low-k dielectric film. As the toluene partial pressure increases, the critical angle for total reflection also increases (moves to higher Q) in response to the toluene condensing in progressively larger pores. The standard uncertainty in log(R) is less than the line width

reflectivity drops significantly. It is clearly evident that this critical point shifts to higher wave vectors (higher angles) as the environment becomes saturated with toluene. Qualitatively one can see that at low partial pressures, there is at first a moderate increase of θ_c for increments of P/P_o of 0.1. Then, at moderate values of P/P_o, the same partial pressure increments lead to much larger changes of θ_c, indicating a larger volume of pores being filled at these intermediate pressures. Finally, at high P/P_o values the changes in θ_c for equal increments of P/P_o become very small, indicating that few additional pores are filled at the highest partial pressures. As we shall see below, the nature of these increments contains detailed information about the pore size distributions.

To extract the critical angle, we fit the reflectivity data using a least-squares recursive multi-layer fitting algorithm [5]. Specifically we model the low-k dielectric film as a single layer of uniform density (through the film thickness) on top of a thick Si wafer. As the partial pressure increases, we assume (to a first approximation) that the density increase from condensing toluene is uniform throughout the film, i.e., the average density of the film increases. This average density comes from the fitting algorithm in terms of Q_c^2. By examining Eq. (3), one can see that θ_c^2, or it's reciprocal space analog Q_c^2, is directly proportional to the electron (and thus mass) density of the film. For more details regarding the experimental protocol, refer to a recent publication on the technique [6].

The premise of X-ray porosimetry is analogous to ellipsometric porosimetry (EP) [7], which also uses the capillary condensation of organic vapors to map out adsorption/desorption isotherms. However, EP tracks condensation/evaporation through changes in porous films index of refraction, not the electron density. EP implicitly assumes that the optical polarizablilities of the adsorbate and adsorbent are additive, which is not immediately evident. It remains to be seen if this genuine concern in the low-k films where the condensed organic liquid is confined to nanometer-sized pores. Confinement is known to affect deviations from bulk-like behavior in many physical properties, bringing the additivity issue into question. However, the number electrons per atom do not change upon confinement, meaning that the values of Q_c^2 provide a direct and clean measure of the toluene uptake. In our view, this is one advantage of X-ray porosimetry over EP.

Effects of varying P or T

Examples of the adsorption/desorption isotherms are shown in Figure 3. In this representation Q_c^2 is plotted as a function of the partial P/P_o. Clearly there are pronounced changes in Q_c^2 as toluene condenses in the porous HSQ film. Specifically there are two sets of adsorption/desorption isotherms corresponding to the isothermal P (squares) and T (circles) variations respectively. The open symbols denote the adsorption branch of the isotherm while the closed symbols indicate the desorption branch. Recall that the non-isothermal T variations have been transformed into their equivalent 20 °C isotherms through Eq. (2). This allows us to directly compare the true isothermal P/P_o variations to their T variation counterpart.

There are several striking features in Figure 3 that are worth discussing in greater detail. First and foremost is the fact that the T and P variation methods are not equivalent. This clearly indicates the simplistic assumptions behind the Kelvin equation are not entirely appropriate. Factors like heats of adsorption or deviations from the bulk physical properties for the condensed toluene are probably significant. However, the exact reason for this failure has yet to be

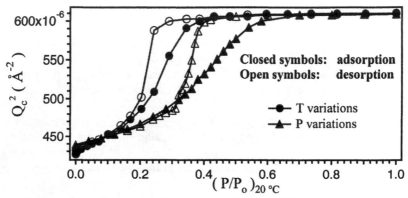

Figure 3. Adsorption/desorption curves generated by the T (circles) and P (triangles) variation techniques (see text). The two techniques do not produce isotherms, indicating that adsorption is not temperature invariant. The standard uncertainty in Q_c^2 is comparable to the size of the data markers.

identified. Of course, the pore size distributions calculated from Eq. (1) will not be unique and clearly depend on the measurement protocol. A more sophisticated analysis is required to fully comprehend these differences. Current aims to understand the differences between these T and P variation data sets.

Comparing low-k dielectric films with different levels of porosity

Regardless of the differences between the T and P variations, it is instructive to compare the isotherms for a several low-k dielectric films, as shown in Figure 4. These include the porous HSQ film discussed in Figures 2 and 3 as well as a series of methylsilsesquioxane (MSQ) films formulated with variable amounts of porogen (1 % , 10 %, 20 % and 30 % by mass fraction). Note that in Figure 4 the T variation method is used to induce the changes in P/P_o. Unlike Figure 3, the Q_c^2 values have been converted into the toluene uptake in terms of a density. This conversion can be done if the atomic composition of adsorbate (film) and adsorbent (toluene) are known. The atomic composition of toluene is trivial (C_7H_8), but additional experiments may be required to determine the composition of the film. Previously we have shown that ion beam scattering can be used to deduce the film composition [8], and the porous HSQ film contains (25 ± 2) % Si, (48 ± 2) % O, (7 ± 2) % C, and (20 ± 2) % H. The uncertainties in the composition are approximate, based on experience with how accurately we can determine the composition of a standard film. If the composition is known from other factors (i.e., stoichiometry), these errors can be greatly reduced.

In-vacuo, Q_c^2 of the porous HSQ film is $(4.25 \pm 0.01) \times 10^{-4}$ Å^{-2}. From Eq. (2) and the composition data (i.e., the number of electrons per mole) the average film density is determined to be (0.990 ± 0.062) g/cm^3 (standard uncertainty indicated). Taking into account that ensuing changes in Q_c^2 are due to toluene condensing in the pores, one can determine the toluene uptake in terms of a density, as shown in Figure 4. At saturation Q_c^2 in the porous HSQ film is $(6.10 \pm 0.01) \times 10^{-4}$ Å^{-2} (standard uncertainty indicated) which corresponds to an overall density of (1.389 ± 0.087) g/cm^3 (standard uncertainty indicated), a toluene uptake of $(.399 \pm 0.106)$ g/cm^3 (standard uncertainty indicated). By assuming the density of the condensed toluene is bulk-like, we calculate that the porosity ϕ of the porous HSQ film is 43 ± 25 %. Note that the composition numbers dominates the large uncertainty. If the composition is accurately known by other means, the uncertainty will be greatly reduced. Similar ϕ calculations for the MSQ films with 1 %, 10 %, 20 %, and 30 % by mass porogen yield respective ϕs of (12 ± 25) %, (18 ± 25) %, (27 ± 25) %, and (34 ± 25) %. Going back to curves in Figure 4, the maximum uptake or plateau at $P/P_o = 1.0$ graphically denotes the overall pore volume fraction or porosity. The porosities quoted above are consistent with the level of the plateau values at $P/P_o = 1.0$; larger porosity corresponds to a higher plateau.

For HSQ and MSQ films, swelling was not observed in the presence of the toluene. However, in some cases appreciable swelling can occur, which affects the porosity calculations. This is tractable though since X-ray reflectivity provides an accurate measurement of the film thickness (in addition to Q_c^2). In fact, matrix swelling may be desirable since it ensures that the toluene has access to all the pores (no blocked pores). By assuming volume conservation as toluene adsorbs/absorbs into the low k film, ϕ can be determined from the changes in Q_c^2 and film thickness as follows:

$$\phi = \frac{\rho_2 t_2 - \rho_1 t_1}{\rho_{tol} t_1} - \frac{t_2 - t_1}{t_1} \tag{4}$$

where t_1 and t_2 are film thickness before and after absorption, ρ_2 and ρ_1 are the corresponding electron densities, and ρ_{tol} is the electron density of liquid toluene.

In addition to the total porosity, Figure 4 reveals qualitative information about the pore structures in the different films. In the MSQ film with 1 % by mass porogen, the uptake saturates at low partial pressures and the adsorption and desorption branches coincide. This is characteristic of the IUPAC definition of micropore filling (pores widths less than 2 nm [9,10]), and consistent with the fact that MSQ materials should be inherently microporous.

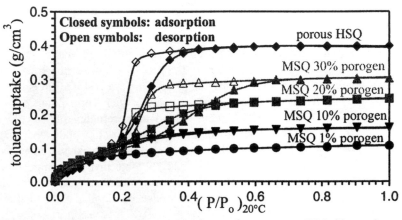

Figure 4. Adsorption/desorption isotherms for a porous HSQ film and a series of MSQ films formulated with varying amounts of porogen. The solid lines are smooth fits using the sum of a sigmoidal and lognormal curve. The standard uncertainty in the uptake is on the order of (6 to 7) %.

As the porogen content increases, a second population of larger pores fills at larger partial pressures. A hysteresis loop typical of mesopores also appears for these larger porogen-created pores. Notice that the adsorption branch is broader than the desorption branch in the vicinity of this hysteresis loop. This suggests a broad distribution of pore sizes with pore blocking or "ink bottle" effects on the desorption branch. The term ink bottle implies there are large pores that would like to desorb at relatively high partial pressures, but can not because the egress of the toluene must pass through narrow passageways, or bottle necks. The larger pores are only free to desorb at lower partial pressures when the smaller blocking pores naturally drain.

Examples of this pore blocking effect on the desorption branch of the isotherm have been well documented in the porosimetry literature. Several years ago Kraemer [11] pointed out that in an ink bottle-type pore, capillary condensation on the adsorption branch would be dominated by the radius of curvature of the larger pore whereas the emptying of the pores through the desorption branch would be governed by the radius of the narrow neck. This results in a sharp drop in the desorption isotherm, similar to that seen in Figure 4, when the small and large pores simultaneously drain. One could easily envision a similar scenario for the porogen loaded MSQ films. Toluene desorption from the isolated large pores created by the porogen has to proceed through regions of inherently microporous MSQ resin.

Interpreting the adsorption/desorption isotherms

The adsorption/desorption isotherms in Figure 4 convey a wealth of qualitative structural information. However, it can be a challenge to quantify these characteristics in terms of an average and distribution of pore sizes. This is equally true for the thin porous films presented here as well as all isotherms in general, and not an inherent limitation of X-ray porosimetry. The adsorption/desorption process is complex and affected by many factors. Consequently there are several ways to interpret isotherms, each focusing on different aspects or based on different assumptions. Generally these analysis schemes can be grouped into three general classes. The first group emphasizes the initial stages of adsorption, while the first monolayer(s) of adsorbate adhere to the surface. Well-known examples include the Henry's Law type analysises, the Hill – de Boer equation [12], Langmuir theory [13], and the widely known Brunauer-Emmett-Teller (BET) analysis [14]. These theories have been developed for adsorption onto non-porous solids and are primarily surface area techniques.

The low partial pressure regions of an isotherm are also where micropore filling occurs. Micropores, with widths less than 2 nm, are easily filled by a few monolayers of most adsorbents, and the second group of equations or theories attempt to extract micropore characteristics from the initial stages of the isotherm. These are similar to the surface area techniques and include Henry's Law based interpretations, the Langmuir-Brunauer equation [15], and the Dubinin-Stoeckli based theories [16,17].

The final group of equations focuses on the latter stages of adsorption, where the mesopores (ca. 2 nm to 50 nm in width according the IUPAC definition [9,10]) are filled. This is the region where capillary condensation occurs and the Kelvin equation is the simplest of these interpretations. There are numerous variations on the Kelvin equation that account for effects like multilayer adsorption prior to capillary condensation (i.e., BJH method [18]), disjoining pressure effects in the condensed liquid (i.e., DBdB method [19]), etc.

However, it is not our intention to ascertain which adaptation is most appropriate. Rather, we apply the simplest form of the Kelvin equation to demonstrate the technique. It would be easy to use an alternate data interpretation scheme if it is deemed more appropriate.

Eq. (1) provides the conversion between P/P_o and the pore sizes. Figure 5 displays the corresponding pore size distributions extracted from data in Figure 4. These pore size distributions now quantify the qualitative discussions in reference to the isotherms of Figure 4. Notice that pore size distributions

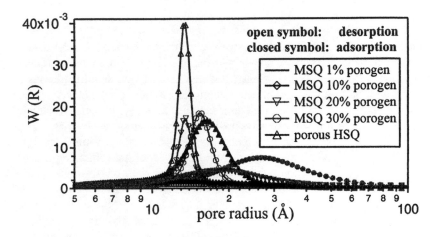

Figure 5. Approximate pore size distributions from the smooth fits through the isotherms in Figure 4 (Eq. (1) used to convert P/P_o into a pore size). The distributions from the adsorption branch (dotted lines) are always broader and shifted to larger pore sizes than the corresponding desorption branch.

have been extracted from both the adsorption (dotted lines) and desorption (solid lines) branches of the isotherm, and that the distributions are always broader and shifted to a higher average pore size for the adsorption branch. This discrepancy is especially evident for the highly porous samples, and consistent with the pore blocking effects discussed earlier. It is sometimes a "general practice" in the porosimetry field to report pore size distributions for the desorption branch of the isotherm. However, in the presence of pore blocking this leads to artificially narrow and smaller distributions. It is crucial to look at both the adsorption and desorption isotherms to obtain a more comprehensive understanding of the pore structure. Once again though, this criticism is applicable to all forms of porosimetry and is not unique for X-ray porosimetry.

The pore size distributions in Figure 5 are approximate and only intended to demonstrate the technique. The averages are well below 20 nm which means that the pore sizes are in danger of being in error by well over 100 % [2-4]. However, this is a matter of interpretation and the raw isotherms in Figure 4 are robust, and the relative differences between the different materials are clear and reliable. Porosimetry in general is very model dependent. To interpret the data one must assume a model and then verify how well it fits the data. This does not ensure that the solutions obtained are unique. To reach a proper interpretation requires supporting information, like the nature of the interactions between the adsorbent and adsorbate, feasible schemes for the pore architecture (i.e., isolated spheres, interconnected channels, fractals), etc. This type of supporting data will help facilitate a correct interpretation of the isotherms.

Generally the physics of capillary condensation in pores smaller than 20 nm is poorly understood. Studies show that more accurate interpretations require computer modeling and/or simulations of the isotherms. Gage Cell Monte-Carlo (GCMC) simulations and non-linear density functional theory (NLDFT) methods [2-4] have recently been developed for helping to interpret adsorption/desorption isotherms from several templated nano-porous materials and reveal that all of the existing interpretations of capillary condensation fail to extract reliable pore sizes. Currently these GCMC and NLDFT techniques are the only reliable methods for understanding capillary condensation in pores significantly smaller than 20 nm, and should therefore be considered when studying low-k dielectric films with exceedingly small pores.

Conclusions

We demonstrate that X-ray reflectivity is a powerful tool for extracting porosity information from highly porous, ultra-thin low-k dielectric films. As the partial pressure of an organic solvent is increased, capillary condensation occurs and this results in a marked increase in the critical angle for total X-ray reflectance. Tracking this critical angle as a function of the partial pressure generates the adsorption/desorption isotherms that are traditionally analyzed in the field of porosimetry. If the atomic composition of the film is known, it is possible to calculate the absolute porosity from the total toluene uptake. If a reliable adsorption/desorption models exists, one can extract pore size distributions from the isotherms. This shown through the simple model of the Kelvin equation and capillary condensation.

Acknowledgements

The authors are grateful for support from the Office of Microelectronics Programs at NIST and International SEMATECH through the J-105 program.

Literature Cited

1. Klein, J.; Kumacheva, E.; *J. Chem. Phys.* **1998**, *108,* 6996.
2. Neimark, A.; Ravikovitch, P. I.; Vishnyakov, A.; *Phys. Rev. E* **2000**, *62*, R1493.
3. Neimark, A. V., Schuth. F.; Unger, K. K.; *Langmuir* **1995**, *11*, 4765.
4. Ravikovitch, P. I.; Vishnyakov, A.; Neimark, A.; *Phys. Rev. E* **2001**, *64*, 011602.
5. Anker, J. F.; Majkrzak, C. J.; *Neutron Optical Devices and Applications*, SPIE Proceedings, SPIE, Bellingham, WA, 1992; Vol. 1738, pp260.
6. Lee, H. J.; Soles C. L.; Liu, D. –W.; Bauer, B. J.; Wu, W. L., *J. Polym. Sci.: Part B: Polym. Phys.*, **2002**, *40*, 2170.
7. Baklanov, M. R.; Mogilnikov, K. P.; Polovinkin, V. G.; Dultsev, F. N., *J. Vac. Sci. Technol. B*, **2000**, *18*, 1385.
8. Wu, W.-l.; Wallace, W. E.; Lin, E. K.; Lynn, G. W.; Glinka, C. J.; Ryan, E. T.; Ho, H.-M., *J. Appl. Phys.* **2000**, *87*, 1193.
9. Evertt, D. H., *Pure Appl. Chem.*, **1972**, *31*, 579.
10. Sing, K. S. W.; Evertt, D. H.; Hual, R. A. W.; Mocsou, L.; Pierotti, R. A.; Rouquerol, J.; Siemieniewska, T., *Pure Appl. Chem.*, **1985**, *57*, 603.
11. Kraemer, E. O., *A Treatise on Physical Chemistry*, Taylor, H. S.; Ed.; Macmillan, New York, 1931; p1661.
12. de Boer, J. H., *The Dynamical Character of Adsorption*, Oxford University Press, London, 1968; p 179.
13. Langmuir, I., *J. Am. Chem. Soc.*, **1916**, *38*, 2221.; ibid, **1918**, *40*, 1361.
14. Brunauer, S.; Emett, P. H.; Teller, E., *J. Am. Chem. Soc.*, **1938**, 60, 309.
15. Brunauer, S., *The Adsorption of Gases and Vapors*, Princeton University Press, Princeton, NJ, 1945.
16. Dubinin, M. M.; Radushkevich, L. V., *Proc. Acad. Sci. USSR*, **1947**, *55*, 331.
17. Stoeckli, H. F., *J. Colloid Interface Sci.*, **1977**, *59*, 184.
18. Barrett, E. P.; Joyner, L. G.; Halenda, P. H., *J. Am. Chem. Soc.*, **1951**, *73*, 373.
19. Broekhoff, J. C. P.; de Boer, J., *Catalysis*, **1967**, *9*, 8.

Chapter 17

Micro- and Nanoporous Materials Developed Using Supercritical CO$_2$

Novel Synthetic Methods for the Development of Micro- and Nanoporous Materials Toward Microelectronic Applications

Sara N. Paisner[1] and Joseph M. DeSimone[1,2]

[1]Department of Chemistry, University of North Carolina at Chapel Hill, Chapel Hill, NC 27599
[2]Department of Chemical Engineering, North Carolina State University, Raleigh, NC 27606

A number of different types of nanoporous materials for use in low dielectric constant applications have been developed in recent years, including nanoporous silica, polyimides, poly(arylethers), and poly(methyl silsesquioxanes). Recently, much research has been done in the field of supercritical carbon dioxide (scCO$_2$) and its use in the synthesis of polymers for microelectronic applications. A variety of different methods using supercritical CO$_2$ to form micro- and nanoporous materials towards applications in the microelectronic industry are described.

Introduction

Demand for low dielectric constant (k) materials in the microelectronics industry has led to extensive efforts to explore the applicability of porous materials, especially nanoporous materials.[1] Nanoporous or mesoporous materials have potential utility in many applications including membranes, sensors, waveguides, dielectrics, and microfluidic channels.[2-7] For these uses, a number of different types of nanoporous materials for microelectronic application have been developed. They include nanoporous silica,[8-14] polyimides,[15,16] poly(arylethers),[17-19] and poly(methyl silsesquioxanes).[5,6,20-22] Most of these ultralow-k materials are prepared by introducing air-filled pores into the film, taking advantage of air's low dielectric constant ($k_{air} = 1$). For microelectronic applications, control over the pore size, shape and distribution is critical to obtain materials with suitable mechanical and electrical properties.

Recently, a large amount of research has been done to develop chemistry in scCO$_2$. The potential benefits of using CO$_2$ to make nanoporous materials are numerous, and in recent years many CO$_2$ based processes have proven their worth in other industries, such as industrial extraction processes,[23] polymer processing,[24] and even consumer CO$_2$-based garment cleaning.[25] The low viscosity and surface tension of CO$_2$ potentially allow it to address certain problems associated with microelectronics industry in general, and the synthesis of micro- and nanoporous materials in particular.[26] Another important advantage of using CO$_2$ as a solvent in material synthesis is that it is relatively environmentally benign. Unlike organic solvents and aqueous solutions used in conventional micro- and nanoporous material synthesis, CO$_2$ is non-toxic and readily recyclable.[27] In an industry where an enormous amount of organic and aqueous waste is produced daily, the introduction of a readily recyclable, environmentally benign solvent has the potential to dramatically reduce the cost, both financially and environmentally, of microchip fabrication.

This review chapter will focus primarily on the synthesis of nanoporous materials using CO$_2$ in one of two main methods. The first is *via* extraction, which takes advantage of the differing solubilities of polymers in scCO$_2$. The second method involves foaming of materials with CO$_2$ as the blowing agent. Both methods have resulted in micro- and nanoporous polymers some of which may have potential applications as low dielectric constant materials.

Formation of Porous Materials *via* scCO$_2$ Extraction

By taking advantage of the differing solubilities of polymers in scCO$_2$, novel nano- and microporous materials can been developed. In particular block and graft copolymers can form micro-phase separated morphologies such as spheres, cylinders and lamellae, which can be used to control the orientation of nanostructures of large areas.[28] The molecular structure and M_w of the second block segment determines the pore size and volume fraction of the dispersed phase.[29] Selective removal of the minority component from the crosslinked material then leaves a matrix filled with nanoscopic voids.

Block copolymers which contain a thermally labile segment have been used to form nanoporous materials.[30] Following this approach, we have focused on the formation of phase separated polystyrene-g-poly(dimethylsiloxane) (PS-g-PDMS) and PS-b-PDMS copolymers (Figure 1) followed by chemical decomposition of the PDMS blocks using acid in scCO$_2$. Once the PDMS decomposed, it was extracted out from the PS matrix using scCO$_2$, in which the oligomeric siloxanes were highly soluble. This resulted in nanoporous materials, as the size of the PDMS spheres within the PS matrix could be controlled by controlling the weight % of the PDMS block. TEM analysis of a film formed by evaporation from chloroform, showed 5-8 nm size spheres of PDMS in a PS matrix (Figure 2). Thin films were formed by spin coating on to a silicon wafer from trifluorotoluene and SEM analysis of the thin films show a film thickness of approximately 500 to 700 nm. By DSC, two T_g's are observed; 86 °C for PS, -106 °C for PDMS as well as a T_m at -46 °C for PDMS. The PDMS grafts (or blocks) of these materials were decomposed into oligomers by trifluoroacetic acid in scCO$_2$. By SEM analysis of cross sections of films, the sizes of the pores were determined to be approximately 5-20 nm (Figure 3).

A second example of scCO$_2$ as an a development solvent for use in the formation of low-k materials was demonstrated using hot-filament chemical vapor deposition (HFCVD, also known as pyrolytic CVD) to form directly patterned low-k films.[31] HFCVD using hexafluoropropylene oxide (HFPO) as the precursor gas has been shown to produce fluorocarbon films spectroscopically similar to polytetrafluoroethylene (PTFE, $k = 2.0$),[32] a material with a very low dielectric constant. Films formed from this process were exposed to e-beam doses followed by scCO$_2$ development to determine contrast curves. The contrast for each sample described correlated qualitatively with the concentration of OH and C=O species, and the lower the OH/CF$_2$ ratio, the higher the contrast under e-beam exposure. Both of these sets of values can be controlled by careful design of the HFCVD process which was described previously.[33,34] By developing low-k materials compatible with the damascene process in scCO$_2$, multiple steps presently required to produce patterned insulators can potentially be removed.

A method which involves the use of scCO$_2$ to extract out oligomers from a polymer matrix has also been described.[35] Plasma enhanced chemical vapor deposition (PECVD) was used to obtain siloxane films from tetravinyltetramethylcyclotetrasiloxane, (TVTMCTS, a liquid source) mixed in

226

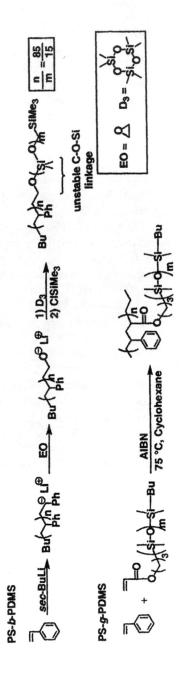

Figure 1. Synthesis of PS-b-PDMS and PS-g-PDMS

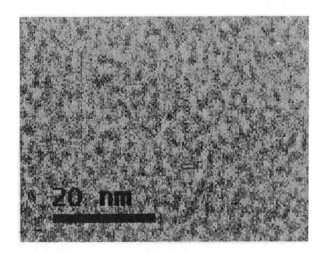

Figure 2. TEM at high magnification of PDMS spheres in a PS matrix.

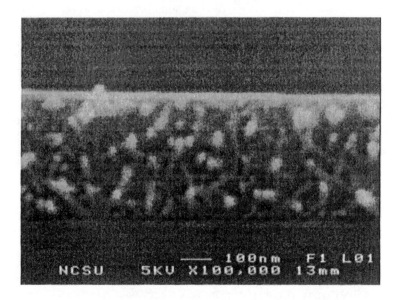

Figure 3. SEM of cross section of PS nanoporous film after PDMS removal in scCO₂ using trifluoroacetic acid.

with a C_2F_8 gas and H_2 plasma. This mixture was deposited on silicon substrates using PECVD to obtain composite films which were stable to 400 °C. The films were then pressurized with $scCO_2$ at 200 °C and 8650 psi for 8h to extract out the low molecular oligomers. This was possible as low molecular weight oligomers of this siloxane were soluble in $scCO_2$, whereas the high molecular weight materials were not. Once the oligomers were removed, a porous film was obtained. Dielectric constants of $k = 2.5$ to 3.3 were observed in the materials extracted with $scCO_2$.

Carbon Dioxide Foaming

Another approach for forming micro- and nanoporous materials is *via* $scCO_2$ foaming. Use of environmentally friendly physical blowing agents in their supercritical or non-supercritical state has become important in the past few decades. For use of foamed materials in any microelectronic features, pore size must be much smaller than the film thickness, as otherwise the foams may collapse.

One physical foaming process involving glassy poly(ether imide)s and poly(ether sulfone)s using CO_2 has been investigated.[36] Two types of porosities were observed, closed microcellular and bicontinuous. In this work, the foaming behaviors of thin (~75μM) extruded poly(ether imide) (PEI) and poly(ether sulfone) (PES) films were studied (Table I).

Porous materials were formed using a discontinuous solid-state microcellular foaming process with CO_2 as the blowing agent. Temperature-concentration conditions ("foam diagrams") that mark the foaming region (*i.e.* where the CO_2-saturated polymer/gas mixture changes into a cellular structure) were determined. Closed-cellular morphologies were no longer observed above certain CO_2 pressures in the mixture and instead, nanoporous bicontinuous foams were observed. For example, closed cell pores were seen until *ca.* 40 and 50 bar for PEI and PES, respectively, after which the structure became bicontinuous. A good guide for the foaming boundaries were determined to be the T_g of the polymer/gas mixture and what the authors called the "upper foaming temperature limit." This temperature was where cells became unstable due to CO_2 loss (due to diffusion) and a strong decrease in viscosity of the polymer was observed (Figure 4).

Interestingly, increasing CO_2 saturations levels to *ca.* 47 cm^3 was found to cause cell diameters to suddenly decrease below 100 nm and cell densities to increase by two orders of magnitude, up to 10^{14} cells/cm^3 (Figure 5). This was accompanied by, and thought to be due to, the change from a closed pore to bicontinuous morphology.

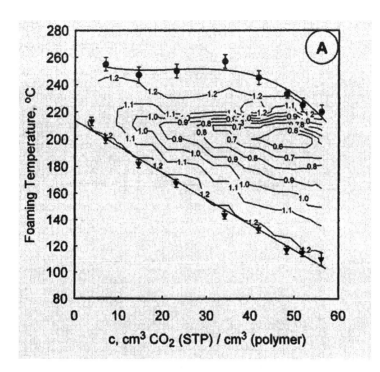

Figure 4. Mass density of PEI as a function of the dissolved amount of CO_2 and the foaming temperature. Numbers written close to the straight lines represent the mass density of the foam in g/cm^3. The mass density contours are constructed by linear interpolation from experimental data series of nine different saturation pressures, equally distributed over the investigated concentration range. The glass transition temperature (T_g) and T_{upper} are presented dependant on the dissolved amount of CO_2. The straight line represents a least-squares fit of the experimental glass transition data. (Reprinted from reference 36. Copyright 2001 American Chemical Society.)

230

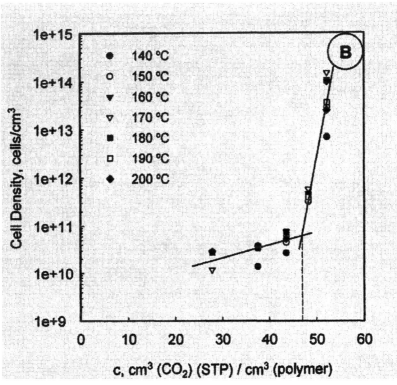

Figure 5. Cell densities of porous PES morphologies vs. the CO₂ concentration of the saturated polymers. Cell densities obtained at different foaming temperatures are included. Lines are included to emphasize the transition at approximately 47 cm³ (STP)/cm³ (polymer). (Reprinted from reference 36. Copyright 2001 American Chemical Society.)

Table I. Chemical Structure, Mass Density, Glass Transition Temperature, and Dual Mode Sorption Parameters of CO_2 for the PEI and PES Films at 25 °C.

	Polyetherimide	Polyethersulfone
T_g, °C	218	230
Mass density, g/cm^3	1.28	136
Kp, cm^3(STP/cm^3 (polymer)/bar	0.428	0.265
C_H, cm^3(STP)/cm^3 (polymer)	40.01	45.11
b, bar^{-1}	0.088	0.126

Source: Reproduced from reference 36. Copyright 2001 American Chemical Society

Another example of foaming, while not immediately applicable to microelectronics due to large pores sizes, was a method by which microcellular materials were formed *via* self-assembly in CO_2.[37] This system used foaming in CO_2 of hydrogen bonded small compounds and polymers to form low density, low thermal conductivity, microfibrillar foams. The precursors were non-polymeric bisurea and trisurea compounds, acrylate copolymers, and fluorinated polyacrylate urea (Figure 6.) The behavior of the various compounds in CO_2 reflected the self-association strength (number of physical cross-links points, strength of the interactions) versus the solute-CO_2 interaction strength. If self-association via hydrogen bonding was too strong, the material either did not dissolve in CO_2 or precipitated as a powder. If self-association was too weak versus interaction with CO_2, the compound dissolved readily but required significant depressurization prior to precipitation, and the resulting foam exhibited relatively large cells (Figure 7.) The trisurea compounds (1) with a symmetric spacer in the hydrogen bond core and three association points per molecule, were found to provide the optimum structure to generate strong microcellular foams.

Conclusions

Nanoporous materials have become important for the next generation of ultralow-*k* dielectrics for the microelectronics industry. Towards this goal a number of different methods have been developed to form such materials. One subset of these involve the use of scCO$_2$ as either a development solvent (i.e. extraction of a component) or as a foaming agent. Various (co-)polymers and additives have been tested to determine the best conditions which lead to nanoporous materials. Some of these materials and methods have resulted in low

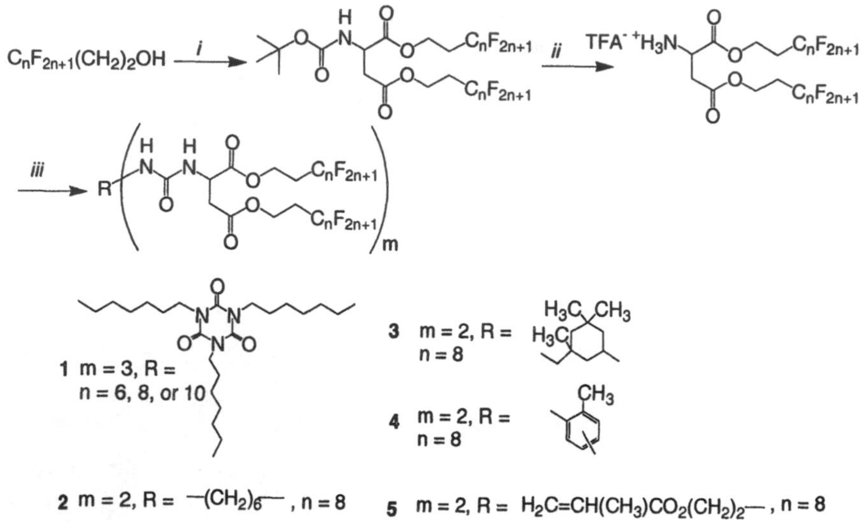

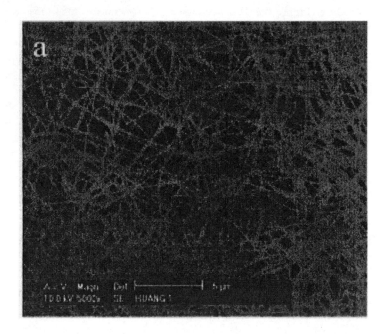

$C_nF_{2n+1}(CH_2)_2OH$ $\xrightarrow{i}$ [BOC-protected Asp diester with C_nF_{2n+1} chains] $\xrightarrow{ii}$ $TFA^-\ {}^+H_3N$ [Asp diester with C_nF_{2n+1} chains]

$\xrightarrow{iii}$ $R\left(\begin{array}{c}\overset{H}{N}\overset{H}{N}\text{—}[\text{Asp diester, } O\text{—}C_nF_{2n+1}, O\text{—}C_nF_{2n+1}]\end{array}\right)_m$

1 m = 3, R = [isocyanurate tris-alkyl structure]
n = 6, 8, or 10

2 m = 2, R = —(CH₂)₆— , n = 8

3 m = 2, R = [trimethyl/ethyl cyclohexyl structure]
n = 8

4 m = 2, R = [methyl-substituted aromatic structure]
n = 8

5 m = 2, R = $H_2C=CH(CH_3)CO_2(CH_2)_2$— , n = 8

*Figure 6. Synthesis of Urea compounds **1-5**. Conditions: (i)EDCI, DMAP N-BOC-Asp; (ii) TFA; (iii) TEA, isocyanate. (Adapted from reference 37)*

*Figure 7. SEM image of the foam from triurea compound **1b** in carbon dioxide from concentration of 0.1 wt %. (Reprinted from reference 37. Copyright 2002 American Chemical Society.)*

dielectric constant materials. Further improvements, however, in reducing pore size and increasing material strength so as to form low-k materials for industrial applications are still necessary in order to replace the currently used porous silicates.

Acknowledgements

Support for this work is provided in part by the Kenan Center for the Utilization of CO_2 in Manufacturing and by the STC Program of the National Science Foundation under Agreement No. CHE-9876674. The authors acknowledge Dr. W. Ambrose and Dr. D. Bachelor for assistance with TEM and SEM, respectively.

References

(1) Miller, R. D. *Science* **1999**, *286*, 421.

(2) Yang, S.; Mirau, P. a.; Pai, C.-S.; Nalamasu, O.; Reichmanis, E.; Pai, J. C.; Obeng, Y. S.; Seputro, J.; Lin, E. K.; Lee, H.-J.; Sun, J.; Gidley, D. W. *Chem. Mater.* **2002**, *14*, 369.

(3) Boyle, T. J.; Brinker, C. J.; Gardner, T. J.; Sault, A. G.; Hughes, R. C. *Comments Inorg. Chem.* **1999**, *20*, 209.

(4) Huo, Q. S.; Zhao, D. Y.; Feng, J. L.; Weston, K.; Buratto, S. K.; Stucky, G. D.; Schacht, S.; Schuth, F. *Adv. Mater.* **1998**, *9*, 974.

(5) Baskaran, S.; Liu, J.; Domansky, K.; Kohler, N.; Li, X. H.; Coyle, C.; Fryxell, G. E.; Thevuthasan, S.; Williford, R. E. *Adv. Mater.* **2000**, *12*, 291.

(6) Lu, Y. F.; Fan, H. Y.; Doke, N.; Loy, D. A.; Assink, R. A.; LaVan, D. A.; Brinker, C. J. *J. Am. Chem. Soc.* **2000**, *122*, 5258.

(7) Hedrick, J. L.; Miller, R. D.; Hawker, C. J.; Carter, K. R.; Volksen, W.; Yoon, D. Y.; Trollsas, M. *Adv. Mater.* **1998**, *10*, 1049.

(8) Nitta, S. V.; Jain, A.; P.C. Wayner, J.; Gill, W. N.; Plawsky, J. L. *J. Applied Phys* **1999**, *86*, 5870.

(9) Domansky, K.; Fryxell, G. E.; Liu, J.; Kohler, N. J.; Baskaran, S.; Li, X.; Thevuthasan, S.; Coyle, C. A.; Birnbaum, J. C. In *World Intellectual Property Organization*: USA, 2000.

(10) Pevzner, S.; Regev, O.; Yerushalmi-Rozen, R. *Current Opinion in Colloid & Interface Science* **2000**, *4*, 420.

(11) MacDougall, J. E.; Heier, K. R.; Weigel, S. J.; Wediman, T. W.; Demos, A. T.; Bekiaris, N.; Lu, Y.; Mandal, R. P.; Nault, M. P. In *European Patent Office*: Germany, 2001.

(12) Cheng, P.; Doyle, B. S.; Chiang, C.; Tran, M. T.-H.; Intel Corporation, Santa Clara, CA: USA, 2001.

(13) Vélez, M. H.; Garrido, O. S.; Barbeyto, R. M. B.; Shmytko, I. M.; Poza, M. M. B.; Burgos, L. V.; Martínez-Duart, J. M.; Ruíz-Hitzky, E. *Thin Solid Films* **2002**, *402*, 111.

(14) Flannery, C. M. *Ultrasonics* **2002**, *40*, 237.

(15) Hedrick, J. L.; Carter, K. R.; Cha, H. J.; Hawker, C. J.; DiPietro, R. A.; J.W.Labadie; Miller, R. D.; Russell, T. P.; Sanchez, M. I.; Volksen, W.; Yoon, D. Y.; Mecerreyes, D.; Jerome, R.; McGrath, J. E. *Reactive & Functional Polymers* **1996**, *30*, 43.

(16) Cha, H. J.; Hedrick, J.; DiPietro, R. A.; Blume, T.; Beyers, R.; Yoon, D. Y. *Appl. Phys. Lett.* **1996**, *68*, 1930.

(17) Xu, Y.; Tsai, Y.-p.; Tu, K. N.; Zhao, B.; Liu, B.-Z.; Brongo, M.; Sheng, G. T. T.; Tung, C. H. *Appl. Phys. Lett.* **1999**, *75*, 853.

(18) Godschalz, J. P.; Bruza, K. J.; Niu, Q. S. J.; Cummins, C. H.; III, P. H. T.; The Dow Chemical Company; 2030 Dow Center, Midland, MI 48674, US: International Patent, 2001, p WO 01/38417 A38411.

(19) Xu, Y.; Zheng, D. W.; Tsai, Y.; Tu, K. N.; Zhao, B.; Liu, Q.-Z.; Brongo, M.; Ong, C. W.; Choy, C. L.; Sheng, G. T. T.; Tung, C. H. *J. Electronic Materials* **2001**, *30*, 309.

(20) Kim, H.-C.; Wilds, J. B.; Hinsberg, W. D.; Johnson, L. R.; Volksen, W.; Magbitang, T.; Lee, V. Y.; Hedrick, J. L.; Hawker, C. J.; Miller, R. D.; Huang, E. *Chem. Mater.* **2002**, *14*, 4628.

(21) Nguyen, C.; Hawker, C. J.; Miller, R. D.; E.Huang; Hedrick, J. L. *Macromolecules* **2000**, *33*, 4281.

(22) Yang, G. Y.; Briber, R. M.; Huang, E.; Kim, H.-C.; Volksen, W.; Miller, R. D.; Shin, K. *NIST Special Publication* **2002**, *977*, 30.

(23) Erkey, C. *Journal of Supercritical Fluids* **2000**, *17*, 259.

(24) Cooper, A. I. *Journal of Materials Chemistry* **2000**, *10*, 207.

(25) DeSimone, J. M. "Cleaning Process Using Carbon Dioxide as a Solvent and Employing Molecularly Engineered Surfactants," The University of North Carolina at Chapel Hill, 1999.

(26) Namatsu, H.; Yamazaki, K.; Kurihara, K. *J. Vac. Sci. Technol. B* **2000**, *18*, 780.

(27) Hashimoto, K. *Materials Science and Engineering A* **2001**, *304*, 88.

(28) Hashimoto, T.; Tsutsumi, K.; Funaki, Y. *Langmuir* **1997**, *13*, 6869 and references therein.

(29) Xu, T.; Kim, H.-C.; DeRouchey, J.; Seney, C.; Levesque, C.; Martin, P.; Stafford, C. M.; Russell, T. P. *Polymer* **2001**, *42*, 9091.

(30) Zalusky, A. S.; Olayo-Valles, R.; Wolf, J. H.; Hillmyer, M. A. *J. Am. Chem. Soc.* **2002**, *124*, 12761.

(31) Pryce-Lewis, H. G.; Weibel, G. L.; Ober, C. K.; Gleason, K. K. *Chem. Vap. Depos.* **2001**, *7*, 195.

(32) Lau, K. K. S.; Gleason, K. K. *J. Fluorine Chem.* **2000**, *104*, 119.

(33) Lau, K. K. S.; Caulfield, J. A.; Gleason, K. K. *Chem. Mater.* **2000**, *12*, 3032.

(34) Lau, K. K. S.; Caulfield, J. A.; Gleason, K. K. *J. Vac. Sci. Technol. A* **2000**, *18*, 2404.

(35) Lubguban, J. A.; Sun, J.; Rajagopalan, T.; Lahlouh, B.; Simon, S. L. *Appl. Phys. Lett.* **2002**, *81*, 4407.

(36) Krause, B.; Sijbesma, H. J. P.; Münüklü, P.; Vegt, N. F. A. V. d.; Wessling, M. *Macromolecules* **2001**, *34*, 8792.

(37) Huang, Z.; Shi, C.; Enick, R.; Beckman, E. *Chem. of Mater.* **2002**, *14*, 4273.

Chapter 18

Thermally Degradable Photocross-Linking Polymers

**Masamitsu Shirai, Satoshi Morishita, Akiya Kawaue,
Haruyuki Okamura, and Masahiro Tsunooka**

Department of Applied Chemistry, Graduate School of Engineering,
Osaka Prefecture University, Sakai, Osaka 599–8531, Japan

A novel monomer (MOBH) which has both epoxy moiety and
thermally cleavable tertiary ester moiety in a molecule was
synthesized and characterized. Homopolymer of MOBH and
copolymers of MOBH with *tert*-butyl methacrylate, *tert*-
butoxystyrene or styrenesulfonates were synthesized. On UV
irradiation the polymer films containing photoacid generators
became insoluble in organic solvents. When the crosslinked
polymer films were baked at 100-220 °C, they became soluble
in methanol. The effective baking temperature was strongly
dependent on polymer structure. The crosslinked polymers
having styrenesulfonic acid ester units became soluble in
water after bake treatments.

Introduction

Polymers which become insoluble in solvents on UV irradiation are used as
photosensitive materials such as photoresists, printing plates, inks, coatings and
photocurable adhesives (*1*). Since photochemically crosslinked polymers are
insoluble and infusible networks, scratching or chemical treatments with strong
acid or base must be applied to remove these networks from substrates.
However, it is difficult or impossible to thoroughly remove crosslinked
polymers without damaging underlying materials. Recently, some thermosets
which are thermally or chemically degradable under a given condition have

been reported.　　Epoxy resins containing disulfide linkages were reported. They could be cleaved by treatment with triphenylphosphine to generate thiols (*2, 3*). Epoxy resins having acetal linkages were also studied (*4*) because acetal linkages can be easily decomposed by acids.　Furthermore, epoxy resins with tertiary ester linkages (*5, 6*) or carbamate linkages (*7*) were reported to undergo network breakdown upon heating.　　Diacrylate and dimethacrylate monomers containing thermally cleavable tertiary ester linkages were synthesized and the networks obtained by photopolymerization were observed to decompose on heating to form partially dehydrated poly(acrylic acid) or poly(methacrylic acid) and volatile alkenes (*6*).　Decomposed products were soluble in basic solutions and could be removed by a simple thermal treatment followed by washing with a basic solution.　　Re-usable polymers based on bicyclic ortho esters and spiro ortho esters were also reported (*8, 9*).　These polymers can be converted to monomers by depolymerization.

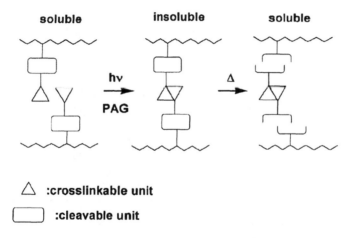

△ :crosslinkable unit

▭ :cleavable unit

Figure1.　Concept of thermally degradable photocrosslinking polymer.

In this paper, we report the synthesis and characterization of photocrosslinkable polymers bearing thermally degradable property (*10, 11*). The concept of the present system is shown in Figure 1. On irradiation network formation takes place by the photoinduced-acid catalyzed reactions of the crosslinkable moieties.　A thermal treatment of the crosslinked polymers induces the cleavage of the network linkages.　Based on this concept, we have synthesized polymers bearing both an epoxy moiety and a tertiary ester linkage in the side chain. These polymers are important as a photocrosslinkable material which can be removed by baking after use.

Experimental

Materials

Benzene, *N,N*-dimethylformamide (DMF), dichloromethane, methyl methacrylate, *tert*-butyl methacrylate (TBMA) and *p-tert*-butoxystyrene (tBOSt) were purchased and distilled before use. 2,2'-Azobisisobutyronitrile (AIBN) was purified by recrystallization from ethanol. 9-Fluorenilideneimino *p*-toluenesulfonate (FITS) (*12*), neopentyl styrenesulfonate (NPSS) (*13*), cyclohexyl styrenesulfonate (CHSS) (*14*) and phenyl styrenesulfonate (PhSS) were prepared according to the literature. Oxone (potassium peroxymonosulfate) (Aldrich) and triphenylsulfonium triflate (TPST) (Midori Kagaku) were used as received.

1-Methyl-1-(4-methyl-cyclohex-3-enyl)ethyl methacrylate (MMCEM)

MMCEM was prepared by the reaction of methacryloyl chloride with α-terpineol (Figure 2) . To a cold (< 5 °C) solution of α -terpineol (34.8 g, 0.226 mol) and 4-dimethylaminopyridine (DMAP) (2.7 g, 0.0221 mol) in anhydrous pyridine (31 mL) was slowly added a solution of 24.0 g (0.230 mol) of methacryloyl chloride in 110 mL of anhydrous dichloromethane. The mixture was stirred at ambient temperature for 40 h and then thoroughly washed with 2N H_2SO_4. The organic phase was separated and washed with saturated $NaHCO_3$ solution and then with water. The organic layer was dried over anhydrous $MgSO_4$. The product was purified by column chromatography; yield 25.6 g (51.0 %). ^{1}H NMR (400 MHz, $CDCl_3$) δ 5.92 (s, 1H, CH_2=C), 5.39 (s, 1H, CH_2=C), 5.30 (s, 1H, -CH=C-), 1.83 (s, 3H, CH_3), 1.57 (s, 3H, CH_3), 1.41 (d, 6H, 2CH_3), 1.20-2.10 (m, 11H, CH, CH_2).

1-Methyl-1-(6-methyl-7-oxabicyclo[4.1.0]hept-3-yl)ethyl methacrylate (MOBH)

MOBH was obtained by epoxidation (*15*) of MMCEM. Into a three-necked round-bottom flask fitted with an efficient magnetic stirrer, a Claisen adapter, two addition funnels, and a pH meter electrode were placed MMCEM (25.6 g, 0.115 mol), dichloromethane (160 mL), acetone(190 mL, 2.64 mol), phosphate buffer (pH = 7.4, 630 mL), and 18-crown-6 (1.26 g, 0.00477 mol). The flask was cooled to 0 – 5 °C using an ice-water bath. Oxone (2KHSO$_5$ · KHSO$_4$ · K$_2$SO$_4$) (107 g, 0.174 mol) in 390 mL of water was added dropwise over the course of 2 h. At the same time, a solution of KOH (40 g, 0.713 mol) in 190 mL of water was also added dropwise to keep the reaction mixture at pH 7.1 ~ 7.5. After the addition of Oxone, the reaction mixture was stirred at 5 °C for an additional 4 h. The resulting mixture was filtered and extracted with three 80 mL aliquots of dichloromethane, and the combined organic layers were washed with water and dried over anhydrous MgSO$_4$. The oily residue was subjected to column chromatography to obtain the pure product; yield 14.3 g (52.1 %). ^{1}H NMR (400 MHz, CDCl$_3$) δ 5.92 (s, 1H, CH$_2$=C), 5.40 (s, 1H, CH$_2$=C), 2.95 (d, 1H, epoxy HCOC), 1.83 (s, 3H, CH$_3$), 1.38 (s, 3H, CH$_3$), 1.25 (d, 6H, 2CH$_3$), 1.20-2.10 (m, 7H, CH, CH$_2$).

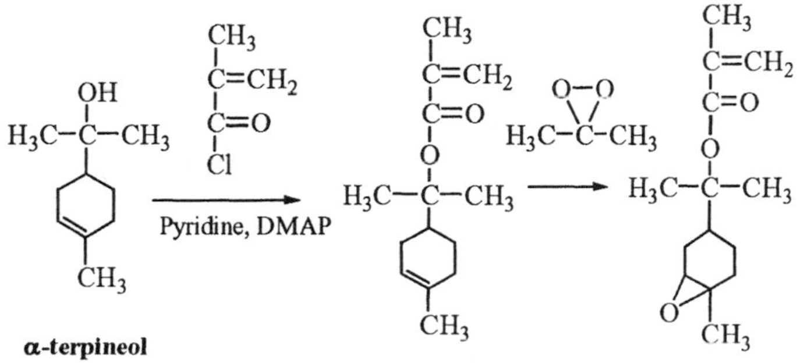

Figure 2. Synthesis of the monomer MOBH.

Preparation of Polymers

Poly(MOBH) (PMOBH), poly(MOBH-*co*-TBMA) (P(MOBH-TBMA)), poly(MOBH-*co*-tBOSt), (P(MOBH-tBOSt)), poly(MOBH-*co*-CHSS) (P(MOBH-CHSS)), poly(MOBH-*co*-NPSS) (P(MOBH-NPSS)) and poly(MOBH-*co*-PhSS) (P(MOBH-PhSS) were prepared by radical polymerization in degassed DMF solution at 30 °C using AIBN (0.12 mol/L) as an initiator with irradiation using a medium-pressure mercury lamp (Toshiba SHL-100UV) with a cutoff filter (Toshiba UV-35). Monomer concentrations were 2.6-6.9 mol/L. The resulting polymers were purified by reprecipitation from chloroform / methanol. Yields of polymers were 37-56%. The fraction of the MOBH incorporated into the polymers was determined from the peak intensity of the ^{1}H NMR spectra. When the conventional radical polymerization of MOBH was carried out at 50-60°C, gel formation frequently occurred due to undesired reaction of the epoxy moiety in MOBH. Thus, homopolymer and copolymers of MOBH were usually prepared by photopolymerization at 30 °C to prevent gel formation. Glass transition temperatures (T_g) of these polymers were 95 – 122 °C which were close to that of poly(methyl methacrylate). Number-average molecular weights of the polymers were 3.0 – 15 x 10^4. Thermal decomposition temperatures (T_d: onset temperature) of the polymers were 140 – 234 °C. The T_d values above 200 °C were due to the decomposition of carboxylic acid ester units. On the other hand, the T_d values at 140 °C for P(MOBH-CHSS) and 174 °C for P(MOBH-NPSS) were due to the decomposition of styenesulfonic acid ester units. Structures and properties of the polymers are shown in Figure 3 and Table I, respectively.

Measurements

All sample films were prepared on silicon wafers by spin-casting from solutions of cyclohexanone / chloroform (1:1, v/v) containing sample polymer and photoacid generators. The sample films were dried on a hot plate at 120 °C for 2 min. The thickness of films was about 0.5 μm except for the sample films for the FT-IR measurements (1.9 μm). Irradiation was performed at 254 nm in air using a low-pressure mercury lamp (Ushio ULO-6DQ, 6 W) without a filter.

Table I. Characteristics of polymers

Polymer	M_n $X 10^{-4}$	M_w/M_n	T_g (°C)	T_d (°C)	F_{ins} (%)
PMOBH	5.1	2.2	118	216	83
P(MOBH(14)-TBMA)	4.6	2.3	118	229	60[a]
P(MOBH(32)-TBMA)	9.3	2.6	117	218	65
P(MOBH(15)-tBOSt)	3.0	1.8	95	234	40[a]
P(MOBH(42)-CHSS)	9.5	5.1	107	140	62
P(MOBH(38)-NPSS)	9.1	5.4	122	174	60
P(MOBH(28)-PhSS)	15	2.5	115	206	90

[a] Post-exposure-bake treatment was done at 90 °C for 30 min.

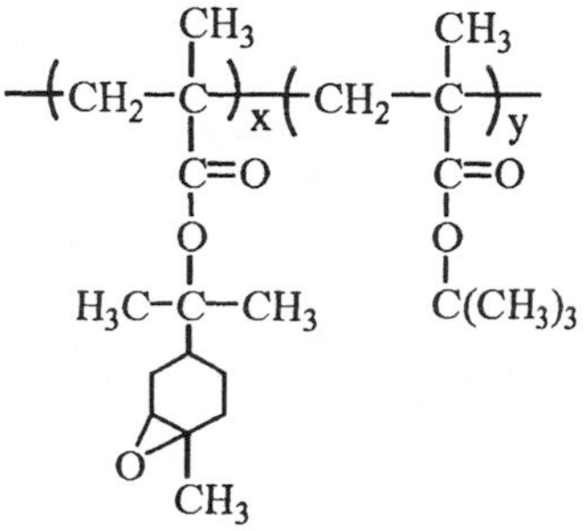

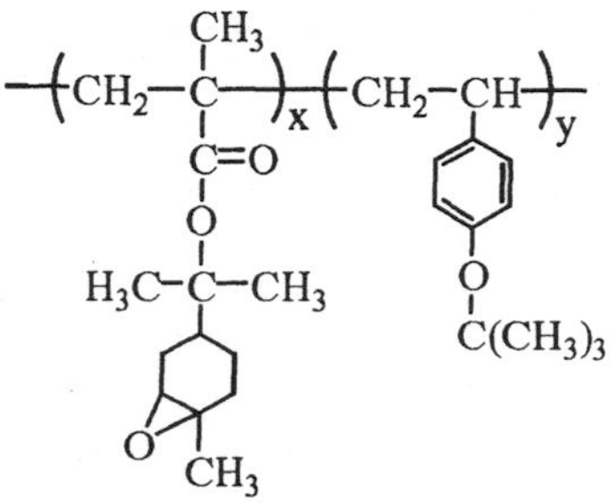

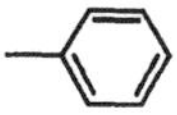

Figure 3. Structures of polymers used.

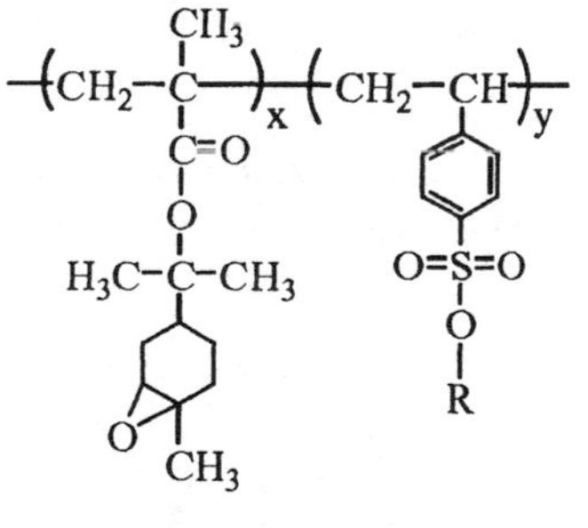

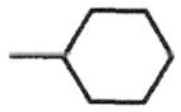

The intensity of the light was measured with an Orc Light Measure UV-M02. Insoluble fraction was determined by comparing the film thickness before and after developments with tetrahydrofuran (THF). Thickness of films was measured by interferometry (Nanometrics Nanospec M3000).

^{1}H NMR spectra were observed at 400 MHz using a JEOL LA-400 or at 270 MHz using a JEOL GX-270 spectrometer. UV-vis spectra were taken on a Shimadzu UV-2400 PC. FT-IR measurements were carried out using a JASCO IR-410. In-situ FT-IR spectroscopy was also performed using Litho Tech PAGA-100. Thermal decomposition behavior was investigated with a Rigaku TAS 100 thermogravimetric analyzer (TGA) and differential scanning calorimeter (DSC) under nitrogen flow. Heating rate was $10°C/min$ for both measurements. Size exclusion chromatography (SEC) was carried out in THF on a JASCO PU-980 chromatograph equipped with polystyrene gel columns (Shodex KF-806M + GMNHR-$_N$; 8.0 mm i.d. x 30 cm each) and a differential refractometer JASCO RI1530. The number-average molecular weight (M_n) and molecular weight distribution (M_w/M_n) were estimated on the basis of a polystyrene calibration.

Results and Discussion

Photocrosslinking

Polymer films containing the photoacid generator FITS were irradiated at 254 nm and insoluble fraction in THF was studied. FITS was photolyzed to generate p-toluenesulfonic acid. The photoinduced-acid initiated cationic polymerization of epoxy units in the side chain to generate networks (Figure 4). Insoluble fractions (F_{ins}) for the polymers were 60-90%, except for P(MOBH(14)-TBMA) and P(MOBH-tBOSt), when irradiated (50 mJ/cm^2) at room temperature (see Table I). The photoinduced insolubilization of the polymers increased with MOBH unit fraction and molecular weight. No significant insolubilization was observed for P(MOBH(14)-TBMA) and P(MOBH-tBOSt) because of low fractions of MOBH units (14 -15 mol%). The insoluble fraction of the present polymers was increased by the post-exposure-baking (PEB) treatment at relatively lower temperatures (60-120 ˚C). Although, after the irradiation (50 mJ/cm^2) at room temperature, the insoluble fractions for P(MOBH(14)-TBMA) and P(MOBH-tBOSt) were negligibly low, the insoluble fraction became 60 and 40%, respectively, after PEB treatment at 90˚C for 30 min. No insolubilization was observed when unirradiated polymer films were baked at 90 °C for 30 min.

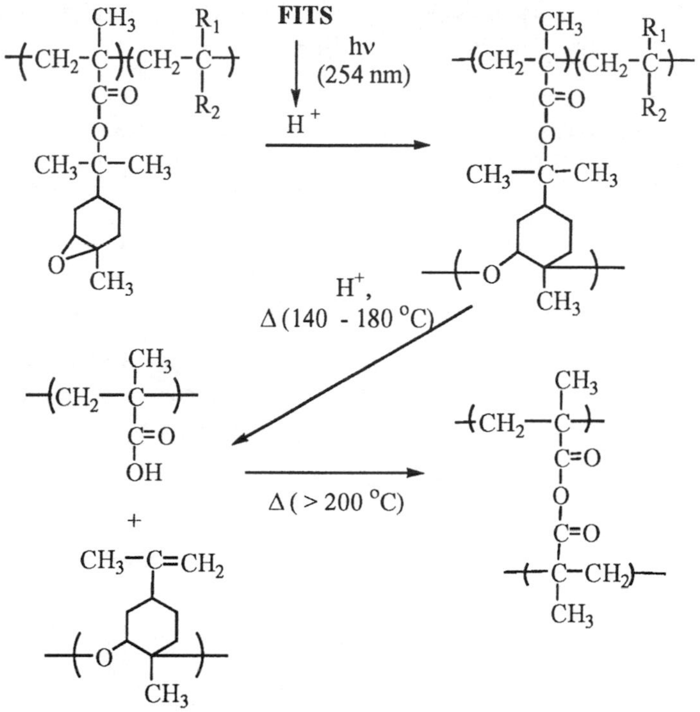

Figure 4. Photoinduced-acid catalyzed crosslinking and thermal degradation of PMOBH.

Thermal Degradation

MOBH Homopolymer

It is known that *tert*-butyl esters of carboxylic acids thermally decompose to form carboxylic acids and isobutene (*16*). The thermal decomposition temperature is lower if strong acids are present (*16*). Figure 5 shows FT-IR spectral changes of PMOBH film containing 3.6 mol% FITS on irradiation. After irradiation of the film with a dose of 180 mJ/cm^2 at room temperature, no significant changes in the FT-IR spectrum were observed though the film became insoluble in THF. When the irradiated film was baked at 90 °C for 10 min, the FT-IR spectrum showed no significant changes except the slight decrease of the peak at 846 cm^{-1} due to epoxy moieties. When the irradiated film was baked at 160 °C for 10 min, the peak at 1740 cm^{-1} due to ester carbonyl observed for the unirradiated film shifted to 1715 cm^{-1} and the peak at 1140 cm^{-1} due to C-O-C disappeared. Furthermore, a broad peak appeared at 2500 ~ 3500 cm^{-1} due to OH of carboxylic acid units. This finding suggests that the cleavage of the tertiary ester moiety in PMOBH occurred to generate poly(methacrylic acid) as shown in Figure 4. The peak at 846 cm^{-1} due to the epoxy moiety disappeared after the bake treatment. Two pathways are possible to explain the disappearance of the peak due to epoxy units: (1) ring-opening reaction completely occurred; (2) the side chain moiety degraded during baking treatment was vaporized. The second pathway was confirmed by TGA analysis of the film as will be discussed below. A new peak appeared at ca. 1800 cm^{-1} when the irradiated film was baked at 200 °C for 36 min. This is due to the formation of carboxylic acid anhydride moieties. It was reported that the dehydration reaction started almost simultaneously with the thermal decomposition of tertiary ester moieties in the side chain of the polymers (6, 16).

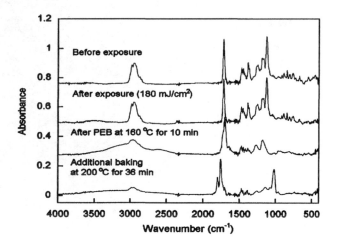

Figure 5. FT-IR spectra of PMOBH film containing 3.6 mol% FITS. Film thickness: 1.9 μm.

TGA analysis was carried out for PMOBH film and irradiated PMOBH film containing 3.6 mol% of FITS. PMOBH started to decompose at 216 °C to generate poly(methacrylic acid). The weight loss at 320 °C was 64% which was consistent with the weight of the ester moiety of MOBH units of PMOBH.

However, FT-IR spectrum of the PMOBH baked at 320 °C showed the formation of anhydride. If anhydride was formed, the weight loss should be 68%. This disagreement may be due to the non-volatile epoxy oligomers thermally formed from MOBH moiety. The PMOBH film containing p-toluenesulfonic acid generated photochemically started to decompose at ca. 150 °C. The weight loss for PMOBH films with and without acids was almost the same. The crosslinked PMOBH film could be thermally decomposed to poly(methacrylic acid) and/or poly(methacrylic acid) anhydride at lower temperatures than PMOBH in the absence of acid.

Decrease of thickness was observed for PMOBH film containing FITS when irradiated and baked at given temperatures. Although, on baking at 140 °C, no significant decrease of film thickness was observed, decrease of the thickness (~ 55%) was observed when baked at 160 and 180 °C for 14 min. This finding was consistent with the data from TGA analysis.

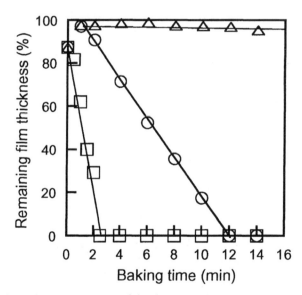

Figure 6. Dissolution properties of the thermally decomposed PMOBH films crosslinked by irradiation. Baking temperature: (△) 140, (○) 160 and (□) 180 °C.

Dissolution properties of the photochemically crosslinked PMOBH are shown in Figure 6. PMOBH film containing 3.6 mol% of FITS was irradiated and followed by baking at 140 – 180 °C. The film baked at 140 °C was not soluble in methanol. When baked at 160 or 180 °C, the crosslinked film became soluble in methanol. Longer bake treatment was necessary if baking temperature was low. It is likely that the degraded polymer does not really dissolve, but breaks off from a substrate due to diminished mechanical and adhesive properties. The irradiated PMOBH film baked at 180 °C was dissolved in methanol and filtered with a membrane filter (pore size: 0.5μm). The filtrate was evaporated to dryness and almost 100% of the polymer was recovered. This finding suggests that the degraded polymer did not break off from a substrate. Furthermore, it was confirmed that the separated polymer was poly(methacrylic acid) by reference to the FT-IR and ^{1}H-NMR spectra of the authentic poly(methacrylic acid). However, the dissolved polymer may have small amounts of anhydride groups which can not be detected by the spectroscopic analysis.

MOBH Copolymers

When P(MOBH(14)-TBMA) film containing 3.6 mol% FITS was irradiated and baked at 160 °C for 10min, the film gave the same FT-IR spectrum as observed for the thermally degraded PMOBH film. This suggests that the thermal degradation of MOBH and TBMA units occurred simultaneously to generate poly(methacrylic acid). Furthermore, after the thermal treatment of the irradiated P(MOBH-tBOSt) film containing 3.6 mol% FITS, formation of poly(methacrylic acid-co-vinylphenol) was confirmed by FT-IR measurements. The peak at 1140 cm^{-1} due to C-O-C disappeared and the peak at 1240 cm^{-1} ascribed to phenolic C-O stretching newly appeared. Poly(t-butoxystyrene) is known to thermally decompose to generate poly(vinylphenol) and isobutene at relatively low temperatures (~ 100 °C) in the presence of strong acids.[17] P(MOBH-tBOSt) started to decompose at ca. 230 °C and the second decomposition started at 270 °C due to the cleavage of the tertiary ether linkages in tBOSt units. However, in the presence of photoinduced-acid, the decomposition of t-butyl ether linkages and tertiary ester linkages in P(MOBH-tBOSt) occurred simultaneously at 160 °C. This was confirmed by the finding that a TGA curve for P(MOBH-tBOSt) showed one-stage decomposition.

The thermal degradation of copolymers of MOBH and CHSS, NPSS, or PhSS was studied by in-situ FT-IR spectroscopy. The spectral changes of the sample film were measured at a constant temperature. When the crosslinked P(MOBH-CHSS) film was baked at 120 °C, the peaks at 1720 (carboxylic acid ester) and 1178 cm^{-1} (sulfonic acid ester) decreased and the peak at 1018 cm^{-1} (sulfonic acid) increased. The carboxylic acid ester moiety due to MOBH unit and sulfonic acid ester unit due to CHSS decreased simultaneously. The increase of the peak at 1018 cm^{-1} corresponded to the decrease of the peak at 1178 cm^{-1}. When the crosslinked P(MOBH-NPSS) film was baked at 150 °C, the peak at 1720 cm^{-1} (carboxylic acid ester) decreased and then followed by the decrease of the peak at 1178 cm^{-1} (sulfonic acid ester). The increase of the peak at 1018 cm^{-1} (sulfonic acid) corresponded to the decrease of the peak at 1178 cm^{-1} as shown in Figure 7. Furthermore, no thermolysis of PhSS units in the crosslinked P(MOBH-PhSS) was observed when baked at 240 °C.

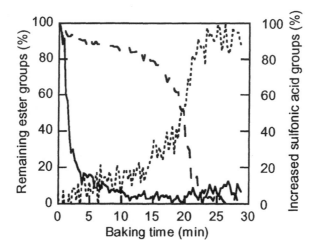

Figure 7. Thermolysis of the crosslinked P(MOBH-NPSS) film measured by in-situ FT-IR spectroscopy. Solid line: carboxylate group (1720 cm^{-1}). Broken line: sulfonate ester group (1178 cm^{-1}). Dotted line: sulfonic acid group (1018 cm^{-1}). (Reproduced with permission from reference 11. Copyright 2002 The Chemical Society of Japan.)

Figure 8 shows the dissolution properties of the crosslinked P(MOBH(32)-TBMA), P(MOBH-tBOSt) and PMOBH. The irradiated films (exposure dose: 60 mJ/cm^2) were baked at 90 °C for 10 min before the bake treatment at 160 °C. The crosslinked PMOBH and P(MOBH(32)-TBMA) became soluble in methanol after baking for 6 – 12 min. Although the crosslinked P(MOBH-tBOSt) became soluble after baking for 2 min, the dissolved fraction decreased when baked for longer period than 2 min. This may be due to the formation of partial network by the esterification reactions between carboxylic acid and phenol units which were generated by the thermolysis of P(MOBH-tBOSt).

Figure 9 shows effect of baking temperature on the dissolution properties of the irradiated (60 mJ/cm^2) P(MOBH-CHSS), P(MOBH-NPSS), and P(MOBH-PhSS) films containing 3.6mol% FITS. When baked below 100°C, these films were insoluble in water. The crosslinked P(MOBH-CHSS) and P(MOBH-NPSS) films became soluble in water after baking at 120-220 and 160-200°C, respectively. No dissolution in water was observed for the crosslinked P(MOBH-PhSS) film after baking at 80-240 °C. Thus, the generation of both carboxylic acid units and sulfonic acid units was necessary to dissolve the crosslinked films into water. Re-insolubilization of P(MOBH-CHSS) and P(MOBH-NPSS) which were soluble in water was observed after baking at above 220-240 °C. This is due to the partial formation of carboxylic acid anhydride moieties.

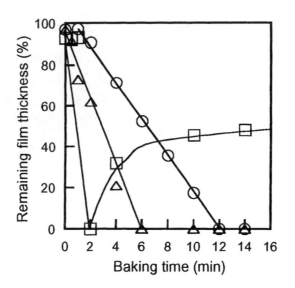

Figure 8. Dissolution properties of the crosslinked PMOBH (○), P(MOBH(32)-TBMA) (△) and P(MOBH-tBOSt) (□) films containing 3.6 mol% FITS.

Baking temperature to dissolve the crosslinked polymers was dependent on the photoacid generator used. In the case of the crosslinked PMOBH film containing 3.6 mol% of TPST, after baking at 90 – 110 °C for 0.5 – 10 min, it became soluble in methanol. TPST can generate triflic acid on irradiation. When FITS was used as a photoacid generator, the baking at 160 – 180 °C for 3 – 12 min was necessary to completely dissolve the crosslinked PMOBH. Thus, the baking temperature to thermally decompose the crosslinked polymers could be selected by changing photoacid generators.

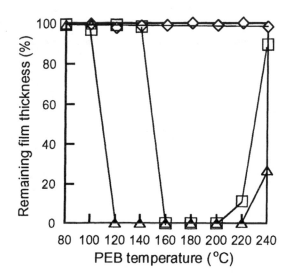

Figure 9. Dissolution properties of the crosslinked P(MOBH-CHSS) (△), P(MOBH-NPSS) (□) and P(MOBH-PhSS) (◇) films after baking for 10 min. (Reproduced with permission from reference 11. Copyright 2002 The Chemical Society of Japan.)

Conclusion

We have synthesized a novel methacrylate monomer (MOBH) which has an epoxy moiety and a tertiary ester linkage in a molecule. Homopolymer of MOBH and copolymers of MOBH with *tert*-butyl methacrylate, *tert*-butoxystyrene, cyclohexyl styrenesulfonate, neopentyl styrenesulfonste or phenyl styrenesulfonate were obtained by the conventional radical photopolymerization. Polymer films containing photoacid generators (PAG) became insoluble in tetrahydrofuran on UV irradiation because of the photoinduced-acid catalyzed crosslinking reaction of epoxy units.

250

Decrosslinking of the irradiated polymer films was performed by thermal treatment at 90 – 200 °C the temperature of which was dependent on the polymer structure. It was confirmed that decrosslinking occurred by the cleavage of tertiary ester, t-butyl ether or sulfonic acid ester linkages. The polymers can be applied to photocrosslinkable materials for temporary use.

Acknowledgment

This work was supported by a Grant-in-Aid for Scientific Research (B)(2) No. 13450382 from Japan Society for the Promotion of Science. We are also grateful to the Ministry of Education, Culture, Sports, Science and Technology for financial support through 21st Century COE Program 24403 E-1.

References

1. Fouassier, J. P.; Rabek, J. F. *Radiation Curing in Polymer Science and Technology*, Elsevier Applied Science, New York, 1993.
2. Tesero, G. C.; Sastri, V. *J. Appl. Polym. Sci.* **1990**, *39*, 1425.
3. Sastri, V. J.; Tesero, G. C. *J. Appl. Polym. Sci.* **1990**, *39*, 1439.
4. Buchwalter, S. L.; Kosber, L. L. *J. Polym. Sci. Part A: Polym. Chem.* **1996**, *34*, 249.
5. Yang, S.; Chen, J.- S.; Korner, H.; Breiner, T.; Ober, C. K. *Chem. Mater.* **1998**, *10*, 1475.
6. Ogino, K.; Chen, J.- S.; Ober, C. K. *Chem. Mater.* **1998**, *10*, 3833.
7. Wang, L.; Wong, C. P. *J. Polym. Sci. Part A: Polym. Chem.* **1999**, *37*, 2991.
8. Hitomi, M.; Sanda, F.; Endo, T. *J. Polym. Sci. Part A: Polym. Chem.* **1998**, *36*, 2823.
9. Hitomi, M.; Sanda, F.; Endo, T. *Macromol. Chem. Phys.* **1999**, *200*, 1268.
10. Shirai, M; Morishita, S.; Okamura, H.; Tsunooka, M. *Chem. Mater.* **2002**, *14*, 334.
11. Shirai, M; Kawaue, A.; Okamura, H.; Tsunooka, M. *Chem. Lett.* **2002**, 940.
12. Shirai, M.; Kinoshita, H.; Tsunooka, M. *Eur. Polym. J.* **1992**, *28*, 379.
13. Okamura, H.; Takatori, Y.; Tsunooka, M.; Shirai, M. *Polymer,* **2002**, *43*, 3155.
14. Shirai, M.; Nakanishi, J.; Tsunooka, M.; Endo, M. *J. Photopolym. Sci. Technol.* **1998**, *11*, 841.
15. Crivello, J. V.; Liu, S. *Chem. Mater.* **1998**, *10*, 3724.
16. Ito, H.; Ueda, M. *Macromolecules* **1988**, *21*, 1475.
17. Conlon, D. A.; Crivello, J. V.; Lee, J. L.; O'Brien, M. J. *Macromolecules* **1989**, *22*, 509.

Chapter 19

Evaluation of Infrared Spectroscopic Techniques to Assess Molecular Interactions

Robert K. Oldak and Raymond A. Pearson

Microelectronic Packaging Laboratory, Lehigh University,
Bethlehem, PA 18015

The task of promoting adhesion cannot be fulfilled without considering the acid-base interactions. Such interactions play a major role in promoting adhesion. However, these interactions are difficult to characterize as well as quantify. In this work, a model underfill resin system based on cycloaliphatic epoxy resin is studied by FTIR spectroscopy to characterize possible molecular interactions. In addition, interactions in molecules with structure corresponding to parts of epoxy resin, as well as simple molecules such as acetone and ethyl acetate have been studied. Two infrared spectroscopic techniques, carbonyl peak shifts and hydroxyl peak shift are used to determine Drago constants. Such constants can be used to quantify the acidic and basic character of molecular interactions. It was found that Drago constants determined by hydroxyl peak shifts agreed well with literature values. In contrast carbonyl peak shifts were interfere plagued by solvent effects and concentration dependence.

Introduction

Intermolecular acid-base interactions are essential for promoting adhesion (1). The quantification of intermolecular interactions is necessary for understanding predicting, and controlling adhesion. However, quantifying molecular interactions is difficult to achieve. Several physical methods such as contact angle measurements and flow microcalorimetry have been evaluated by our group in the past in an effort to quantify site-specific interactions. In this work, spectroscopic methods (FTIR) are applied to characterize the molecular interactions of a model underfill resin system based on a cycloaliphatic epoxy resin. In addition, a series of other molecules with structures corresponding to parts of the epoxy resin have been studied. As a control, ethyl acetate and acetone have been used. The model underfill system contains an epoxy resin, Araldite CY179, which is cycloaliphatic epoxy resin with ester linkage. The strong absorption by C=O from ester group has been utilized. In this study the shift in frequency of carbonyl peak upon introduction of acceptor is used to detect molecular interactions. To complement our studies shifts in frequency of OH peaks upon interactions with various donor molecules were used. Such IR shifts can be calibrated with heat data and used to determine Drago constants, which quantify the acidic and basic character of the epoxy resin as well as allow to predict strength of interactions for molecules with known constants.

The direct measurement of heats of acid-base complex formation ΔH^{AB} was pioneered by Fowkes (1-3), who used them to characterize polymer surfaces. Fowkes applied Drago's (4) acid-base theory to interpret his calorimetric results. In contrast to Fowkes, Drago performed calorimetric studies on simple molecules such as organic and inorganic acids interacting with organic bases. In his work, Drago used dilute solutions with neutral organic solvents to keep van der Waals interactions unchanged. Thus, the heat of interaction represents only acid-base interactions. The empirical relation was obtained from analysis of such data (4,5):

$$-\Delta H^{AB} = E_A E_B + C_A C_B \qquad (1)$$

The constant C represents the covalent contribution and constant E represents the electrostatic contribution, with the subscripts referring to either acid (acceptor) or base (donor). The advantage of the Drago approach it that it can be very quantitative (6-8).

Infrared (IR) peak shifts

IR spectroscopy has been shown to be useful for estimating the heats of complexation by many researchers. In order to obtain the heats of complexation IR spectra must be correlated with ΔH^{AB}. For example, the shift in the OH stretching frequency has been used to calculate the enthalpy of acid-base interactions, ΔH^{AB} by Drago (5, 7, 8). Fowkes et al. (2, 3, 9) were amongst the first researchers to study IR spectral shift in polymers. They determined acid-base shift in esters, and associated it with calorimetrically determined heats of formation. Then, followed this pathway to study the acid-base shift in polyesters. The carbonyl oxygens of esters are electron donor sites, which can form acid-base bonds with the electron-accepting sites of acidic molecules such as phenol. The stretching frequency of such bonded carbonyl oxygen is increased by an amount Δv^{AB} proportional to the enthalpy of bonding ΔH^{AB}.

Based on Fowkes experimental data (9) an empirical equation was derived for carbonyl peak shifts:

$$\Delta H^{AB} = 0.976 \Delta v^{AB} + 3.31 \; kJ*mol^{-1}*cm^{-1} \tag{2}$$

Where, Δv^{AB} is the amount of peak shift, and ΔH^{AB} is again the acid-base heat of complexation.

The similar relation can be found for hydroxyl peak shifts. The equations for used acids are based on work by Drago (5, 10-12). We can obtain relation between shift and heat of complexation by following formula:

HFIP

$$\Delta H^{AB} = 0.0481 \Delta v^{AB} + 15.06 \qquad kJ*mol^{-1}*cm^{-1} \tag{3}$$

Phenol

$$\Delta H^{AB} = 0.0439 \Delta v^{AB} + 12.55 \qquad kJ*mol^{-1}*cm^{-1} \tag{4}$$

Infrared shifts for acid complexes with molecules containing carbonyl can be used to determine their Drago E_B and C_B constants. If the test acid of known C_A and E_A gives a Δv^{AB} shift, which leads to determination of ΔH^{AB} from Equations 2-4.

Objective

The purpose of this study is to evaluate the use of IR techniques to quantify molecular interactions in a model underfill resin. Such interactions should provide an insight into strengthening the adhesion between underfill resin and flip chip passivation coatings.

Experimental Approach

Materials

The major focus was on a cycloaliphatic epoxy resin, Araldite CY 179, which is common component in some commercial underfill systems. Additional epoxies studied included: a cyclohexane oxide, Bisphenol F and Bisphenol A based epoxy resins. The structure of cycloaliphatic epoxy resin (shown in Figure 1a) indicates presence of ester group, which contains a carbonyl unit. Methylcyclohexyl cyclohexanecarboxylate was also used, since its structure corresponds to the part of the resin (Figure 1b). To complete the study, dicyclohexyl ketone contain only a carbonyl group between cyclohexane rings was also examined. Ethyl acetate and acetone were used as reference molecules. For OH shift study two oxygen Lewis acids were chosen with higher possible range of hardness (C/E ratio), phenol and HFIP (Figure 1i). The OH peak shift has advantage over the carbonyl one that any basic molecule can be studied. We were able to examine various resins showed in Figure 1 d-f.

All solvents and probe molecules were carefully dried, except the anhydrous ones. The resin and ketone were used as received; the ester was synthesized and purified by double fractional vacuum distillation.

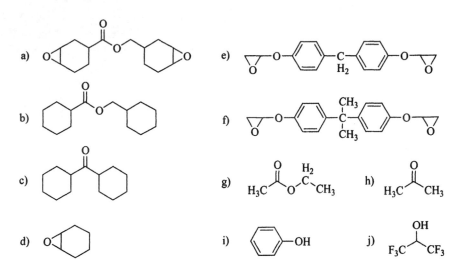

Figure 1. Chemical structures of used model molecules: a) Cycloaliphatic epoxy resin (Araldite CY 179), b) Methylcyclohexyl cyclohexanecarboxylate (Ester), c) Dicyclohexyl ketone d) cyclohexane oxide e) Bisphenol F epoxy resin (Epon 862), f) Bisphenol A epoxy resin (DER 331), g) Ethyl acetate, h) acetone, i) Phenol, j) 1,1,1,3,3,3 Hexafluoro-2-propanol (HFIP)

FTIR spectroscopy

The carbonyls undergo stretching vibrations under an IR beam (13), which is detectable as strong absorption at about 1720 cm^{-1}. The hydroxyls shows a sharp strong peak at about 3623 cm^{-1}, and it was chosen as reference in DDC solution. For solution in CCl_4 peak at 3616 cm^{-1} was the reference. FTIR experiments were carried out on a computer driven Perkin-Elmer model 1605 FTIR spectrometer with MIR source. A strong apodization, 4cm^{-1} resolution and usually 16-64 scans in ratio mode were used. The background was measured prior to the each series of measurement.

Preparation of dilute solutions

The samples were dissolved for times long enough to ensure complete dissolution of probe molecules. The solutions were made in optimized concentrations, low enough to avoid self- association and high enough to give readable signals. For carbonyl peak shift study cyclohexane was used, while in case of solutions for hydroxyl peak shift dodecane and carbon tetrachloride were used. As the neutral solvent in previous study (14) we used cyclohexane, as neutral solvent (15), which was conform by Chehimi et, al. (16). However, the transparency in hydroxyl region of interest, 3650-3200 cm^{-1} is not satisfactory. Cyclohexane shows some overtone peaks in OH region, thus we used dodecane instead. Dodecane has dispersion component of surface tension similar to that of cyclohexane, this indicates that it exhibits the same level of interactions due to van der Waals forces. The dispersion components for cyclohexane and dodecane are 25.24 and 25.35 mJ/m^2 (15) respectively. Such proximity makes the interactions with these solvents comparable, for example carbonyl peak of ethyl acetate in dodecane is 1748.4 cm^{-1} versus 1748.6 cm^{-1} in cyclohexane. The difference is well below measurement error.

Typically the concentration of model molecules was about 40 mM, and the ratio to the probe acids ranged from 1:2 to 1:5 in carbonyl shift study and 1:1 to 1:3 acid/molecule in hydroxyl peak shift study. The samples were measured using a liquid cell between two NaCl or KCl windows, and BaF_2 in OH shift, equipped with Teflon spacer. Spectra were analyzed using PE Spectrum 2.0 software. In effort to increase clarity of spectra some were treated with Fourier Transformation deconvolution. For two probe acids, $SbCl_5$ and Me_3Al quite different procedure was used. Because of precipitation of resin and ketone in contact with probe instead of liquid cell Perkin-Elmer HATR accessory was used to analyze the precipitate.

Determination of Drago's constants

To compare results from carbonyl shift and hydroxyl peak shift we used a method that utilized hardness of base to determine Drago constants. This method has been presented before (*14*). The method is based on assumption that the hardness, i.e. ratio C/E of bases with similar structure are equal. Thus, we assumed hardness of ester and cycloaliphatic epoxy model molecules to be equal hardness of ethyl acetate, $C^*_B/E^*_B = 0.61$, hardness of dicyclohexyl ketone to be equal to that of acetone, $C^*_B/E^*_B = 0.72$ and hardness of Bisphenol A and F epoxies to be equal to that of cyclohexane oxide, $C^*_B/E^*_B = 1.5$ (*17-20*). Thus, relation (5) can be used to find E_B:

$$E_B = -\Delta H^{AB}/C_A * (C^*_B/E^*_B) + E_A \tag{5}$$

Results and Discussion

Carbonyl peak shift

The analysis of spectra clearly indicates interactions between probe molecules and resin, based on red IR shift (Figure 2), and results are shown in Table I.

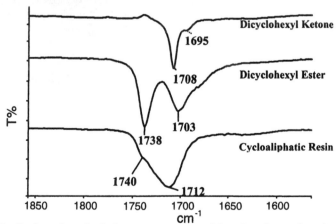

Figure 2. Carbonyl peak shift in spectra of model molecules with iodine monobromide

The IR shift depends on the strength of interactions (*9,16*). It can be said that our dicyclohexyl ester showed stronger C=O peak shifts than the other model molecules. The C=O peak shifts in the cycloaliphatic epoxy were more often greater then these in the ketone. Therefore, the basicity of the carbonyl group can be rated as:

$$C=O_{ester} > C=O_{epoxy} > C=O_{ketone}$$

Table I. Collective results for model molecules in dilute cyclohexane solutions with Drago constants for acids (*19*).

Acid	Ea (kJ/mole)$^{1/2}$	Ca (kJ/mole)$^{1/2}$	C=O IR shifts, Δv (cm^{-1})		
			ketone	ester	epoxy
I$_2$	4.09	1.02	-7	-22	10
CH$_2$Cl$_2$	1.76	0.23	-12	-17	-14
CHCl$_3$	3.19	0.90	-12	-18	-15
t-BuOH	2.19	1.41	-8	-15	-20
ICl	5.97	3.40	-3	-26	-22
PhOH	4.64	2.19	-37	-29	-26
CF$_3$CH$_2$OH	4.23	2.17	-19	-34	-27
IBr	2.45	6.73	-15	-35	-29
SbCl$_5$	7.45	21.31	-41	-153	-77
Me$_3$Al	17.71	7.53	-54	-84	-86

Hydroxyl peak shift

Our study showed significant OH peak shifts with all model molecules. Among the probe acids, HFIP exhibited significantly larger shifts then the phenol. Small molecules tend to show more shift then larger ones The largest shift was caused by interactions with acetone and cyclohexane oxide. The resulting shifts are showed in Table II.

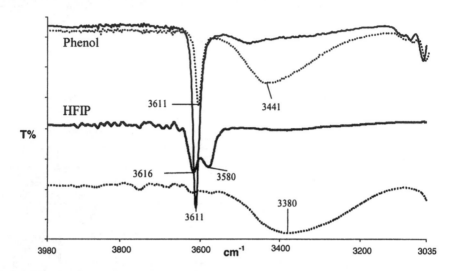

Figure 3. Hydroxyl peak shift due to interaction with cycloaliphatic epoxy. Solid lines -probe acids in solvent, doted lines - after resin was introduced.

It can be said that our dicyclohexyl ester and cycloaliphatic epoxy resin showed similar peak shifts to ethyl acetate. Bisphenol based epoxy resins showed smaller shifts then cyclohexyl epoxy resin. That may indicate contribution of ester group to total interaction. In general, smaller molecules tend to show more shift probably due to more conformational freedom. The arrangement allowing interaction of strongest sites is easier to achieve. Also, there is possible that solvent molecules prevent bigger molecules from stronger interactions. Therefore, the basicity of the small molecules interacting with the OH group can be rated as:

oxirane> ketone> ester

Acid-base constants from C=O peak shifts

Drago constants determined from C=O peak shift are shown in Table III. The results for acetone are with literature data see Table IV.

Table II. Collective hydroxyl shift results for model molecules in dilute solutions, shifts are in cm^{-1}

Molecule	Phenol		HFIP	
	Dodecane	CCl$_4$	Dodecane	CCl$_4$
acetone	-223	-226	-267	-286
dicyclohexyl ketone	-201	-187	-255	-236
ethyl acetate	-195	-187	-244	-234
dicyclohexyl ester	-193	-185	-245	-237
cycloaliphatic epoxy	-192	-169	-245	-239
cyclohexane oxide	-232	-194	-267	-300
bis A epoxy	-	-161	-	-230
bis F epoxy	-	-161	-	-233

Table III. Drago constants for model molecules determined in this study from carbonyl peak shift in cyclohexane

Molecule	E_B [kJ/mol]$^{1/2}$	C_B [kJ/mol]$^{1/2}$
acetone	3.56	2.58
ethyl acetate	5.67	3.46
dicyclohexyl ketone	3.54	2.55
dicyclohexyl ester	5.89	3.59
cycloaliphatic epoxy	4.29	2.62

Table IV. Drago constants for model molecules obtained from literature
(12)

Molecule	E_B $[kJ/mol]^{1/2}$	C_B $[kJ/mol]^{1/2}$
acetone	3.56	2.58
ethyl acetate	3.31	2.00
cyclohexane oxide	2.97	4.38

Discrepancies arise in case of ethyl acetate. The results for ethyl acetate obtained from carbonyl peak shift are significantly higher from those from literature. Interestingly, there is no strict agreement on the Drago constants values for ethyl acetate. Drago (19) listed E_B and C_B values as 3.31 and 2.00 respectively, while Fowkes reported 3.80 and 2.23 (from reference 9 after transformation of constants to new scale using procedure from reference 17). It is worth mentioning that our constants relay on Fowkes calorimetric data, thus are higher as well. Also, when one explores results from carbonyl shifts for ethyl acetate reported by various groups, it can be find that results vary (9, 20).

The discrepancies may be related to the difference in solvent used or concentration dependence for some molecular probes. For weak Lewis acids, carbonyl peak showed to be concentration dependent, see Figure 4. The molar ratio of 250 corresponds to solvent free solution, i.e. only acetone and $CHCl_2$ or pyridine. The dependence is due to acid-base interactions with solvent or self-associated donor, or donor-solvent complexation. The shifts presented in Figure 4 are significantly larger then those induced by change in van der Waals interactions. Such large shifts are related to acid-base interactions. Since pyridine and acetone are considered to be basic, one of them must actually act as an acid. This shows that interactions even in solutions of simple molecules can be very complex. The constants determined for dicyclohexyl ketone are similar to that of acetone, also constants determined for dicyclohexyl ester are close to those of ethyl acetate. The constants obtained for cycloaliphatic epoxy lies between. Thus, the results obtained for our ketone appear to be consistent. There some uncertainty of values for our the epoxy, since only C=O peak shift were monitored.

Acid-base constants from OH peak shifts

Drago constants for our model molecules are shown in Table V. In general, there is only slight difference between results obtain for HFIP and phenol and different solvents. In contrary to carbonyl peak shift data, OH peak shifts were large and did not display a concentration or solvent dependence. The comparison with literature data in Table IV with average constants presented in Table VI shows that values obtained for ketones and esters match literature data very well.

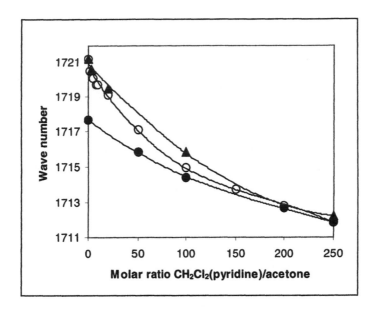

Figure 4. *Acetone concentration dependence, cyclohexane solutions:* ▲-*pyridine,* ○-*methylene chloride;* ●-*methylene chloride in CCl₄*

The constants determined for cycloaliphatic epoxy are close to those of esters. The data for epoxies (excluding cycloaliphatic one) are self-consistant, On the other hand, the Drago constants for cyclohexane oxide are 14% lower from those published in literature, however we have no knowledge about the method used to determine them. Upon these circumstances, we can consider our data to be very good. The hydroxyl peak shift seems to be valuable tool to assess molecular interactions and to quantify them.

Conclusions and Future Work

IR hydroxyl peak shifts have proven to be a convenient method for evaluating site-specific interactions. IR peak shifts showed to be useful tool to detect acid-base interactions between probe acids and model molecules containing carbonyl and other basic groups. The calculated Drago constants based on OH peak shifts showed to be more consistent with literature data then constants derived from carbonyl peak shifts. Such constants allow the prediction of the strength of acid-base interaction.

In the future, the acid-base interactions at epoxy-polyimide interfaces will be studied.

Table V. Drago constants for model molecules obtained from hydroxyl peak shift

Molecule	Phenol				HFIP			
	Dodecane		CCl_4		Dodecane		CCl_4	
	E_B	C_B	E_B	C_B	E_B	C_B	E_B	C_B
acetone	3.56	2.58	3.60	2.58	3.25	2.33	3.40	2.45
dicyclohexyl ketone	3.38	2.43	3.27	2.35	3.15	2.27	3.01	2.17
ethyl acetate	3.50	2.15	3.44	2.09	3.21	1.96	3.13	1.90
dicyclohexyl ester	3.48	2.13	3.42	2.09	3.23	1.96	3.15	1.92
cycloaliphatic epoxy	3.48	2.13	3.27	2.00	3.23	1.96	3.17	1.94
cyclohexane oxide	2.70	4.05	2.48	3.70	2.43	3.64	2.62	3.93
bis A epoxy	I/S	I/S	2.27	3.40	I/S	I/S	2.21	3.31
bis F epoxy	I/S	I/S	2.27	3.40	I/S	I/S	2.23	3.33

Constants are in $kJ*mol^{-1}$, I/S- insoluble

Table VI. Drago constants determined in this study

Molecule	Average		STD	
	E_B	C_B	E_B	C_B
acetone	3.45	2.49	0.16	0.12
dicyclohexyl ketone	3.20	2.31	0.16	0.11
ethyl acetate	3.32	2.03	0.18	0.12
dicyclohexyl ester	3.32	2.03	0.16	0.10
cycloaliphatic epoxy	3.29	2.01	0.13	0.09
cyclohexane oxide	2.56	3.83	0.12	0.19
bis A epoxy	2.24	3.36	0.04	0.06
bis F epoxy	2.25	3.37	0.03	0.05

Acknowledgments

This work was funded by the Semiconductor Research Corporation (SRC contract No. 658.001).

References:

1. Fowkes, F. M.; Mosafa, M. A. *Ind. Eng. Chem. Prod. Res. Dev.* **1978,** *17,* 3.
2. Fowkes, F. M. *J. Adhesion Sci. Technol.* **1987,** *1,* 7.
3. Fowkes, F. M. *J. Adhesion Sci. Technol.* **1990,** *4,* 669.
4. Drago, R. S.; Wayland, B. B. *J. of Am. Chem. Soc.* **1965,** *87,* 3571.
5. Drago, R. S.; Vogel, G. C.; Needham, T. E. *J. of Am. Chem. Soc.* **1971,** *93,* 6014.
6. Jensen, W. B. *The Lewis Acid-Base concepts*; John Willey & Sons: New York, NY, 1980 p 247.
7. Drago, R. S.; David, E. T. *J. of Am. Chem. Soc.* **1969,** *91,* 2883.
8. Drago, R. S. *Coord. Chem. Rev.* **1980,** *33,* 251.
9. Fowkes, F. M.; Tischler, D. O.; Wolfe, J. A.; Lanningan, L. A.; Ademu-John, C. M.; Halliwell, M. J. *J. of Polym. Sci., Polym. Chem.* **1984,** *22,* 547.
10. Drago, R. S.; O'Bryan, N; Vogel, G. C. *J. of Am. Chem. Soc.* **1970,** *92,* 3924
11. Nozari, M. S.; Drago, R. S. *J. of Am. Chem. Soc.* **1970,** *92,* 7086.
12. R. S. Drago *Structure and Bonding* **1973,** *15,* 73.
13. Cooper, J. W. *Spectroscopic techniques for organic chemists*; John Willey & Sons: New York, NY, 1980 p 43.
14. Pearson, R. A. Oldak, R. K *Proc. of the 24th Ann. Meet. of Adhes. Soc.,* Williamsburg, VA, 2001, p 125.
15. van Oss, C. J. *Interfacial Forces in Aqueous Media*; Marcel Dekker, Inc.: New York, NY, 1994, p 171.
16. Chehimi, M. M.; Pigois-Landureau, E.; Delmar, M.; Watts, J. F.; Jenkins, S. S.; Gibson, E. M. *Bull. Soc. Chi. Fr.* **1992,** *129,* 137.
17. Drago, R. S.; Ferris D. C.; Wong, N. *J. of Am. Chem. Soc.* **1990** *112,* 8953.
18. Drago, R. S.; Vogel, G. C. *J. of Chem. Ed.* **1996,** *73,* 701.
19. Drago, R. S.; Dadmun A. P.; Vogel, G. C. *Inorg. Chem.* **1993,** *32,* 2473.
20. Drago, R. S.; Vogel, G. C. *J. of Am. Chem. Soc.* **1992,** *114,* 9527.

Chapter 20

Study on Metal Chelates as Catalysts of Epoxy and Anhydride Cure Reactions for No-Flow Underfill Applications

Zhuqing Zhang and C. P. Wong

School of Materials Science and Engineering, Georgia Institute of Technology, Atlanta, GA 30332

For no-flow underfill applications, high curing latency for the epoxy resins is required. Metal chelates are effective latent catalysts for epoxy and anhydride cure reactions. A screen test on the catalytic behavior of metal acetylacetonates shows that this system offers a wide range of cure latency and material properties. An isothermal kinetic study on the catalytic behavior of metal chelates with first row transition metal ions is conducted and analyzed using an autocatalytic model. It is found that the activation energies of the systems containing divalent metal chelates follow the Irving and Williams rule. The activation energies of reaction obtained in the kinetics study are compared with the dissociation energies of the metal/ligand bond and the results are discussed. The effect of ligand chemistry is also investigated and similar behavior is observed.

INTRODUCTION

Flip-chip has advantages over other interconnection methods including high I/O counts, better electrical performance, high throughput, low profile, etc. [1], and has been practiced in industry for four decades. Recently, the desire for low cost, mass production has resulted in the increasing use of organic substrates such as FR-4 printed wiring boards (PWB) instead of ceramic substrates. In order to alleviate the thermal stress on the solder joint caused by the difference between the coefficients of thermal expansion (CTE) of the silicon chip and the organic substrate, underfill was invented and its application in flip-chip has greatly enhanced the reliability of the package by redistributing stress among the chip, substrate, underfill and all the solder joints [2,3]. However, the current underfill process encounters various problems. Conventional underfill is drawn into the gap between the chip and the substrate by capillary flow, which is usually slow and can be incomplete, resulting in voids. It also produces non-homogeneity in the resin/filler system. In addition, cure of the underfill takes hours in the oven [4]. These problems are further aggravated by increasing chip dimensions and I/O counts, and decreasing gap distances and pitch sizes.

In order to address the problems associated with conventional underfill, the no-flow underfill process was invented and the first successful no-flow underfill material was developed by Wong and Shi [5]. In the no-flow underfill process, the underfill is dispensed onto the substrate prior to chip placement. The underfill usually has the fluxing capability to facilitate the solder to wet the contact pads of the substrate during solder reflow. This technology simplifies the underfill process by eliminating the flux application and cleaning, underfill dispensing and capillary flow, and combining the solder reflow and underfill cure into one step [6]. In order to develop a successful no-flow underfill material, high cure latency of the underfill is required to maintain its low viscosity until the solder joints are formed. Otherwise, gelled underfill would prevent the melting solder from collapsing onto the contact pads, resulting in low yield of solder joints.

Organometallic compounds have been explored as the latent catalysts for various epoxy resin systems [7,8,9]. Metal acetylacetonates (AcAc's), in particular, are found to be effective latent accelerators for epoxy and anhydride cure reactions [10,11,12]. Based on the epoxy/ anhydride/ metal AcAc system, underfill materials have been developed for flip-chip applications [5,6,13]. Metal AcAc's are unique as catalysts for epoxy cure reactions in that they not only provide high cure latency, but also offer a wide range of cure temperatures.

Despite the fact that metal AcAc's have been used extensively as catalysts for epoxy resins, there have been few systematic studies on their catalytic behavior. P.V. Reddy et al. focused on the cure kinetics of the epoxy/ anhydride/ Co (III) AcAc system [11]. Their results showed that it followed first order kinetics. J.D.B Smith studied the cure behavior of the epoxy/anhydride/metal AcAc's system for a number of different metal AcAc's [10]. He suggested that the decomposition products of the metal AcAc's might be the active species responsible for initiating polymerization in epoxy/ anhydride resin systems. A systematic study on metal chelate cured epoxy resins was conducted by A.V. Kurnoskin [14,15,16,17,18]. The study included the reaction mechanisms, the structures, and various properties of epoxy/ metal chelate matrices. In this work, a variety of metal AcAc's are investigated as the catalysts of epoxy/ anhydride systems. The cure kinetic study on this system can provide more insight into the nature of the reactions.

EXPERIMENTAL

Materials

Two types of epoxy resins were used in this study. They were bisphenol A type epoxy resins (EPON 828 from Shell Chemicals), and cycloaliphatic epoxy resins (ERL 4221 from Union Carbide). The epoxy equivalent weight (EEW) of ERL 4221 was 134 g/mol, and the EEW of EPON 828 was 188 g/mol. The curing agent used in this study was hexahydro-4-methylphthalic anhydride (HMPA), purchased from Lindau Chemicals. The molecular weight of HMPA was 168.2 g/mol. The chemical structures of the epoxy resins and HMPA are shown in Figure 1.

The metal AcAc's studied are included in Table 1. Most of the metal AcAc's were purchased from Research Organic/Inorganic Chemical Corp. However, Co (II) AcAc, Co (III) AcAc, Fe (II) AcAc, Fe (III) AcAc, and Ni (II) AcAc, and Co (II) haxafluoroacetylacetonate (HFAcAc) hydrate were purchased from Aldrich Chemicals, Inc.; Cu (II) AcAc and Cr (III) AcAc were purchased from Avocado Research Chemicals Ltd.; Mn (II) AcAc and Mn (III) AcAc were purchased from TCI. The melting points of the metal AcAc's were obtained either from literature or from DSC experiments and are listed in Table 1.

For formulations based on cycloaliphatic epoxy resin, the mixing weight ratio of ERL 4221: HMPA: metal AcAc was 100:100:0.80. The metal AcAc's were dissolved in the epoxy resins at elevated temperature before the addition of HMPA. The dissolving temperature depended on the solubility of the metal AcAc's in ERL 4221. The dissolving behavior of the metal AcAc's was recorded and included in Table 1. For formulations based on bisphenol A resin, the weight ratio of EPON 828: HMPA was 100:71.2. A molar concentration of 0.005 mole metal ion/ 1 mole epoxide was used in this case.

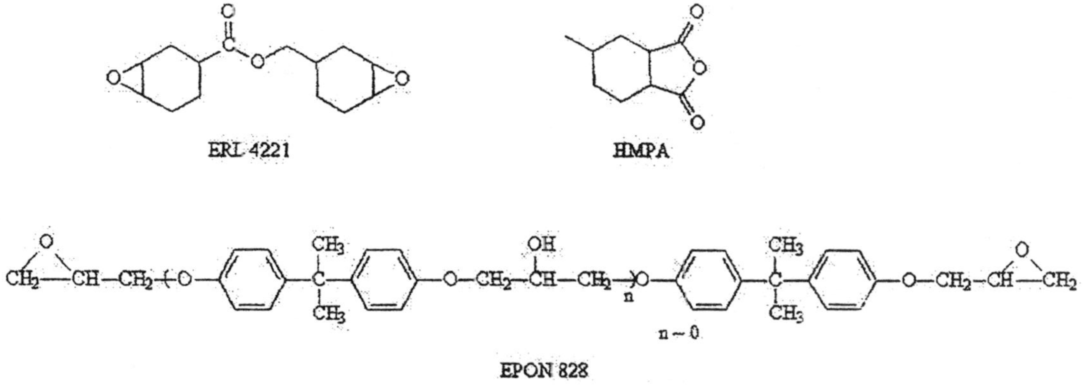

Figure 1. Chemical Structures of Epoxy Resins and Hardener in This Study

Instruments

The thermal stability of each metal AcAc was investigated using a Thermal Gravimetric Analyzer (TGA), Model 2050 by TA Instruments. A sample of ~20 mg was placed in a platinum pan and heated to 400 °C at a heating rate of 5 °C/min under N2 purge (77 ml/min in the vertical direction and 12 ml/min in the horizontal direction). A curve of weight loss versus temperature was obtained.

The dynamic cure profile of a formulation was obtained using a modulated Differential Scanning Calorimeter (DSC), Model 2920 by TA Instruments. A sample of ~10 mg was placed into a hermetic DSC sample pan and heated in the DSC cell from room temperature to 350 °C at a heating rate of 5 °C/min under N2 purge (40 ml/min). The heat of reaction was recorded as a function of temperature. The onset temperature and the peak temperature of the cure reaction were recorded. For kinetic study, isothermal DSC experiments were conducted. The sample was placed into a hermetic DSC sample pan and kept at a certain temperature for a sufficient amount of time. For each formulation, four isothermal temperatures were chosen within ± 20 °C of the onset cure temperature. The heat of reaction was recorded as a function of time.

RESULTS AND DISCUSSION

Screen Tests

The screening of the catalytic behavior of metal AcAc's was performed on the ERL 4221/HMPA/metal AcAc system. Altogether, 43 formulations were investigated in terms of the visually determined solubility of metal AcAc in the epoxy resin, the DSC cure profile at a heating rate of 5 °C/min, and the qualitative properties of cured resins after curing at 170 °C for 30 min and then at 230 °C for another 30 min. These results are included in Table 1. As can be observed in the table, metal AcAc's offer a wide range of cure temperatures and material properties for the ERL 4221/HMPA system.

For selection of the no-flow underfill formulation for flip-chip application, four basic criteria are considered: (1) The metal ion does not have any potential to form mobile ions. (2) The metal acetylacetonate can be totally dissolved in the epoxy resin at a temperature lower than 150 °C. The existence of metal acetylacetonate particles may cause the variation of the cured material properties. (3) The curing peak temperature must be high enough to ensure the required curing latency for the solder reflow (higher than 170 °C for eutectic SnPb solder bumped flip-chip application and higher than 200 °C for lead-free solder bumped flip-chip application). (4) The material should be cured at 230 °C without any cracks after cooling. High temperature post-cure introduces high residual stress and causes secondary reflow of the solder interconnects. According to these criteria, no-flow underfill formulations were developed based on metal chelates catalyzed epoxy system for both eutectic SnPb solder and lead-free solder bumped flip-chp applications [6, 13]. In this study, we are mainly interested in understanding what causes the difference in the catalytic behavior among these metal AcAc's.

As mentioned in Smith's work [10], low solubility of some metal AcAc's may have been a contributing factor to increasing the catalytic latency. However, some lanthanide ions, e.g., Nd, Sm, Gd, etc., did not dissolve into the epoxy resins but showed the catalytic effect at a low temperature, indicated by their onset cure temperatures. Also according to Smith, the mechanism of reaction may be related to the tendency of some metal AcAc's to decompose or dissociate at elevated temperature. Hence TGA studies on the decomposition of some metal AcAc's were conducted. TGA results on these metal AcAc's did not show a strong correlation between the decomposition temperature of the metal AcAc's and the cure temperature of epoxy/anhydride/metal AcAc system as shown in Table 2. For instance, Co (II) AcAc and Fe (II) AcAc displayed similar decomposition temperatures, but their catalytic reactivities were quite different.

However, considering that the dissociation of metal AcAc's as solids in the nitrogen atmosphere might differ from that in a solution of epoxy resins, it is possible that the metal-ligand bonding strength or energy is responsible for their reactivity in catalyzing the cure reaction.

According to the Crystal Field Theory (CFT), metal complexes are stabilized through the Crystal Field Stabilization Energy (CFSE) generated by the deformation of the d electron cloud when the metal ion is surrounded by ligands [19]. If the catalytic behavior of metal AcAc's is related to the dissociation of the bond between the metal ion and the ligand, the cure reactivity of these formulations should correlate with the CFSE of the metal complexes. Most of the metal AcAc's investigated here can be classified into three categories in terms of their catalytic behavior.

The first group consists of chelates with alkali and alkaline earth metal ions. They typically give a low onset cure temperature and a high cure peak temperature with relatively poor properties of the cured resins. As these ions either do not have d orbitals or only filled d orbitals, crystal field stabilization cannot occur. There is only electrostatic bonding between the ion and the ligand and the bonding strength is poor. So the metal chelates easily dissociate in the epoxy resins and start the reaction at low temperatures. However, since the metal ions cannot form strong bonds with epoxy or HMPA due to the nature of their electron configuration, they do not show an effective catalytic behavior, resulting in high cure peak temperatures and poor material properties. Exceptions in this group include Li (I) AcAc and Be (II) AcAc, which showed sufficient catalytic effects for the ERL 4221/ HMPA system. These two ions have low atomic number and small radius. Hence the bonding of the ions to the ligand possesses high portion of valence bond rather than pure electrostatic bond. So the bonding

Table 1. Basic Cure Properties of ERL 4221/HMPA/Metal AcAc System

Metal Chelate	Melting Point (°C)	Solubility in ERL 4221	Onset Temperature (°C)	Peak Temperature (°C)	Post Curing Material Property	
					30 min. @ 170°C	30 min. @ 170°C + 30 min 230°C
Li(I) AcAc	230°C	130°C	168°C	219°C, 322°C	cured, rigid	-
Na(I) AcAc	210°C	130°C	132°C	176°C	air bubble, brittle	cured, rigid
Be(II) AcAc	108°C	60-70°C	162°C	198°C, 239°C	tacky	cured, rigid
Mg(II) AcAc	259°C	not dissolved	-	315°C	tacky	brittle, crack
Ca(II) AcAc	175°C	130°C	112°C	257°C	tacky	brittle, crack
Sr(II) AcAc	220°C	130°C	101°C	227°C	tacky	brittle, crack
Ba(II) AcAc	320°C	130°C	165°C	282°C	tacky	brittle, crack
Ga(III) AcAc	194°C	130°C	100°C	181°C	cured, rigid	cured, rigid
In(III) AcAc	186°C	not dissolved	76°C	132°C	soft material	brittle, crack
Y(III) AcAc	160-165°C	not dissolved	75°C	135°C	cured, rigid	cured, rigid
Pb(II) AcAc	141-144°C	not dissolved	83°C	179°C	cured, rigid	cured, rigid
La(III) AcAc	130°C	110°C	168°C	209°C	cured, rigid	cured, rigid
Pr(III) AcAc	139°C	not dissolved	98°C	209°C	cured, brittle	cured, brittle
Nd(III) AcAc	140°C	not dissolved	88°C	204°C	cured, brittle	cured, brittle
Sm(III) AcAc	138°C	not dissolved	94°C	203°C	cured, brittle	cured, brittle
Gd(III) AcAc	100°C	not dissolved	88°C	195°C	cured, brittle	cured, brittle
Tb(III) AcAc	101°C	not dissolved	92°C	205°C	cured, brittle	cured, brittle
Dy(III) AcAc	103°C	not dissolved	152°C	199°C	cured, rigid	cured, rigid
Ho(III) AcAc	104°C	110°C	148°C	189°C	cured, rigid	cured, rigid
Er(III) AcAc	105°C	not dissolved	97°C	187°C	cured, brittle	cured, brittle
Tm(III) AcAc	97°C	not dissolved	165°C	202°C	tacky	cured, rigid
Lu(III) AcAc	-	not dissolved	169°C	195°C	cured, brittle	cured, rigid
U(II) AcAc	225°C	cured	-	-	-	-
Th(II) AcAc	171°C	130°C	120°C	169°C	tacky	cured, rigid

Cd(II) AcAc	235°C	not dissolved	95°C	200°C	cured, brittle	cured, rigid
Sc(III) AcAc	187-190°C	90°C	121°C	160°C	cured, brittle	cured, rigid
Y(III) AcAc	102°C	90°C	97°C	138°C	cured, brittle	cured, rigid
Cu(II) AcAc	284°C	90°C	199°C	232°C	cured, rigid	cured, rigid
Ti(IV) AcAc	subl. 200°C	not dissolved	140°C	266°C	tacky	cured, rigid
Zr(IV) AcAc	-	Cured	-	-	-	-
Hf(IV) AcAc	195°C	130°C	200°C	288°C	tacky	cured, rigid
Co (II) AcAc	166°C	60°C	156°C	207°C	tacky	cured, rigid
Co (III) AcAc	210-213°C	60°C	180°C	217°C	tacky	cured, rigid
Fe (II) AcAc	175°C	60°C	124°C	176°C	cured, rigid	cured, rigid
Fe (III) AcAc	182-185°C	60°C	50°C	149°C	cured, rigid	cured, rigid
V(III) AcAc	184°C	90°C	148°C	195°C	cured, rigid	cured, rigid
Mo(III) AcAc	179°C	not dissolved	167°C	212°C	cured, rigid	cured, rigid
Mn(II) AcAc	261°C	130°C	166°C	205°C	cured, rigid	cured, rigid
Ru(III) AcAc	260°C	90°C	199°C	263°C	tacky	crack
Rh(III) AcAc	265°C	90°C	174°C	235°C	cured, rigid	cured, rigid
Pd(II) AcAc	132-205°C	not dissolved	94°C	205°C	cured, brittle	cured, brittle
Pt(II) AcAc	230°C	130°C	220°C	248°C	tacky	crack
Ni(II) AcAc	240°C	90°C	185°C	249°C	tacky	cured, rigid

Table 2. Onset temperatures of decomposition of some metal AcAc's and their catalytic behavior in ERL 4221/HMPA system

Metal Ion	Onset Decomp. Temp.	Onset Cure Temp.	Peak Cure Temp.
Co (II)	170 °C	156 °C	207 °C
Fe (II)	172 °C	124 °C	176 °C
Mn (II)	215 °C	166 °C	205 °C
Ni (II)	212 °C	185 °C	249 °C
Cu (II)	200 °C	199 °C	232 °C

strength of Li (I) and Be (II) to the ligand is substantially higher than that of the other ions in this group, which makes their catalytic behavior different from other alkali and alkaline earth metal ions.

The second group consisted of chelates with lanthanide and actinide ions. Although these ions possess partially filled f orbitals, the f electrons are not effective in forming coordination bond. So they showed similar behavior to the first group, namely, low cure onset temperature, high cure peak temperature, and poor material properties of cured resins.

The third group contained chelates with transition metal ions with partially filled d orbitals. They displayed a variety of cure reactivities and usually better material properties after curing as indicated in Table 1. In order to take a close look at these metal AcAc's, studies were conducted on the EPON 828/ HMPA/ metal AcAc system. Bis-A type epoxy was chosen for the kinetic study because it possesses a symmetric structure, unlike ERL 4221 whose two epoxide groups might have different reactivity which introduced complexity in cure kinetics. The metal chelates containing the first row of transition metal ions were investigated, including Cr, Mn, Fe, Co, Ni, Cu, and Zn. In order to keep a consistent metal ion to epoxide group ratio, a molar concentration of 0.005 mole / mole epoxide was adopted for all the formulations.

Cure Kinetics Study

In order to further investigate the catalytic behavior of some transition metal AcAc's, kinetic studies were conducted on EPON 828/HMPA/metal AcAc system. First, a DSC dynamic cure profile at a heating rate of 5 °C/min was obtained for each EPON 828/ HMPA/ metal AcAc formulation. Following the dynamic cure studies, isothermal cure behavior of EPON 828/ HMPA/ metal

AcAc system was investigated at different temperatures. At each temperature, an autocatalytic kinetic model was utilized to obtain the reaction order and the rate constant. The model can be expressed as [20]:

$$\frac{dC}{dt} = kC^m(1-C)^n$$

where C is the fractional conversion, n is the reaction order, m is the independent order, k is the rate constant. According to the kinetic theory, reaction constant k is related to temperature and activation energy of the reaction by Arrhenius equation:

$$k = k_0 \exp(-\frac{E_a}{RT})$$

where E_a is the activation energy and k_0 is the frequency factor. From the slope of the plot of lnk versus $-1/T$, E_a for the cure reaction can be calculated.

Isothermal kinetic parameters of seven formulations were obtained in the same way and listed in Table 3. The onset and peak temperatures of the dynamic cure profiles of the EPON 828/ HMPA/ metal AcAc system, together with the activation energy obtained from the isothermal studies are illustrated in Figure 2.

Table 3. Isothermal Kinetic Parameters of EPON 828 /HMPA/ Metal AcAc System

Catalyst	Ea (kJ/mol)	Log k	n	m
Mn (II) AcAc	73.2	6.59	0.339	0.230
Fe (II) AcAc	74.8	7.95	0.821	0.260
Fe (III) AcAc	103	11.0	0.823	0.195
Co (II) AcAc	98	10.0	1.42	0.925
Co (III) AcAc	122	12.7	1.46	0.913
Ni (II) AcAc	101	8.88	0.657	0.045
Cu (II) AcAc	133	12.8	0.89	0.47

Since these systems possess different reaction orders, it is difficult to find a general trend of catalytic activity from the onset and peak cure temperatures of EPON 828/ HMPA/ Metal AcAc. However, the activation energy of these reactions provides a better understanding of the catalytic behavior of metal AcAc's. Theoretical studies have shown that the charge and radius of the ion,

274

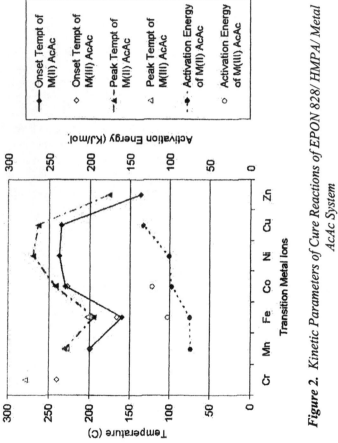

Figure 2. *Kinetic Parameters of Cure Reactions of EPON 828/ HMPA/ Metal AcAc System*

and the stability due to orbital splitting are the important factors determining the stability of the metal complexes [21]. As can be observed in Figure 2, trivalent metal AcAc's have higher activation energies in catalyzing the epoxy cure reaction than the divalent metal AcAc's since the covalency of the bond and the stability of the metal complexes formed with the same ligand generally increases with increasing oxidation number of the metal ion. For divalent metal AcAc's, the activation energy follows the rule discovered by Irving and Williams [22] that the order of complex-forming ability in the series of similar, divalent metal ions is as follow:

Mn < Fe < Co < Ni < Cu > Zn

The order can be explained by the decrease in metal-ion radius from Mn to Zn, and by the increase of crystal-field stabilization energy from Fe to Cu. The d orbitals of Zn (II) are full, so no stabilization energy is gained through the complex. That is the main reason that Zn (II) AcAc behaved similarly to that of the alkaline earth metal ions and the lanthanide and actinide ions, initiating curing at low temperature but giving poor mechanical properties after cure.

The activation energy reflects the energy barrier the reactants need to overcome to start the reaction. It is closely related to the dissociation energy of metal AcAc in a solution of epoxy resin and HMPA. The metal-ligand bond energies have been studied using a number of methods [23] including theoretical calculation [24], ion molecule reactions, ion beam, photo-dissociation, mass spectrometry [25], and collision-induced dissociation (CID) [26] etc. M.T. Rodgers and P.B. Armentrout did a systematic study on the metal-ligand bond energies for a wide variety of metal ions and ligands using the CID technique [26]. Although metal AcAc's were not included in their study, the general trend for the first row transition metal ions was proven to be a decrease from Ti to Mn and an increase from Mn to Cu. However, in each particular case, the change in the bond energies depended largely on the ligands. The bond energies in their study varied from 50 to 250 kJ/mol, which is of the same order of magnitude as those from the current isothermal curing study. The study by C. Reichert and J.B. Westmore was performed on metal (II) AcAc's using mass spectrometry [25]. The dissociation energies from their study were 9 kcal/mol, 30 kcal/mol, and 51 Kcal/mol for Mn (II) AcAc, Fe (II) AcAc, and Co (II) AcAc, respectively. Although they followed the same general trend, the differences between these metal AcAc's were greater than what was obtained from the current study. However, noticing that in the current study, the metal AcAc's were in a solution of epoxy resin and HMPA, the energy states of the metal ions were more complicated, as compared to most literature studies where only pure metal complexes were investigated. The transition state of the reaction was considered to be as followed:

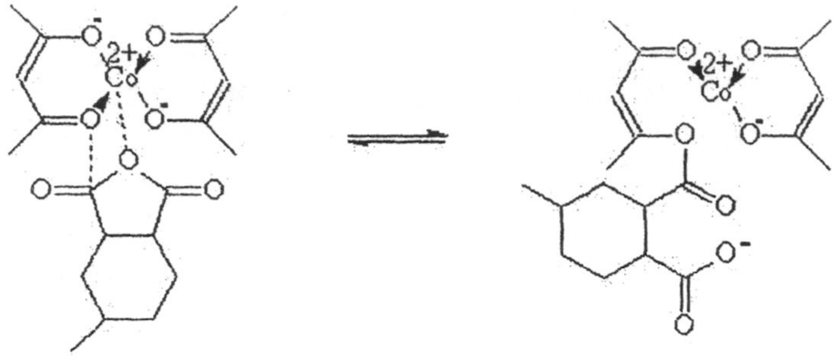

where both the breakage of the coordinate bond between the metal ion and AcAc and the establishment of the bond between AcAc and HMPA should be taken into account in addition to the solvation effect from the epoxy resins. Despite all these complications, the activation energies obtained from isothermal curing study of epoxy/HMPA/metal AcAc can provide an insight into the coordinate bond energy of metal AcAc's and the cure reaction mechanism of this system. These insights can be used as a guideline for designing epoxy curing systems in a wide variety of applications including the design of no-flow underfill materials.

Ligand Effect

In order to further understand the relationship between the activation energy of the cure reaction and the stability of the ligand bond of the metal chelates, Cobalt (II) chelate with a different ligand was used as the catalyst for the epoxy/ anhydride cure reaction. Figure 3 shows the chemical structure of Co(II) AcAc and Co(II) HFAcAc.

Co (II) acetylacetonate   Co (II) hexafluoroacetylacetonate

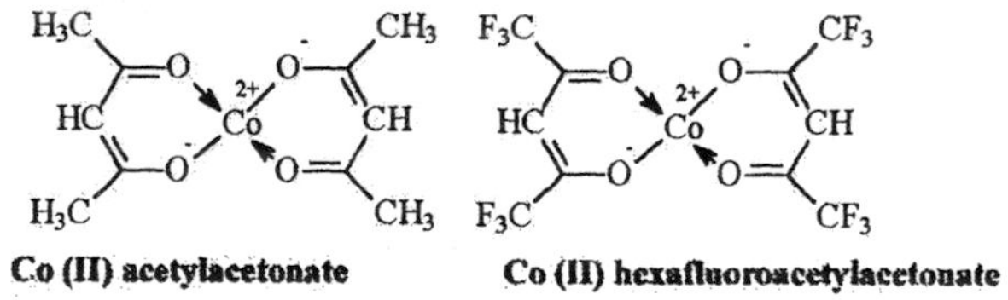

Figure 3. Chemical Structure of Co(II) AcAc and Co(II) HFAcAc

It is expected that with the substitution to fluorinated groups, there is a steric hindrance effect that would interfere with the planar configuration of the chelating ring and thereby lower the stability of the metal complex [27]. On the other hand, the electron-withdrawing effect from fluorinated groups also weakens the ligand bond. The kinetic parameters of the cure reaction of EPON 828/ HMPA/ Co(II) HFAcAc compared with that of EPON 828/ HMPA/ Co(II) AcAc are illustrated in Table 4. The result shows that the activation energy of

EPON 828/ HMPA/ Co(II) HFAcAc is lower than that of the formulation with Co(II) AcAc, which confirms that the stability of the metal-ligand bond is directly related to the catalytic latency.

Table 4. Comparison of Kinetic Parameters of Co(II) AcAc and Co(II) HFAcAc

Catalyst	Ea (kJ/mol)	Log k	n	m
Co (II) HFAcAc	85	7.78	1.03	0.699
Co (II) AcAc	98	10.0	1.42	0.925

CONCLUSIONS

Metal acetylacetonates offer a wide range of catalytic latency and cure behavior for epoxy/anhydride system and are used in no-flow underfill materials for flip-chip applications. The DSC dynamic cure profiles at a heating rate of 5 °C/min were obtained for 43 formulations of ERL 4221/ HMPA/ metal AcAc system. The results revealed that the cure latency was closely related to the bonding strength of the metal ion to the ligand. The DSC isothermal curing study at different temperatures was performed on EPON 828/ HMPA/ metal AcAc containing the first row transition metal ions. Most cure reactions can be analyzed using the autocatalytic kinetic model. The activation energy of the system containing the divalent metal AcAc's followed the Irving and Williams rule. The activation energies were comparable with the dissociation energies of the metal/ligand bond from the literature. Since the metal AcAc's were in a solution of epoxy resin and HMPA, the energy states of the metal ions were more complicated. Different ligands effect was also investigated. In summary, our isothermal cure kinetics study provided insights into the coordinate bond energy of metal chelates and cure reaction mechanism of this system.

REFERENCES

1. R. R. Tummala, E. J. Rymaszewski, and A. G. Klopfenstein, Ed., *Microelectronics Packaging Handbook;* Chapman & Hall, New York, NY, 1997.
2. F. Nakano, T. Soga, and S. Amagi, *Proceedings of International Society of Hybrid Microelectronics Conference,* **1987,** pp. 536.

3. D. Suryanarayana, R. Hsiao, T. P. Gall, and J. M. McCreary, *Proceedings of IEEE 40th ECTC*, **1990,** pp. 338.

4. S. Han and K. K. Wang, *IEEE Trans. On CPMT, Part B*, **1997**, Vol. 20(4), pp. 424.

5. C.P. Wong and S.H. Shi, *U.S. Patent* 6,180,696, **2001.**

6. C.P. Wong, S. H. Shi, and G. Jefferson, *IEEE Trans. On CPMT, Part A*, **1998**, Vol. 21(3), pp. 450.

7. H. Starck, and F. Schlenker, *US Patent* 2,801,228, **1957.**

8. M. Naps, *US Patent* 2,876,208, **1959.**

9. M. Markovitz, *US Patent* 3,812,214, **1974.**

10. J.D.B. Smith, *Journal of Applied Polymer Science* **1981**, 26, pp. 979.

11. P.V. Reddy, R. Thiagarajan, and M.C. Ratra, *Journal of Applied Polymer Science* **1990**, 41, pp. 39.

12. D.M. Stoakley and St. Clair, *Journal of Applied Polymer Science* **1986**, 31, pp. 225.

13. Z. Zhang, S.H. Shi, and C.P. Wong, *IEEE Trans. on Components, and Packaging Technologies*, **2000**, 24 (1), pp. 59.

14. A.V. Kurnoskin, *Journal of Applied Polymer Science* **1992**, 46, pp.1509.

15. A.V. Kurnoskin, *Journal of Applied Polymer Science* **1993**, 48, pp. 639.

16. A.V. Kurnoskin, *Polymer* **1993**, 34 (5), pp. 1060.

17. A.V. Kurnoskin, *Polymer* **1993**, 34 (5), pp. 1068.

18. A.V. Kurnoskin, *Polymer* **1993**, 34 (5), pp. 1077.

19. F. Basolo, and R. Johnson, *Coordination Chemistry*, W.A Benjamin, Inc., New York, **1964.**

20. A.A. Skordos, and I.K. Partridge, *Polymer Engineering and Science* **2001**, 41 (5), pp. 793.

21. J. Inczedy, and J. Tyson, *Analytical Applications of Complex Equilibria, Chapter 1: Complexes and Their Properties*, John Wiley & Sons Inc., New York, **1976.**

22. H. Irving, and R.J.P. Williams, *Journal of the Chemical Society* **1953**, pp. 3192.

23. B.S. Freiser (edited), *Organometallic Ion Chemistry*, Kluwer Academic Publishers, the Netherlands, **1996.**

24. M. Rosi, and C. W. Bauschicher, Jr., *Journal of Chemical Physics* **1990**, 92 (3), pp. 1876.

25. C. Reichert, and J.B. Westmore, *Inorganic Chemistry* **1969**, 8 (4), pp. 1012.

26. M.T. Rodhers, and P.B. Armentrout, *Mass Spectrometry Reviews* **2000**, 19, pp. 215.

27. S. Martell, *Chemistry of Metal Chelate Compounds*, Prentice Hall, Inc, Englewood cliffs, N.J., **1962.**

Chapter 21

Benzocyclobutene-Based Polymers for Microelectronic Applications

Ying-Hung So, Phil E. Garrou, Jang-Hi Im, and Karou Ohba

Advanced Electronic Materials Department, The Dow Chemical Company, Midland, MI 48640

Benzocyclobutene polymerization chemistry is unique and attractive in microelectronics because the thermally activated BCB ring-opening reaction does not produce any volatiles such as water, and products from BCB reactions are nonpolar hydrocarbon moieties. The excellent film-forming properties of divinyltetramethylsiloxane bisbenzocyclobutene (DVS-bisBCB) oligomer solutions and the outstanding electrical, thermal, and planarization properties of the cured final product make these polymers the dielectric materials of choice in a variety of microelectronic applications. The ductility of DVS-bisBCB-based polymer is enhanced with the incorporation of an elastomer in the resin matrix. This paper discusses the chemistry of polymers based on DVS-bisBCB and recent advances in product development.

Introduction

The continuous evolution of microelectronic systems towards higher frequency, better performance, lower cost, and robust reliability requires dielectric materials with demanding properties. Some of the notable requirements are low dielectric constant, low dissipation factor, high glass-transition temperature, minimum moisture uptake, good mechanical properties, and good thermal stability. No evolution of volatiles during fabrication is also a highly desirable feature (*1-5*).

Two key advantages of polymers based on benzocyclobutene (BCB) are (a) the curing process does not emit any volatiles and (b) the products from the BCB ring-opening reaction during polymerization are nonpolar hydrocarbon moieties. This paper focuses on the chemistry of polymers based on divinyl-tetramethylsiloxane bisbenzocyclobutene (DVS-bisBCB) and how recent advances continue to make these polymers important dielectric materials in microelectronic applications.

Benzocyclobutene Ring-Opening

The benzocyclobutene ring opens thermally to produce *o*-quinodimethane. This extremely reactive species readily reacts with dienophiles via a Diels-Alder reaction. In the absence of a dienophile, *o*-quinodimethane reacts with itself to give a dimer, 1,2,5,6-dibenzocyclooctadiene, or undergoes a polymerization reaction similar to that of a 1,3-diene to give poly(*o*-xylene) (*6, 7*).

Polymerization based on the BCB ring-opening reaction has become a popular topic since the mid-1980s. Reaction of the BCB ring-opening intermediate, *o*-quinodimethane, has been used to build macromolecules (*8-11*).

Researchers have also made polymers with pendant BCB moieties and used them as crosslinking agents to transform a linear polymer into a thermoset polymer (*12-14*).

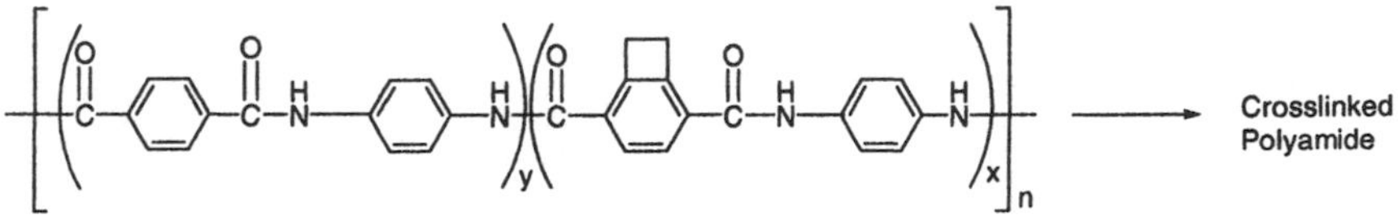

Crosslinked Polyamide

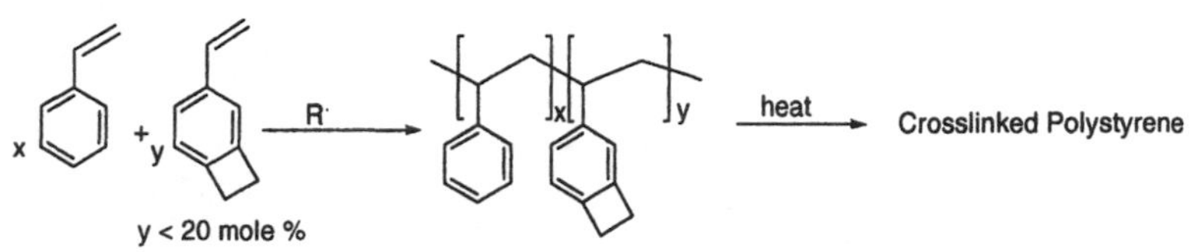

heat → Crosslinked Polystyrene

y < 20 mole %

DVS-bisBCB Monomer Synthesis and B-Staging

The scheme shown below describes the synthesis of divinyltetramethylsiloxane bisbenzocyclobutene (DVS-bisBCB) monomer. The BCB hydrocarbon (*15*) may be made from the pyrolysis of α-chloro-*o*-xylene (*16*). Treatment of the hydrocarbon with bromine provides 4-bromo-BCB with excellent yield. Palladium-catalyzed coupling of 4-bromo-BCB with divinyltetramethylsiloxane produces the monomer DVS-bisBCB. Distillation of the monomer can lower ionic impurities to the ppb level if required (*17*). The viscosity of DVS-bisBCB monomer is 100 cP at room temperature as measured with a Brookfield viscometer at 20 rpm.

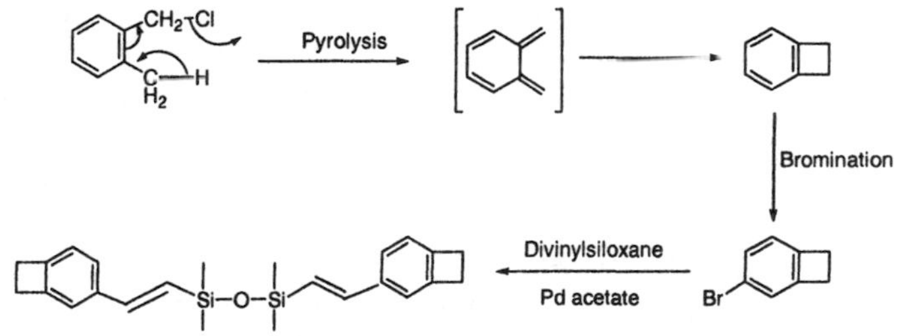

The Diels-Alder reaction is the predominant reaction in B-staging of DVS-bisBCB and in the subsequent curing step to generate the cured product. B-staged DVS-bisBCB solution is formulated in mesitylene and was commercialized by The Dow Chemical Company as CYCLOTENE 3022 electronic resin. (CYCLOTENE is a trademark of The Dow Chemical Company.) These products are dry-etchable. Products from the CYCLOTENE 4000 electronic resin series contain a diazo crosslinker, 2,6-bis(4-azidobenzylidene)-4-alkyl cyclohexanone. These products are photosensitive.

Curing of DVS-bisBCB Polymer Thin Film

The BCB ring starts to open at approximately 160°C and the reaction intensifies at or above 200°C. The curing kinetics of DVS-bisBCB polymer has been studied in detail (*18*). The infrared (IR) absorption peak at 1475 cm^{-1} is assigned to the BCB four-member ring (Figure 1). The peak progressively decreases during cure and virtually disappears at higher than 95% conversion. The absorption peak at 1500 cm^{-1} is attributed to the tetrahydronaphthalene group being formed in the product. This peak is absent in the monomer and progressively develops into a strong absorbance. The peak at 1251 cm^{-1} is the rocking mode of the methyl groups attached to the silicon atoms. The intensity of this peak is unaffected by the polymerization reaction. The degree of BCB conversion can be determined by the ratio of peaks at 1475 cm^{-1} and 1251 cm^{-1}.

Curing can be completed in minutes at 285°C or in a few hours at 205°C. DVS-bisBCB polymer films are cured in nitrogen. If more than 100 ppm of oxygen is present in the DVS-bisBCB polymer curing process, IR spectroscopy of the DVS-bisBCB polymer film shows absorption peaks in the carbonyl region, which suggests possible formation of ketones, carboxylic acids, and carboxylic acid anhydrides. The resulting film is a darker color.

DVS-bisBCB Properties

The polymer made from DVS-bisBCB has a low dielectric constant (Figure 2) and dissipation factor, a high T_g, very low moisture uptake, and a high degree of planarization (Figure 3). The electrical, thermal, and mechanical properties of thin films from photosensitive products are summarized in Table I (*19*). Polymers from the dry-etchable products have similar properties.

Dry-etchable DVS-bisBCB polymer film has >90% transmittance across the visible spectrum (Figure 4). Figure 5 shows the coefficient of thermal expansion for photosensitive DVS-bisBCB polymer as a function of temperature.

Thin Film Patterning Processes

A thin film generally has to be patterned to serve its function in microelectronics. The patterning of a non-photosensitive polymer typically requires the

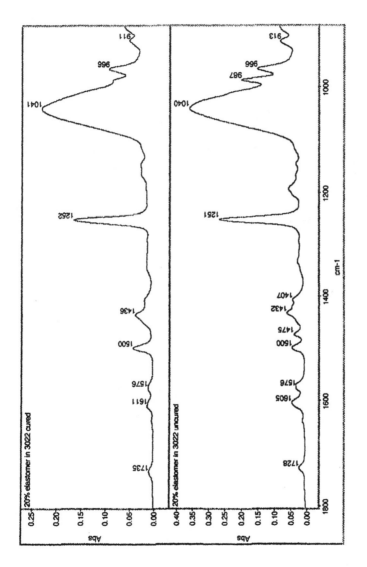

Figure 1. IR spectra of cured and uncured dry-etchable DVS-bisBCB resin containing 20% elastomer.

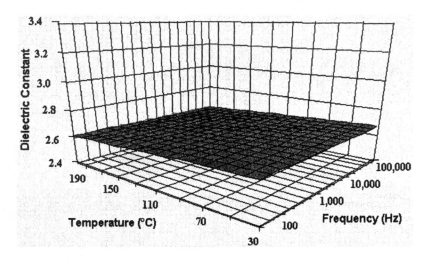

Figure 2. Dielectric constant of DVS-bisBCB resin as a function of temperature and frequency.

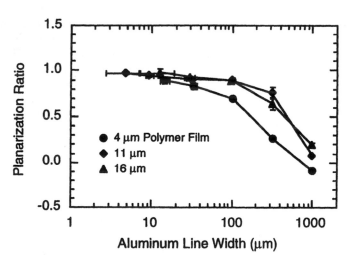

Figure 3. DVS-bisBCB resin degree of planarization on 4-μm-high Al lines.

Table I. Properties of Thin Films from the Photosensitive Products

Property	Value
Dielectric constant (1 kHz-20 GHz)	2.65
Dissipation factor (1kHz-1MHz)	0.0008
Breakdown voltage (V/cm)	3×10^6
Volume resistivity (Ω-cm)	1×10^{19}
Thermal conductivity (W/m°K) at 25°C	0.29
CTE (ppm/°C) 50-150°C	52
Tg (°C)	>350
Water uptake at 84% relative humidity and 23°C (%)	0.14
Tensile modulus (GPa)	2.9
Tensile strength (MPa)	87
Elongation at break (%)	8
Poisson's ratio	0.34
Residual stress on Si at 25°C (MPa)	28

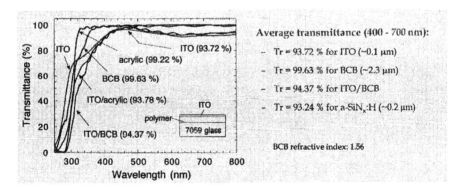

Figure 4. UV-Vis spectra of dry-etchable DVS-bisBCB resin and other materials.

286

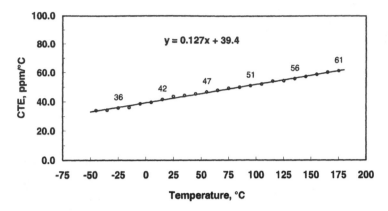

$$y = 0.127x + 39.4$$

Figure 5. Coefficient of thermal expansion for photosensitive DVS-bisBCB resin as a function of temperature.

use of a photoresist and more processing steps than for a photosensitive material. Photosensitive DVS-bisBCB formulation is negative acting and is sensitive to G- and I-line. The patterning of dry-etchable DVS-bisBCB resins and photosensitive DVS-bisBCB resins has been described previously (*17*).

Elastomer-Modified DVS-bisBCB

A limitation of DVS-bisBCB polymer is its restriction in building thick (>50 μm) stacks because of its moderate fracture toughness. The ductility of DVS-bisBCB polymer is significantly enhanced by the addition of an elastomer. Elastomer-toughened thermosets and thermoplastics are widely used in many diverse industries. Recent advances in this area can be found in three American Chemical Society Symposium Series (*20-22*).

The following illustrates recent developments in elastomer-modified DVS-bisBCB formulations. A given amount of an elastomer was added to B-staged DVS-bisBCB. Thin films were spin-coated and cured in a nitrogen-purged oven. Transmission electron micrographs (TEM) of DVS-bisBCB polymer films with 10% and 15% elastomer did not show discrete domains of the additive. In a sample with 20 wt % elastomer, some domains with approximate diameter of 0.1 μm were observed. The area fraction of the observed elastomer was considerably less than 20%. As the amount of the elastomer increased, the amount and size of the domains increased. At 30 wt % the size of the domains was approximately 0.5 μm, and at 40%, the size was approximately 1 μm (*23*). At lower concentrations, the elastomer is soluble in DVS-bisBCB resin, and as the solubility limit is reached, domains of the elastomer are formed.

Chemistry of the Elastomer During DVS-bisBCB Curing

Possible chemical reactions of double bonds in the elastomer, which is a poor nucleophile, with o-quinodimethane from DVS-bisBCB were investigated. IR spectra of DVS-bisBCB resin with elastomer before and after cure are shown in Figure 1. Peaks at 987 cm^{-1} and 966 cm^{-1} are due to the double bonds in DVS-bisBCB resin and the elastomer, respectively. The after-cure spectrum shows the obvious intensity reduction of the DVS-bisBCB double bond. The peak for the elastomer double bond did not change noticeably. The result suggests that the elastomer did not react extensively with DVS-bisBCB during cure.

Chemical bonding between the elastomer and DVS-bisBCB resin was, however, demonstrated by gel permeation chromatography (GPC) coupled with ultraviolet (UV) and refractive index detectors. A mixture of the elastomer and DVS-bisBCB monomer was heated for a short period of time. Only DVS-bisBCB absorbs in the UV range. As an aliphatic compound, the elastomer does not absorb in UV. The reacted mixture was injected into a GPC column, and the eluded components were monitored by UV and refractive index detectors. The eluded elastomer was UV-sensitive, which suggests grafting of DVS-bisBCB on the aliphatic chain.

Electrical Properties

Incorporation of the elastomer did not increase the dielectric constant and dissipation factor of DVS-bisBCB resin. At 1 MHz, dielectric constant and dissipation factor were 2.65 and 0.0008, respectively.

Thermal Analysis

DVS-bisBCB resin with 10, 15, and 18% elastomer was studied by dynamic mechanical spectroscopy (DMS), and the result for DVS-bisBCB with 18% elastomer is shown in Figure 6. Only a slight decrease of storage modulus was observed. Glass transition was not detected in the temperature range of $-100°C$ to $300°C$.

Thermal stability of DVS-bisBCB resin as free-standing films of about 11 μm thick and with 0, 5, 10, 15, and 20 wt % elastomer was studied. Samples were held isothermally at 350°C for 2.5 h. Weight loss for all samples was less then 2% per hour.

Table II lists coefficient of thermal expansion (CTE) values of DVS-bisBCB resin formulation with different amounts of elastomer. The incorporation of an elastomer increases the CTE of DVS-bisBCB resin.

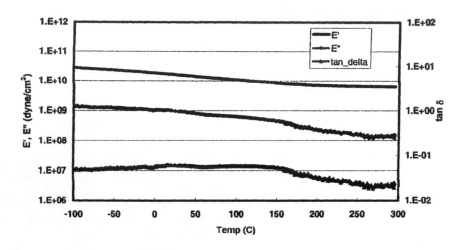

Figure 6. DMS of DVS-bisBCB resin with 18 wt % elastomer.

Table II. CTE Values of DVS-bisBCB Resin with Different Amounts of Elastomer

Elastomer %	CTE (ppm/°C)
0	52
5	57
10	65
15	80

Table III. Tensile Properties of DVS-bisBCB Resin with Elastomer

Additive Content (%)	Tensile Strength (MPa)	Strain (%)	Modulus (GPa)
0	87	8	2.9
5	94	13.5	2.7
10	90	18	2.5
18	95	35	2.4

Mechanical Properties

Thin film ductility, as measured by strain at break, was used to determine the toughness of elastomer-modified DVS-bisBCB resin. Tensile properties of DVS-bisBCB resin with different amounts of elastomer are listed in Table III. Strain at break of DVS-bisBCB resin was significantly enhanced with the elastomer, and tensile modulus was slightly reduced. Figure 7 is the stress versus strain curve of DVS-bisBCB resin with 18 wt % elastomer. There have been reports in the past that formation of discrete domains is important in polymer toughness enhancement (*19-22, 24*). However, formation of discrete elastomer domains does not seem to be necessary for DVS-bisBCB resin toughness enhancement as TEM of the polymer film with 15% elastomer did not show discrete domains of the additive. The mechanism for improved mechanical properties of the matrix is being investigated.

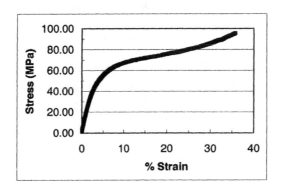

Figure 7. Strain-stress curve of DVS-bisBCB resin with 18 wt % elastomer.

Moisture Absorption

The low moisture absorption of DVS-bisBCB resin is maintained in modified DVS-bisBCB resin. At 85% relative humidity, for DVS-bisBCB resin samples with 5 to 20 wt % elastomer, moisture uptake was less than 0.2 wt %.

Adhesion Properties

A DVS-bisBCB resin formulation with 5 and 10 wt % elastomer was applied on wafers with SiO_2, Si_3N_4, aluminum, and copper coatings. AP3000 was used as an adhesion promoter. The wafers were placed in a pressure vessel at 125°C and 2 atm for 96 h. The samples were evaluated by the tape peel test, and

Table IV. Adhesion Performance of DVS-bisBCB Resin with Elastomer

| | Tape Peel Test Results | | |
Wafer Type	0% Elastomer	5% Elastomer	10% Elastomer
Si_3N_4	5B	5B	5B
SiO_2	5B	5B	5B
Al	5B	5B	5B
Cu	5B	5B	5B

no detachment of DVS-bisBCB resin from any of the surfaces was observed (Table IV).

Applications

DVS-bisBCB resins are notable for their low dielectric constant, low loss factor at high frequency, low moisture absorption, low cure temperature, high degree of planarization, low level of ionics, high optical clarity, good thermal stability, excellent chemical resistance, and good compatibility with various metallization systems. These properties make DVS-bisBCB resins the dielectric materials of choice for many applications in the electronics industry.

Dry-etchable DVS-bisBCB resin is commercially used by LG Phillips in ToPixel technology in which indium tin oxide is deposited on a completely flat DVS-bisBCB polymer surface. A high degree of planarization of DVS-bisBCB polymer is essential to this "high-aperture" display technology. A low curing temperature (210°C), low water absorption, rapid curing, and low viscosity make dry-etchable DVS-bisBCB resin an excellent choice for a variety of applications in gallium arsenide (GaAs) integrated circuits. Triquint Semiconductor, Nortel Networks Corporation, and others use DVS-bisBCB to form multilayer interconnects on their GaAs chips. Others, such as Anadigics, use DVS-bisBCB to encapsulate and thus protect air bridges during plastic packaging.

In bumping/redistribution/wafer level chip scale packaging, polymer layers are used for rerouting of perimeter bond pads to an internal array or for reconfiguring pads for solder bumping. Photosensitive DVS-bisBCB resin is used extensively for this application in the U.S., Taiwan, Korea, and Europe.

Extremely low dielectric constant and low loss at GHz frequencies, low moisture uptake, and copper compatibility make DVS-bisBCB resin an excellent dielectric material for integration of passive RF components. Waveguide structures, which use photosensitive DVS-bisBCB resin as the core material and dry-etchable DVS-bisBCB resin for cladding, have been demonstrated by Ericsson and others. The major advantage of DVS-bisBCB resin is its low loss of <0.2 dB/cm at 1.3 μm. Illustrative figures and references of the above applications can be found in reference (17).

The commercialization of DVS-bisBCB-resin-coated copper foil was recently announced in Japan for use in third-generation (3G) high-frequency telecommunication printed wiring boards (*23*). Elastomer-modified DVS-bisBCB-coated Cu foil satisfies the requirements of high flexibility, flaking resistance, and resin flow control in the hot-press process. The modified polymer maintains the advantages associated with DVS-bisBCB polymer as a highly desirable dielectric material for high-density chip carriers, daughter boards, and GHz telecommunication devices. Figure 8 shows a 1.8-mm FR-4 board with 60-μm-thick, Cu-coated DVS-bisBCB build-up layers on both sides and a 0.35-mm-diameter, electrodeless, Cu-plated, drill-through hole. Illustrations of a CO_2 laser via on a DVS-bisBCB build-up layer after Cu foil was stripped and a Cu-plated laser via can be found in reference (*22*).

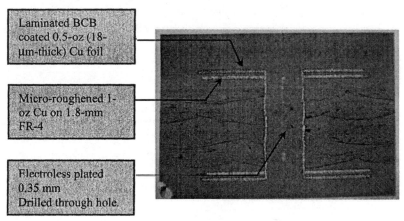

Figure 8. A 1.8-mm FR-4 board with 60-μm-thick, Cu-coated DVS-bisBCB resin build-up layers on both sides.

Conclusion

Benzocyclobutene polymerization chemistry is unique and attractive in microelectronics because the thermally activated BCB ring-opening reaction does not produce any volatiles such as water, and products from BCB reactions are nonpolar hydrocarbon moieties. The excellent film-forming properties of DVS-bisBCB oligomer solutions and the outstanding electrical, thermal, and planarization properties of the cured final product make these polymers the dielectric materials of choice in a variety of microelectronic applications. The ductility of DVS-bisBCB-based polymer is enhanced with the incorporation of an elastomer in the resin matrix. We expect to see the volume of existing DVS-bisBCB applications increase and new applications to emerge in the future.

Acknowledgements

We thank our colleagues at Dow for their contributions to the BCB program, Britton Romain for preparation of the manuscript, and Ann Birch of Editech for editing.

Literature Cited

1. *Polymers for Electronic and Photonic Applications*; Wong, C. P., Ed.; Academic Press: San Diego, CA, 1993.
2. *Fundamentals of Microsystems Packaging*; Tummala, R. R., Ed.; McGraw-Hill: New York, NY, 2001.
3. Garrou, P. Chapter 9 Thin Film Packaging and Interconnect. In *Thin Film Technology Handbook*, Elshabini-Riad, A., Barlow, F. Eds., McGraw Hill: New York, 1997.
4. Garrou, P. Polymer Dielectrics for Multichip Module Packaging. *Proceedings of IEEE*, Vol. 80, 1992, 1942–1954.
5. Czornyj, G.; Asano, M.; Beliveau, R.L.; Garrou, P.; Hiramoto, H.; Ikeda, A.; Kruez, J.A.; Rohde, O. Chapter 11 Polymers in Packaging. In *Microelectronics Packaging Handbook Semiconductor Packaging*, 2nd Ed., Tummala, R. R., Rymaszewski, E. J., Klopfenstein, A. G., Eds.; Chapman & Hall: New York, 1997, 509–623.
6. Rondan, N. G.; Bruza, K. J.; Kirchoff, R. R. Mechanism of Electrocyclization of Benzocyclobutene to o-Xylene. A Theoretical Study. *Preprints, Am. Chem. Soc. Div. Polym. Chem.* **1991**, *32* (1), 385–386 and references therein.
7. Tan, L.-S.; Arnold, F. E. *J. Poly. Sc.: Part A: Poly. Chem.* **1988**, *26*, 1819–1834.
8. Kirchhoff, R. A.; Bruza, K. J. *Prog. Polym. Sci.* **1993**, *18*, 85–185.
9. Kirchhoff, R. A.; Bruza, K. J. *Chemtech* **1993**, 22–25.
10. Kirchhoff, R. A.; Bruza, K. J. *Adv. Polym. Sciences*, **1993**, 117, 1–66.
11. Farona, M.F. *Prog. Polym. Sci.* **1996**, *21*, 505–555.
12. Walker, K. A.; Markoski, L. J.; Gary, A.; Spilman, G E.; Martin, D. C.; Moore, J. S. Crosslinking Chemistry for High-Performance Polymer Networks. *Polymer*, **1994**, *35* (23), 5012–17.
13. Wong, P.K. Reactive Styrene Polymers. U.S. Patent 4,698,394, 1987.
14. Hawker, C. J.; Bosman, A. W.; Harth, E.; Russell, T. P.; Van Horn, B.; Frechet. J. M. J. Approaches to Nanostructures for Advanced Microelectronics Using Well-defined Polymeric Materials. Abstracts of Papers, 222nd ACS National Meeting, Chicago, IL, August 26-30, 2001.

15. The CAS name for the BCB hydrocarbon is bicyclo[4.2.0]octa-1,3,5-triene, and it is also known as 1,2-dihydrobenzocyclobutene and benzocyclobutene.

16. For references of the synthetic steps, see So, Y. H.; Garrou, P. E.; Im, J. H.; Scheck, D. M. *Chemical Innovation*, **2001**, *31*, 40-47.

17. Stokich, T. M.; Lee, W. M.; Peters, R. A. Real Time FT-IR Studies of the Reaction Kinetics for the Polymerization of Divinylsiloxane bis-Benzocyclotutene Monomers, *Mat. Res. Soc. Symp. Proc.* 1991, 227, 103–114.

18. Im, J.; Shaffer II, E. O.; Peters, R.; Rey, T.; Murlick, C.; Sammler, R. L. Physical and Mechanical Properties Determination of Photo-BCB-Based Thin Films. *Proc. Int. Soc. Hybrid Microelectronics*, Minneapolis, MN, 1996, 168–175.

19. Toughened Plastics I–Science and Engineering. Riew, C. K., Kinloch, A. J., Eds.; ACS Advances in Chemistry Series 233; American Chemical Society: Washington, DC, **1993**.

20. Toughened Plastics II–Novel Approaches in Science and Engineering. Riew, C. K., Kinloch, A. J., Eds.; ACS Advances in Chemistry Series 252; American Chemical Society: Washington, DC, **1996**.

21. Toughening of Plastics–Advances in Modeling and Experiments. Pearson, R. A.; Sue, H.-J.; Yee, A. F., Eds.; ACS Symposium Series 759; American Chemical Society: Washington, DC, **2000**.

22. Ohba, K.; Akimoto, H.; Kohno, M.; So, Y. H.; Im, J. H.; Garrou, P. E.; Shimoto, T.; Matsui, K.; Shimada, Y. Development of CYCLOTENE Polymer Coated Cu Foil for Build-up Board Application. 2000 IEMT/IMC Symposium, Tokyo, Japan, April 19–21, 2000, 41–46.

23. So, Y. H., et al. In *Proceedings of the International Microelectronics and Packaging Society*, Boston, 2000; International Microelectronics and Package Society: Washington, DC; 80-86.

24. *Polymer Toughening*; Arends, C. B., Ed.; Marcel Dekker: New York, NY, 1996.

Chapter 22

Formation of Nanocomposites by In Situ Intercalative Polymerization of 2-Ethynylpyrdine in Layered Aluminosilicates: A Solid-State NMR Study

A. L. Cholli[1], S. K. Sahoo[1], D. W. Kim[1], J. Kumar[1], and A. Blumstein[2]

[1]Center for Advanced Materials and [2]Department of Chemistry, University of Massachusetts at Lowell, Lowell, MA 01854

Novel polymer nanocomposites having potential anisotropic properties have been synthesized by *in-situ* intercalative polymerization of 2-ethynylpyridine in the lamellar galleries of Ca^{2+}-montmorillonite. In this paper, we used solid-state ^{13}C CP/MAS NMR spectroscopy and X-ray diffraction technique to demonstrate the spontaneous intercalative polymerization of 2-ethynylpyridine in the galleries of montmorillonite without the aid of external initiator. Nanocomposites were synthesized with varying amounts of polymers. X-ray diffraction data indicate the increase in the d- spacing of Ca^{2+}-montmorillonite with the polymer content. The ^{13}C NMR chemical shifts of various carbons in the *in-situ* polymerized nanocomposite samples were assigned and compared with the bulk synthesized polymer of poly (2-ethynylpyridine) (Poly (2Epy)) and the nanocomposite obtained after intercalation of the bulk polymer into the Ca^{2+}-montmorillonite. Analysis of preliminary solid-state NMR data suggests that the structure of the *in-situ* polymerized sample (Poly (2EPy)) is structurally similar to that of bulk polymerized sample (Poly (2Epy)). *In-situ* and bulk polymerized sample show different ^{13}C NMR chemical shifts for various carbons mostly due to the variation in residual charge density on the pyridine nitrogen.

Introduction

In the past few years there is resurgence of research activity in the area of polymer-clay nanocomposites even though the first reported literature data on these materials goes back to late 50's and early 60's by Blumstein (*1*). Today's fast growing activity in inorganic/organic nanocomposites arises from the realization that these materials show exceptional properties when compared to their macroscopic counterparts. Because of their exceptional and improved thermal, mechanical, and physical properties (*2*), these polymer-based nanocomposites are being considered for a wide range of applications in electronics and automobile industries.

Formation of lamellar polymer-clay nanocomposites had opened new avenues for polymer research in a number of areas (*3,4*) Inorganic layered materials possess ordered interlayer spacings potentially accessible for polymer chains to occupy this space resulting in nanocomposite materials (Figure 1a). The clays used in the formation of polymer nanocomposites are aluminosilicates with a sheet-like structures consisting of silica SiO_4 tetrahedra bonded to alumina AlO_6 octahedra. In the case of naturally occurring clay, montmorillonite (MMT), the ratio of tetrahedra to octahedra is 2:1. The thickness of these sheet-like layered structures consisting of tetrahedra and octahedra structures are in the order of 1 nm and the aspect ratios typically are in the range of 100-1500. These sheets and their edges bear a charge on their surfaces depending on the chemical composition. The charge on the surface is normally balanced by the counter ions. These counter ions normally reside in the gallery space between the layered sheets. The gallery space between the silicate layers is normally ca. 1-2 nm. These are truly nanoparticultaes.

The molecular interaction leading to nano length scale structures provides generally two classes of nanomaterial in the form of intercalated or exfoliated structures (*2*). In the case of intercalated structures, the polymer chains are inserted between layers of ordered multilayered aluminosilicates such that gallery space is expanded to accommodate the presence of polymer chains in way that the layers still maintain well-defined spatial relationship to each other (Figure 1b). In the case of exfoliated structures, the layers do not maintain their order among themselves; the layers are separated as well as dispersed throughout the polymer matrix (Figure 1c).

Our recent interest is, however, on the in-situ intercalative polymerization of pyridine substituted acetylenic monomer (2-ethynylpyridine (2EPy)) within the galleries of Ca^{2+}-montmorillonite (Ca^{2+}-MMT) (*5*). In this process, it is believed that monomer first intercalates into the galleries of montmorillonite, and then the interaction with clay surface initiates the polymerization (*5*). The X-ray

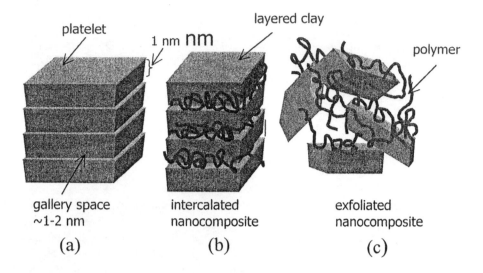

Figure 1: *Schematic representation of layered structure of (a) clay (montmorillonite, MMT), (b) intercalated, and (c) exfoliated polymer nanocomposites (not to scale).*

diffraction data shows a corresponding increase in the d-spacing with adsorption yield, and a trans-transoidal conformational model of the pyridine ring inside the clay lamella is proposed (5). The nature of polymer-clay interaction so far has not been very clear. Recent work using multi nuclei solid-state NMR techniques provides information regarding the structure and dynamics of the polymer chains and the nature of interaction of clay surface with the polymer chains (6,7). To the best of our knowledge, a limited literature deals with the structure of the intercalated substituted imine polyacetylene polymer and interaction of the clay surface with the polymer chains in the in-situ intercalative polymerization (5). The goal of this preliminary work is to investigate the nanocomposites formed as a result of in-situ intercalative polymerization of 2-ethynylpyridine by solid-state ^{13}C CP/MAS NMR and X-ray diffraction techniques.

Experimental

Materials: Ca^{2+}-Montmorillonite (Ca-MMT) was purchased from Clay Minerals Repository with cation exchange capacity of 1.20 mequiv/g. 2-ethynylpyridine (2EPy) was purchased from Farchan Laboratory and used as received.

Synthesis of Poly (2-ethynylpyridine)/Clay nanocomposites: 0.395 g of 2EPy was added to a suspension of 1.0 g of driedCa^{2+}-MMT in 25 ml of benzene and then mixture was stirred at 65±3 °C for 24 hrs. The dark brown precipitate was separated from suspension by centrifugation and washed several times with benzene to exclude the monomer trapped between the aggregates of clay particles and finally dried at 110 °C for 24 hrs. The adsorption yield of the monomer per unit weight of dried Ca^{2+}-MMT was obtained by measuring the concentration in the benzene solution by UV/Vis absorption spectroscopy. The chemical synthesis of bulk-P2EPy, poly (2-ethnylpyridinium hydrochloride) was described elsewhere (8,9)

NMR Spectroscopy: Solid-state ^{13}C NMR experiments were carried out using a dedicated Bruker DMX 300 MHz NMR instrument. Zirconium oxide (ZrO_2) 4mm rotors were used with Kel-F caps for all the measurements. Cross-polarization with magic angle spinning (CP/MAS) and high power dipolar decoupling techniques were used to study these materials. All spectra were recorded using a rotor spinning speed of 10 kHz. The typical parameters for ^{13}C CP/MAS NMR experiments were as follows: spin lock and decoupling field of 70 kHz, a 3 s recycle time, a contact time of 1 ms and a sweep width of 31 kHz. The total number of free induction decays (FIDs) co-added per spectrum ranged from ~10,000 for bulk polymer to 160,000 for intercalated polymer-clay complexes. All the FIDs were processed by an exponential apodization function and with a line broadening of 20-50 Hz. The ^{13}C CP/MAS NMR spectra were externally referenced to glycine by assigning the carbonyl signal at 176.03 ppm with respect to tetramethylsilane (TMS).

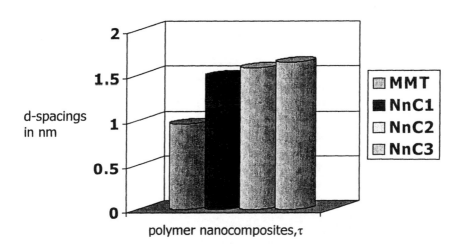

Figure 2. *Graph showing an increase in the basal d-spacing (XRD) of in-situ polymerized nanocomposites NnC1, NnC2, and NnC3, containing 8,15.3 and 21.3% of polymers, respectively.*

Results and Discussion

The in-situ polymerization of 2EPy in the lamellar galleries of Ca^{2+}-MMT resulted an increase in the d-spacing of Ca^{2+}-MMT, as measured by X-ray diffraction at different adsorption yield (τ) (Figure 2). The increased d-spacing of Ca^{2+}-MMT indicates the successful intercalation of 2EPy into the lamellar galleries. The d-spacing increased from 1.49 nm at low adsorption yield (8.0%) to 1.64 nm at an adsorption yield of 21.3%. XRD data suggests that polymers are formed in the gallery space between the aluminosilicate layers maintaining a well-defined spatial relationship to each other.

The solid-state ^{13}C CP/MAS NMR spectrum of a nanocomposite, Poly (2Epy) / Ca^{2+}-MMT with an adsorption yield of 21% is shown in Figure 3 together with the chemical shift values. The solution-state ^{13}C NMR spectrum of the monomer (2EPy) is shown in Figure 4. Comparing the NMR spectra of the monomer and Poly (2Epy)/Ca^{2+}-MMT nanocomposites, it is clear that the peaks of the acetylenic carbons (82.9 and 77.7 ppm resonances in Figure 4) are not present in the solid-state NMR spectra shown in Figure 3. As there are *no* contributions from montmorillonite, the ^{13}C NMR spectrum of the nanocomposite is a characteristic of the adsorbed, intercalated (based on XRD data), and polymerized species within the aluminosilicate matrix. The disappearance of the resonances for the acetylenic carbons and the appearance of broad distribution of resonances in the aromatic/olefinic region suggest the spontaneous in-situ intercalative polymerization of the 2EPy. The structural information obtained from solid-state NMR and an increased d-spacing (from X-ray diffraction) suggests a spontaneous in-situ intercalative polymerization of 2EPy between lamella without any pre-treatment of MMT and without use of external initiator. Significantly the MMT not only acts as a matrix for the preparation of nanocomposites, but acts also as a solid acid catalyst for the polymerization. Resonances due to the ethylenic carbons in the Poly (2Epy) backbone are overlapped with the aromatic ring carbon of the pyridine ring, giving a broad resonance extending from 75 to 250 ppm (Figure 3).

The tentative assignments of various resonances in Figure 3 are as follows: the most intense ~125 ppm peak corresponds to the ethylenic–CH (C7) as well as the C2 (β) and C4 (β') carbons of the pyridine ring (Figure 5). The peak at 138.5 ppm is assigned to the ethylenic C (C6) and C3 (γ) carbons of the pyridine ring, and the most down field peak at 148 ppm is due to C1 (α) and C5 (α'), both being bonded to the nitrogen atom. These assignments are consistent with the chemical shift values of the monomer spectrum (Figure 4) and agree with the work of Maciel et al (*10*). The ^{13}C CP/MAS NMR data were also collected (not shown) on nanocomposites containing varying amount of polymers (adsorption yield). As the adsorption yield increased the relative intensity of the 125.5 ppm peak increased compared to the 147.5 ppm peak. At the low adsorption yield of 8.0%, the solid-state ^{13}C CP/MAS NMR spectra are broad and not well resolved. The strong peak in Figure 3 appears at 125.5 ppm, with a weak shoulder at 147.5

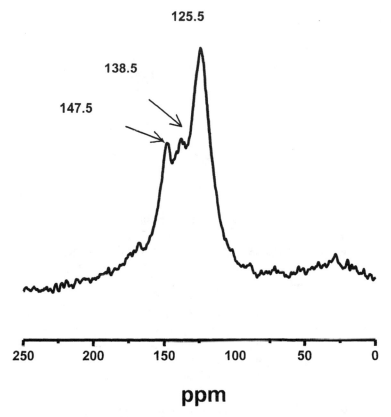

Figure 3. Solid-state ^{13}C CP/MAS NMR spectrum of in-situ polymerized polymer-clay nanocomposite, Poly (2Epy)/Ca^{2+}-MMT, with an adsorption yield of 21.3% of polymer.

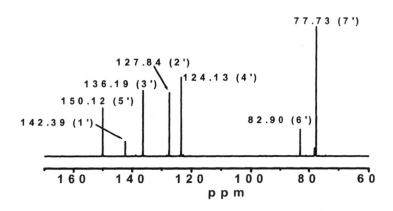

Figure 4. Solution-state ^{13}C NMR spectrum for 2EPy. For labeling of carbons, please see structures in Figure 5.

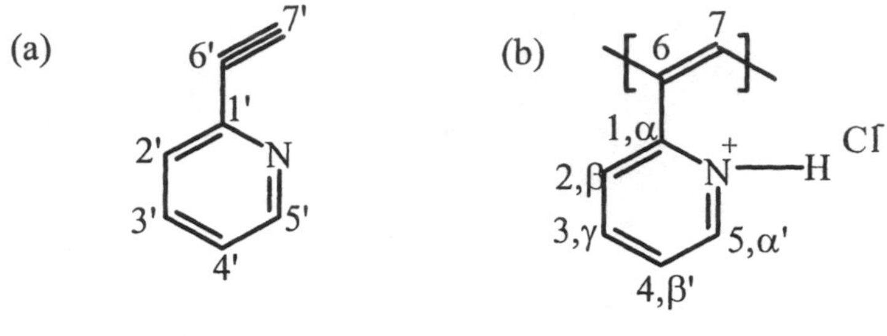

(a)

(b)

Figure 5. *Structure for (a) 2EPy, and (b) P2Epy*

ppm. The chemical shifts of some resonances changed with the adsorption yield suggesting that the local environments of these carbons differ with the amount of polymer in the nanocomposites. The solid state NMR data suggests that the total carbon resonance intensities increased with the polymer content in the nanocomposites (Table 1).

Nature of Polymer Environments in *in-situ* Polymerized Nanocomposites
The ^{13}C NMR chemical shifts of α, β and γ carbons of poly (2Epy) are sensitive to the surroundings of polymer chains in the nanocomposites. The changes in chemical shifts provide both the structural and physical nature of polymers. For this purpose, we synthesized the bulk poly (2Epy) polymer and compared its NMR spectral features with the polymer that is intercalated in the nanocomposites formed by in-situ polymerization. The solid-state ^{13}C CP/MAS

Table 1
Adsorption yield (τ), area of aromatic resonance, and *d*-spacing of the various nanocomposite samples prepared by in-situ intercalative polymerization of 2-ethynylpyridine in the galleries of Ca^{2+}-MMT

Nanocomposite Sample	Adsorption yield (τ)[a] (%)	Total area of aromatic resonance[b]	d-spacing[c] (nm)
1	8.0	1.00	1.49
2	15.3	1.35	1.58
3	21.3	1.50	1.64

[a]The adsorption yield of polymer per unit weight of dehydrated Ca^{2+}-MMT; [b]The relative area determined from solid-state ^{13}C CP/MAS NMR data; [c]The error range of the *d*-spacing is ±0.05 nm.

NMR spectrum of the bulk synthesized Poly (2Epy) is shown in Figure 6. The bulk-Poly (2Epy), prepared in a strong acidic medium (in presence of HCl) is protonated. This shows well-resolved two resonance peaks at 126.5 and 147.0 ppm along with a shoulder at 118.0 ppm. Compared to the NMR spectrum of the polymer in the in-situ polymerized nanocomposite (Figure 4), the spectral features in Figure 6 are different and peaks are narrower. The significant changes are the following: (a) the resonance peak for the carbon C3 (γ) is shifted downfield and merged with the peak at 147 ppm. The resonance peaks for β carbons (2 and 4) and C7 (double bond carbon of the polymer backbone at 118 ppm) are partially resolved due to a downfield shift by 1.5 ppm for β carbons. Even though the bulk synthesized polymer and the polymers formed in the inner layer spacings of MMT in the nanocomposites are protonated, the extent of protonation can be qualitatively assessed based on the chemical shift of γ carbons. The solid state NMR spectrum of the bulk-synthesized polymer in the non-protonated state is helpful to gauge the extent of protonation of the polymer chains that are intercalated with MMT in the nanocomposite.

The solid state NMR spectrum of the non-protonated bulk poly (2Epy) is presented in Figure 7. The spectral features of this polymer are distinctly different compared to that of the protonated bulk polymer shown in Figure 6. The line widths of these resonances are narrower compared to the resonances of the protonated polymer. The removal of the positive charge on the nitrogen atom of the pyridine ring shifted the resonance peak for C3 (γ) to 136 ppm. In a similar way, the peaks for β carbons (2 and 4) are shifted down field to 121.5 ppm. The resonance peaks for α carbons (1 and 5) are resolved and appear at 149 and 155 ppm.

The comparison of chemicals shifts in Figures 3, 6 and 7 allows to understand qualitatively the state of polymer chains in the nanocomposites. The presence of γ carbon resonance at ca. 138 ppm in the solid state NMR spectrum of in-situ polymerized nanocomposite indicates that the nitrogen atom of the pyridine ring in the intercalated polymer is not fully protonated. This is likely due to the interaction of polymer chains with the surface of MMT resulting in the neutralization of charge density on the nitrogen. In a similar way, the chemical shifts for the α and β carbons of the polymer in the nanocomposite are also between the chemical shifts values for the protonated and non-protonated polymer systems (Figures 6 and 7) suggesting that there are some polymers containing both the protonated and non-protonated moieties in the in-situ polymerized nanocomposites.

Nanocomposites Formed by Intercalation of Bulk Polymer poly (2Epy) with MMT It is also possible to form nanocomposites by molecular level incorporation of polymers into MMT by solution blending (2). It is of interest to compare physical nature of poly (2Epy) polymers in these nanocomposites with in-situ polymerized nanocomposites.

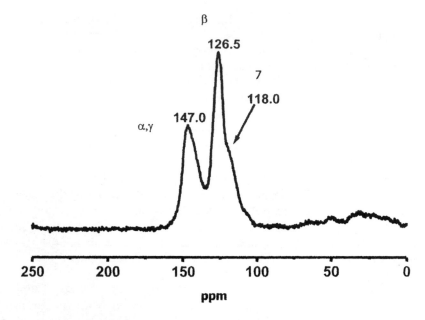

Figure 6. Solid-state ^{13}C CP/MAS NMR Spectrum of bulk-synthesized poly (2Epy) in protonated state

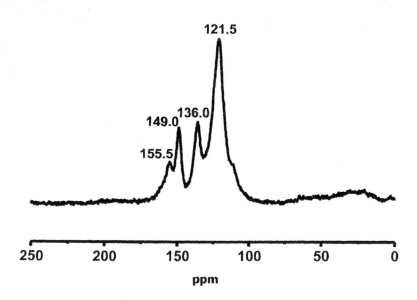

Figure 7. Solid state ^{13}C CP/MAS NMR Spectrum of bulk-synthesized poly (2Epy) in non-protonated state

The X-ray diffraction data of intercalated bulk-Poly (2Epy) shows that d-spacing of the Ca^{2+}-MMT increases to 1.89 nm (9). This value is greater than the d-spacing of Sample 3 (Table 1). The solid-state ^{13}C CP/MAS NMR spectrum of intercalated bulk-P2EPy/Ca^{2+}-MMT nanocomposite is shown in Figure 8b. For comparison, the solid-state ^{13}C NMR spectra of bulk-poly (2Epy) (protonated) and in-situ polymerized nanocomposites are also presented in Figure 8. The results suggest that the polymer obtained by *in-situ* polymerization in the galleries of the Ca^{2+}-MMT is either different in structure (variation of charges on the nitrogen) or having pyridine ring endowed with less mobility which is different from bulk amorphous polymer. We performed variable contact time experiments on the bulk-Poly (2Epy) and the *in-situ* polymerized Poly (2Epy) (Sample 3) (9). The bulk-Poly (2Epy) shows a maximum intensity with a 2ms contact time. This suggests a fair amount of mobility in the bulk amorphous polymer. In contrast, the *in-situ* polymerized Poly (2Epy) sample displays maximum intensity at very short contact time (<1 ms). This data suggests that the polymer chains in the nanocomposite (i.e. *in-situ* polymerized intercalated polymer) have less mobility than in its bulk amorphous state.

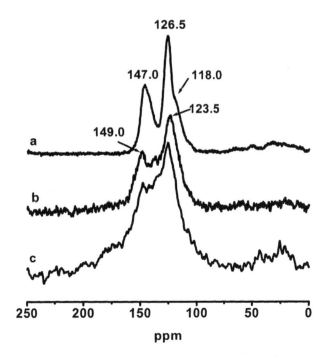

Figure 8. Solid-state ^{13}C CP/MAS NMR spectrum for (a) bulk P2EPy; (b) intercalated bulk-Poly (2Epy)/Ca^{2+}-MMT, and (c) in-situ polymerized Poly (2Epy)/Ca^{2+}-MMT.

Investigation of the nanocomposites formed by bulk-Poly (2Epy) intercalating into the galleries of the Ca^{2+}-MMT (Figure 8b) by solid-state NMR shows spectral features that are similar to the in-situ polymerized poly (2Epy) nanocomposite sample (Figure 8c). This suggests the structural similarity between in-situ and bulk polymerized polymer nanocomposites. The intercalation of the bulk-Poly (2Epy) into the galleries resulted in the electrostatic interaction of the quaternary nitrogen of the polymer with the negatively charged surface of montmorillonite. In the case of the in-situ polymerized polymer nanocomposite sample, as the adsorption yield increased, the difference between the chemical shifts of α–β and α–γ ring carbons also increased. This is because at low adsorption yield, the molecules are tightly bound to the montmorillonite surface by Lewis acid-base interactions. At higher adsorption yield, Brøensted acidity plays a more important role where polymers are either hydrogen bonded or loosely adsorbed (9).

Conclusions

Solid-state ^{13}C CP/MAS NMR has been used to study the polymer-MMT nanocomposites prepared by *in-situ* intercalative polymerization of 2EPy. The NMR results suggest that the structure of the *in-situ* polymerized sample (Poly (2EPy)) is structurally similar to that of bulk polymerized sample (Poly (2Epy)) in the protonated state. *In-situ* and bulk polymerized sample shows different ^{13}C NMR chemical shifts mostly due to the variation in the residual charge density on the pyridine nitrogen atom.

Acknowledgements

The authors are indebted to the National Science Foundation (NSF) for the research grant (DMR-9986644) and also to the Petroleum Research Fund (PRF).

References

1. (a) Blumstn, A., Herz, J., Sinn, V., Sadron,C., C.r. **1958**, 246,1856.
    (b) A. Blumstein, A. *Bull.Chim.Soc.* **1961**, 899.
2. Gilman, J.W., Morgan, A.B., Harris,Jr. R., Manias, E., Giannelis, E.P., Wuthenow,M., In *New Advances in Flame Retardant Technology. Proceedings. Fire Retardant Chemical Association. October 24-27,1999, Tuscon, AZ.* Fire Retardant Chemical Assoc., Lanacaster, PA, **1999**, pp 9-22.

3. Kojima, Y.; Usuki, A.; Kawasumi, M.; Okada, A.; Kurauchi, T.; Kamigaito, O. *J. Polym. Sci., Part A: Polym. Chem.* **1993**, *31*, 1755.
4. Alexandre, M.; Dubois, P. *Mater. Sci. Eng. R* **2000**, *28*, 1.
5. (a) Liu, H.; Kim, D.W.; Blumstein, A.; Kumar, J.; Tripathy, S.K. *Chem. Mater.* **2001**, *13*, 2756. (b) Kim, D.W.; Blumstein, A.; Liu, H.; Downey, M.J.; Kumar, J.; Tripathy, S.K. *J. Macromol. Sci.: Pure & Appl. Chem.* **2001**, *A38(12)*, 1405, (c) Sahoo, S.K., Kim, D.W., Kumar, J., Blumstein, A., Cholli, A.L. *Polym. Materials: Sci. and Technology*, **2002**, 87,394.
6. VanderHart, D.L.; Asano, A.; Gilman, J.W. *Macromolecules* **2001**, *34*, 3819.
7. Mathias, L.J.; Davis, R.D.; Jarrett, W.L. *Macromolecules* **1999**, *32*, 7958.
8. Subramanyam, S.; Blumstein, A. *Macromolecules* **1992**, *25*, 4058.
9. Sahoo, S.K., Kim, D.W., Kumar, J., Blumstein, A., Cholli, A.L., Macromolecules, **2003**, 36, 2777.
10. Maciel, G.E.; Haw, J.F.; Chuang, I-S.; Hawkins, B.L.; Early, T.A.; McKay, D.R.; Petrakis, L. *J. Am. Chem. Soc.* **1983**, *105(17)*, 5529.

Indexes

Author Index

Allard, Dirk, 129
Angelopoulos, Marie, 86, 98
Bao, Zhenan, 1
Bauer, Barry J., 209
Blumstein, A., 294
Bouck, Kevin J., 187
Braun, F., 118
Brehmer, Martin, 129
Cabral, A., 54
Cholli, A. L., 294
Cohen, S., 161
Connor, E., 144
Conrad, Lars, 129
Curtin, J., 54
DeSimone, Joseph M., 223
Eng, L. M., 118
Fedynyshyn, T. H., 54
Fischer, Daniel A., 98
Frank, C. W., 144
Funk, Lutz, 129
Garrou, Phil E., 279
Gates, S., 161
Godschalx, James P., 187, 199
Goldfarb, Darío L., 86, 98
Gorbunoff, A., 118
Hawker, C. J., 144
Hedden, Ronald C., 209
Hedrick, J. L., 144, 161
Helfer, Anke, 129
Huang, E., 144, 161
Huang, Q. R., 144
Im, Jang-Hi, 279
Ishikawa, Seiichi, 72
Itani, Toshiro, 72

Jang, Sung-Yeon, 44
Jin, Jung-Il, 15
Jones, Ronald L., 86, 98
Kawaue, Akiya, 236
Kern, Brandon J., 187
Kim, D. W., 294
Kim, Dong Woo, 15
Kim, H. C., 144, 161
Kim, Kyungkon, 15
Kumar, J., 294
Kunz, R. R., 54
Lee, Hae-Jeong, 209
Lee, Seoung Hyun, 15
Lee, V. Y., 144
Leff, Mary, 187
Lenhart, Joseph L., 86, 98
Lin, Eric K., 86, 98
Lin, Qinghuang, xi
Liu, Da-wei, 209
Loppacher, Ch., 118
Lurio, L., 161
Lyons, John W., 187
Magbitang, T., 144
Malone, K., 161
Marks, T. J., 30
Marquez, Manuel, 44
Marshall, Joan G., 187
Martin, Steven J., 199
Mertig, M., 118
Miller, R. D., 144, 161
Morishita, Satoshi, 236
Mowers, W. A., 54
Moyer, Eric S., 173
Niu, Q. Jason, 187, 199

Ohba, Karou, 279
Okamura, Haruyuki, 236
Oldak, Robert K., 251
Paisner, Sara N., 223
Park, Jung Ho, 15
Park, Yung Woo, 15
Pearson, Raymond A., 251
Pompe, W., 118
Prabhu, Vivek M., 86
Radler, Michael J., 187
Rice, P., 161
Sahoo, S. K., 294
Sambasivan, Sharadha, 98
Sankarapandian, M., 161
Seidel, R., 118
Shida, Naomi, 72
Shirai, Masamitsu, 236
Silvis, H. Craig, 187
Simonyi, E., 161
Sinta, R. F., 54
So, Ying-Hung, 279
Soles, Christopher L., 86, 98, 209
Sotzing, Gregory A., 44
Stokich, Ted M., 187

Sundberg, L., 144
Sworin, M., 54
Syverud, Karin, 187
Théato, Patrick, 129
Toney, M., 144, 161
Toriumi, Minoru, 72
Townsend, Paul H., 199
Trogisch, S., 118
Tsunooka, Masahiro, 236
Tyberg, C., 161
van der Boom, M. E., 30
Voit, B., 118
Volksen, W., 144, 161
Watanabe, Hiroyuki, 72
Wickland, H., 161
Wong, C. P., 264
Wu, Wen-li, 86, 98, 209
Yamazaki, Tamio, 72
Yi, Whikun, 15
Zafran, R., 144
Zhang, Zhuqing, 264
Zhong, Bianxiao, 173
Zhong, Guolun, 15

Subject Index

A

Absorption coefficient
  dependence on fluorine content, 75, 78*f*
  dependence on tetrafluoroethylene (TFE) content, 75, 77*f*
  fluoropolymers, 76*f*, 77*f*
Acetic acid 2-[2-(4-vinyl-phenoxy)-ethoxy]-ethyl ester
  monomer synthesis, 131–132
  synthesis of polymers, 132–133
  See also Block copolymers
Acetone. *See* Infrared (IR) spectroscopy
Acetylacetonates (AcAc). *See* Metal acetylacetonates (AcAc)
Acid-base interactions. *See* Molecular interactions
Acrylamides, atom transfer radical polymerization (ATRP), 139–140
Actinide ions, chelates, 272
Adhesion, divinyltetramethylsiloxane bisbenzocyclobutene (DVS–bisBCB) with elastomer, 289–290
Adsorption/desorption isotherms
  hydrogen silsesquioxane (HSQ) and methyl silsesquioxane (MSQ) films, 218*f*
  interpretation, 219–221
  temperature and pressure variation, 216*f*
Alkali and alkaline earth metal ions, chelates, 269, 272
Aluminosilicates, polymer nanocomposites, 295
Amine polymers

characterization methods, 119–120
  imaging experiments, 120
  organic films containing labile protected amino groups, 123
  selective deprotection, 125–126
  selective labeling with fluoresceine isothiocyanate, 126–127
  structure, 125
  synthesis, 119
  See also Photolabile terpolymers
Amino groups, selective labeling of free, 126–127
Anionic polymerization, functionalized star polymers, 156–157
Arrays, stepwise assembly of organized, 31–32
Atomic force microscopy (AFM)
  method for surface morphology, 89–90
  patterning, 45
  patterns of poly(*p*-phenylenevinylene) (PPV), 26*f*
  surface morphology, 87
  surface roughness of conducting polymers, 50–51
Atom transfer radical polymerization (ATRP)
  ATRP of macromonomers, 204–205
  functionalized star polymers, 156–157
  macromonomer synthesis, 203
  procedure, 201–202
  surface modification via grafting, 139–140
  synthesis of functional initiators, 202
Azobenzene, zirconium phosphonate multilayers, 33

311

312

B

Back-end-of-the-line (BEOL)
  interconnects
  low k dielectrics, 162, 164
  semiconductor terminology, 145
  SiLK™ semiconductor dielectric,
    162
  wiring structure, 163*f*
  See also Porous dielectrics
Benzocyclobutene (BCB)-based
  polymers
  adhesion properties of
    divinyltetramethylsiloxane
    bisbenzocyclobutene (DVS–
    bisBCB), 289–290
  applications of DVS–bisBCB, 290–
    291
  BCB ring-opening, 280–281
  bumping/redistribution/wafer level
    chip scale packaging, 290
  chemistry of elastomer during DVS–
    bisBCB curing, 287
  coefficient of thermal expansion
    (CTE) values of DVS–bisBCB
    resin with varying elastomer, 288*t*
  coefficient of thermal expansion for
    DVS–bisBCB resin vs.
    temperature, 286*f*
  commercialization of DVS–bisBCB-
    resin-coated copper foil, 290–291
  Cu-coated DVS–bisBCB building up
    layers, 291
  curing of DVS–bisBCB polymer thin
    film, 282
  CYCLOTENE™ of The Dow
    Chemical Company, 282
  dielectric constant of DVS–bisBCB
    resin vs. temperature and
    frequency, 284*f*
  Diels–Alder in B-staging, 281
  dry-etchable DVS–bisBCB resin,
    290
  DVS–bisBCB monomer synthesis
    and B-staging, 281–282
  DVS–bisBCB properties, 282

  DVS–bisBCB resin degree of
    planarization, 284*f*
  dynamic mechanical spectroscopy
    (DMS) of DVS–bisBCB resin with
    elastomer, 287, 288*f*
  elastomer-modified DVS–bisBCB,
    286–290
  electrical properties of
    elastomer/DVS–bisBCB, 287
  infrared (IR) spectra of cured and
    uncured dry-etchable DVS–
    bisBCB resin with 20% elastomer,
    283*f*
  integration of passive RF
    components, 290
  mechanical properties of
    elastomer/DVS–bisBCB, 289
  moisture absorption of DVS–
    bisBCB, 289
  properties of thin films from
    photosensitive products, 285*t*
  strain-stress curve of DVS–bisBCB
    resin with elastomer, 289*f*
  tensile properties of DVS–bisBCB
    resin with elastomer, 288*t*
  thermal analysis of elastomer/DVS–
    bisBCB, 287, 288*f*
  thin film patterning processes, 282, 286
  transmission electron micrographs
    (TEM) of DVS–bisBCB, 286
  UV–vis spectra of dry-etchable
    DVS–bisBCB resin, 285*f*
Blending approach, ultra low k
  dielectrics, 174–176
Block copolymers
  atomic force microscopy (AFM)
    images of hydrophilic 5μm lines,
    137*f*
  characterization, 134–135
  collapse of poly(*N*-
    isopropylacrylamide) vs.
    temperature, 141*f*
  contact angle measurements, 134–
    135
  creation of closed structures, 137–
    138

esterification scheme and atom
transfer radical polymerization
(ATRP)of acrylamides, 140*f*
experimental, 131–133
functional groups for surface
modification, 130
glass transition temperatures, 134
grafting via ATRP, 139–140
hydrazinolysis, 133
hydrophilic-hydrophobic pattern
generation, 136–138
lower critical solution temperature,
140, 141*f*
micro-phase separation, 129–130
morphologies, 130
pattern creation, 136
pattern of hydrophilic and
hydrophobic interfaces, 136–137
poly(dimethylsiloxane) (PDMS)
stamp, 136–138
poly(*N*-isopropylacrylamide), 140,
141*f*
polyelectrolytes application, 142
polystyrene-*g*-PDMS (PS-*g*-PDMS)
and PS-*b*-PDMS, 225, 226*f*, 227*f*
polythiophene deposition, 140,
141*f*
reagents, 131
scanning electron microscopy (SEM)
image of 5μm titanium dioxide
lines, 138*f*
scheme of surface patterning, 136*f*
selective deposition of copper, 139
soft lithography, 130
squared deposition of
polyelectrolyte, 142*f*
surface modification, 138–142
swelling of hydrophilic segments,
135
synthesis of monomer acetic acid 2-
[2-(4-vinyl-phenoxy)-ethoxy]-
ethyl ester, 131–132
synthesis of polymers, 132–133
synthesis route to, 133*f*
time dependence of swelling, 135*f*
titanium dioxide deposition, 138

See also Nanoporous and
microporous materials
Brushes. *See* Molecular brushes
Burning out sacrificial spacers
(BOSS) resins
approach, 176–177
compatibility with metal barrier, 182
cure temperature, 180
dielectric constants and modulus by
nanoindentation, 182*f*
dielectric constants of cured films
from, 179*t*
dielectric constants of films from,
180*f*
evaluation of BOSS powder samples,
177, 179
evaluation of thin films, 179–180
N_2 adsorption isotherms for cured,
178*f*
pore characteristics, 181
study of proto-type resin, 181–182
synthesis, 177
thin film properties, 181*t*

C

Capillary condensation
critical radius, 211–212
physics, 221
Carbon
field emission properties of
nanotubes, 26–27
nanopatterning, 27
See also Poly(*p*-phenylenevinylene)
(PPV)
Carbonization, poly(*p*-
phenylenevinylene) (PPV) for
graphitic structure, 25
Casting, soluble conjugated polymers,
8
Catalysts. *See* Metal acetylacetonates
(AcAc)
Chemical oxidants, oxidative solid-
state crosslinking, 47, 49*f*
Chips. *See* Semiconductor industry

Chromophores
  acentric building block with
    phosphonic acid and phosphonate
    ester, 33f
  assembly process of siloxane-based
    multilayers, 36–37
  examples of high-β building blocks,
    38f
  integration into polar films, 41
  NLO (nonlinear optical)-active, in
    poled polymers, 30–31
  NLO responses, 39
  push-pull building blocks, 38
  twisted, with outstanding NLO
    responses, 39f
  zirconium and hafnium phosphonate
    multilayers, 32–33
Clay. See Polymer-clay
    nanocomposites
Coefficient of thermal expansion
  divinyltetramethylsiloxane
    bisbenzocyclobutene (DVS–
    bisBCB) with elastomer,
    288t
  DVS–bisBCB, 286f
Composites. See Polymer-clay
    nanocomposites
Conducting polymers
  electrode materials, 3–4
  See also Intrinsically conducting
    polymers (ICP)
Conjugated polymers, techniques for
    soluble, 8
Contact angle, block copolymers,
    134–135
Copolymers. See Block copolymers;
    Thermally degradable
    photocrosslinking polymers
Copper, selective deposition for
    surface modification, 139
Critical radius, capillary condensation,
    211–212
Crosslinked polymers
  removal from substrates, 236–237
  See also Intrinsically conducting
    polymers (ICP); Thermally

degradable photocrosslinking
    polymers
Crystal field stabilization energy
    (CFSE), metal complexes, 269
Curing chemistry
  divinyltetramethylsiloxane
    bisbenzocyclobutene (DVS–
    bisBCB) thin film, 282
  polyarylene, 191
  See also Benzocyclobutene (BCB)-
    based polymers
Cycloaliphatic epoxy. See Infrared
    (IR) spectroscopy
Cyclohexane oxide. See Infrared (IR)
    spectroscopy
Cyclotene™ polymer, self-assembled
    electro-optical devices, 39–40
Cytop™ polymer, self-assembled
    electro-optical devices, 39–40

D

Degradation. See Thermally
    degradable photocrosslinking
    polymers
Deposition. See Surface modification
Deprotection, amino polymer film,
    125–126
Depth profiling. See Surface depth
    profiling
Design, photolabile polymers, 121
Detector bias, near edge X-ray
    absorption fine structure (NEXAFS)
    spectra, 113, 114f, 115f, 116
Diazosulfonate terpolymer
  characterization methods, 119–120
  decomposition with UV irradiation,
    121–122
  formation of spatially defined gold
    metal layers, 123, 124f
  imaging experiments, 120
  structure, 122
  synthesis, 119
  UV imaging, 123f
  See also Photolabile terpolymers

Dicyclohexyl ester. *See* Infrared (IR) spectroscopy
Dicyclohexyl ketone. *See* Infrared (IR) spectroscopy
Dielectric materials
electronic devices, 9, 11
See also Low k dielectrics; Porous dielectrics
Diels–Alder reaction
divinyltetramethylsiloxane bisbenzocyclobutene (DVS–bisBCB), 281
polyarylene curing chemistry, 191
Divinyltetramethylsiloxane bisbenzocyclobutene (DVS–bisBCB). *See* Benzocyclobutene (BCB)-based polymers
Dynamic mechanical spectroscopy (DMS), divinyltetramethylsiloxane bisbenzocyclobutene (DVS–bisBCB) with elastomer, 287, 288*f*

E

Elastomer modification
divinyltetramethylsiloxane bisbenzocyclobutene (DVS–bisBCB), 286–290
See also Benzocyclobutene (BCB)-based polymers
Electrical conductivity, graphitic films, 25–26
Electrically induced pattern transfer (ECIPT)
patterning, 45, 49, 50*f*
See also Intrinsically conducting polymers (ICP)
Electrical properties, divinyltetramethylsiloxane bisbenzocyclobutene (DVS–bisBCB) with elastomer, 287
Electrochemical Dip-Pen Nanolithography (e-DPN), patterning, 45

Electrochemically modulated surface plasmon resonance (EM–SPR), zirconium phosphonate-azobenzene multilayers, 33
Electronics
size of devices, 145
See also Organic electronics
Electron yield. *See* Near edge X-ray absorption fine structure (NEXAFS)
Electro-optical applications
acentric chromophore building block, 33*f*
all-organic self-assembled electro-optic devices, 39–40
bulky nonlinear optical (NLO)-active chromophores in poled-polymers, 30–31
capping step, 35
channel-type ridged waveguide patterns, 40
covalent siloxane-based self-assembly scheme, 35*f*
criteria, 30
electrically modulated surface plasmon resonance (EM–SPR), 33
examples of acentric siloxane-based monolayers, 36*f*
examples of high-β chromophore building blocks for layer-by-layer formation of siloxane-based superlattices, 38*f*
film stability, 34–35
future outlook, 41
incorporation of refractive index modifying metal oxide layers, 37*f*
layer-by-layer assembly, 31–32
layer-by-layer formation of siloxane-based multilayer, 34*f*
moisture and stability, 34
NLO responses, 39
Pockel effects, 30
push-pull chromophore building blocks, 38
push-pull modulators, 30–31
replacing C–H with C–F and/or C–D groups, 41

316

SAS (self-assembled superlattices),
35
scanning electron microscopy (SEM)
image showing hybrid
Cyclotene™/SAS waveguides, 40*f*
schematic of polar zirconium
phosphate/phosphonate multilayer
film, 32*f*
self-assembled electro-optic
modulator design, 40*f*
self-assembled siloxane-based
multilayers, 34–39
stepwise assembly of organized
arrays, 31–32
streamlined assembly method, 36–37
twisted chromophores with
outstanding NLO responses, 39*f*
two-step layer-by-layer SAS self-
assembly sequence, 37*f*
zirconium and hafnium phosphonate
multilayer assemblies, 32–33
Ellipsometric porosimetry (EP)
pores of burning out sacrificial
spacers (BOSS) resins, 181
porous dielectrics, 169
Epoxy cure reactions
bisphenol A type epoxy resins
(EPON 828), 266
chemical structures of epoxy resins
and hardener, 267*f*
cure agent hexahydro-4-
methylphthalic anhydride
(HMPA), 266, 267*f*
cure kinetics study, 272–273, 275–
276
cure properties of ERL
4221/HMPA/metal acetylacetonate
(AcAc) system, 270*t*, 271*t*
cycloaliphatic epoxy resins (ERL
4221), 266
dynamic cure profile, 268
instruments, 268
isothermal kinetic parameters of
EPON 828/HMPA/metal AcAc,
273*t*

kinetic parameters of EPON
828/HMPA/metal AcAc, 274*f*
metal AcAc as catalysts, 265–266
See also Metal acetylacetonates
(AcAc)
Etching
dependence of dry-etching rate of
fluoropolymers on fluorine
content, 81*f*
fluoropolymer rates, 79–80
See also Plasma etching
Ethyl acetate. *See* Infrared (IR)
spectroscopy
2-Ethynylpyridine (2EPy)
adsorption yield, area of aromatic
resonance, and d-spacing of
nanocomposite, 300*t*
in situ intercalative polymerization,
295–296, 298
solid state ^{13}C nuclear magnetic
resonance (NMR) spectra of
poly(2EPy), 302*f*
solution-state ^{13}C NMR spectrum,
299*f*
structure, 300*f*
synthesis of poly(2EPy)/clay
composites, 296–297
tentative assignments of resonances,
298, 300
See also Polymer-clay
nanocomposites
Exfoliated nanocomposite, schematic,
296*f*

F

Field-effect transistor
device operation, 2–3
schematic structure, 2*f*
Field emission, carbon nanotubes, 26–
27
Film thickness, surface plasmon
resonance (SPR), 135
Flip-chip, advantages, 265

Fluoresceine isothiocyanate (FITC), selective labeling of free amino groups, 126–127

Fluorescence yield. *See* Near edge X-ray absorption fine structure (NEXAFS)

Fluorine content, transparency, 73

Fluorocarbon polymer-based photoresists. *See* Photolithography

Fluoropolymers
  157-nm lithography, 72
  absorption coefficient histogram of, and non-fluoropolymer, 76*f*
  absorption coefficient histogram of backbone and other fluorination, 77*f*
  characterization of fundamental properties, 75, 79–80
  dependence of absorption coefficient on fluorine content, 75, 78*f*
  dependence of absorption coefficient on tetrafluoroethylene (TFE) content, 75, 77*f*
  dependence of dry-etching rate on fluorine atom content, 79–80, 81*f*
  dissolution behavior, 80, 81*f*
  experimental, 73–75
  imaging performances, 82, 83*f*, 84*f*
  lithography performances, 80, 83*f*
  materials, 73–74
  measurements, 74–75
  molecular structures, 76
  QCM (quartz crystal microbalance) method, 74
  QCM traces, 80, 81*f*
  vacuum ultraviolet (VUV) absorption, 74
  VUV spectrum of perfluoropolymer Cytop©, 78*f*, 79

Fluorosilicate glass (FSG), replaying silicon dioxide, 162

Functional groups, surface modification, 130

Functionalized star polymers, combining anionic and atom transfer radical polymerization, 156–157

Functionalized star porogens. *See* Polystyrene star porogens

G

Gallery space
  nanoparticulates, 295
  schematic of clay, 296*f*

Gel permeation chromatography (GPC)
  methyl silsesquioxane (MSSQ) homo- and copolymers, 149, 150*f*
  molecular weight determination of star porogens, 190

Glass transition temperatures, block copolymers, 134

Glassy polymers, self-assembled electro-optical devices, 39 40

Gold, electrode material for organic electronics, 3

Grafting
  atom transfer radical polymerization (ATRP), 139–140
  polyarylene to star porogens, 194–195

Graphitic films
  electrical properties, 25–26
  formation by carbonization of poly(*p*-phenylenevinylene) (PPV), 25

Graphitization, degree, 23–24

H

Hafnium phosphonate, multilayer assemblies, 32–33

4-Hexafluoroisopropanol styrene (HFIP)
  157-nm lithography, 55
  effect of group on absorbance, 59, 60*f*

polymer properties, 58*t*
synthesis, 55–57
thermal stability of copolymers, 61, 65*f*
See also Photolithography
Hydrazinolysis, block copolymers, 133
Hydrogen silsesquioxane (HSQ)
adsorption/desorption isotherms, 218*f*
blending for ultra low k dielectrics, 174–176
nitrogen adsorption isotherms for cured, 175*f,* 176*f*
toluene and porosity, 217
X-ray reflectivity curves as function of toluene partial pressure, 214*f*

I

Imaging. *See* Photolabile terpolymers
Imaging performance, fluoropolymers, 82, 83*f,* 84*f*
Infrared (IR) spectroscopy
acid-base constants for C=O peak shifts, 258, 260
acid-base constants from OH peak shifts, 260–261
carbonyl peak shift indicating interactions, 256–257
concentration dependence of carbonyl peak, 260, 261*f*
determination of Drago constants, 256
Drago constants for model molecules, 259*t*
Drago constants for model molecules from hydroxyl peak shift, 262*t*
Drago constants for model molecules from literature, 260*t*
estimating heats of complexation, 252–253
Fourier transform IR (FT–IR) method, 255

hydroxyl peak shift, 257–258
intermolecular interactions, 251–252
peak shifts, 252–253
See also Molecular interactions
Interactions, molecular. *See* Molecular interactions
Intercalated nanocomposite, schematic, 296*f*
Interconnects. *See* Back-end-of-the-line (BEOL) interconnects
Interfacial issues. *See* Near edge X-ray absorption fine structure (NEXAFS)
Intermetal dielectric (IMD), 162
Intrinsically conducting polymers (ICP)
applications, 45
atomic force microscopy (AFM), 45
chemical oxidants for solid-state crosslinking, 47
conducting pattern of polymer by electrochemically induced pattern transfer, 50*f*
electrochemical oxidative crosslinking, 48*f*
oxidative electrochemistry for solid-state crosslinking, 46–47
oxidative solid-state crosslinking, 48*f*
patterning, 45
patterning via solid-state oxidative electrochemistry, 49
precursor polymer preparation, 46
ring opening metathesis polymerization, 46
strategy for preparation, 45
surface roughness, 50–51
tapping mode AFM images, 51*f*
UV–vis before and after chemical oxidative crosslinking, 49*f*

K

Kinetics, epoxy cure with metal acetylacetonates, 272–273, 275–276

L

Labeling, selective, of free amino
    groups, 126–127
Lanthanide ions, chelates, 272
Laser scanning microscope,
    diazosulfonate terpolymer film after
    UV irradiation, 124*f*
Layer-by-layer assembly
  fine tuning properties, 31–32
  See also Electro-optical applications
Ligand effect. *See* Metal
    acetylacetonates (AcAc)
Line-edge roughness (LER), influence
    of photogenerated acid diffusion, 87
Lithography
  imaging performances of
    fluoropolymers, 82, 83*f*, 84*f*
  patterning, 45
  performance of fluoropolymers, 80,
    83*f*
  transparency requirement for
    materials, 73
  See also Fluoropolymers; Near edge
    X-ray absorption fine structure
    (NEXAFS); Photolabile
    terpolymers; Photolithography
Lower critical solution temperature
    (LCST), poly(*N*-
    isopropylacrylamide), 140, 141*f*
Low k dielectrics
  comparing films of different porosity
    level, 216–219
  comparison, 162, 164
  comparison to silicon dioxide, 163*t*
  extendibility with porous dielectrics,
    164–165
  microelectronics industry, 210, 224
  supercritical CO_2 as development
    solvent, 225
  X-ray reflectivity curves as function
    of toluene partial pressure, 214*f*
  See also Dielectric materials;
    Polyarylene polymers; Porous
    dielectrics; Ultra low k (ULK)
    dielectrics

M

Mechanical properties,
    divinyltetramethylsiloxane
    bisbenzocyclobutene (DVS–
    bisBCB) with elastomer, 289
Mechanism, nucleation and growth,
    147, 154
Metal acetylacetonates (AcAc)
  activation energy, 275–276
  autocatalytic kinetic model, 272–
    273
  bisphenol A type epoxy resins
    (EPON 828), 266
  catalytic activity trend, 273, 275
  chelates with alkali and alkaline earth
    metal ions, 269, 272
  chelates with lanthanide and actinide
    ions, 272
  chelates with transition metal ions,
    272
  chemical structure of Co(II)AcAc
    and Co(II)HFAcAc, 276*f*
  chemical structures of epoxy resins
    and hardener in study, 267*f*
  comparison of kinetic parameters of
    Co(II)AcAc and Co(II)HFAcAc,
    277*t*
  crystal field stabilization energy
    (CFSE), 269
  cure agent hexahydro-4-
    methylphthalic anhydride
    (HMPA), 266, 267*f*
  cure kinetics study, 272–273, 275–
    276
  cure properties of ERL
    4221/HMPA/metal AcAc system,
    270*t*, 271*t*
  cycloaliphatic epoxy resins (ERL
    4221), 266
  dissociation energies, 275
  experimental, 266–268
  instruments, 268
  isothermal kinetic parameters of
    EPON 828/HMPA/metal AcAc,
    273*t*

kinetic parameters of cure reactions of EPON 828/HMPA/metal AcAc, 274*f*

ligand effect, 276–277

materials, 266

onset temperature of decomposition of, and catalytic behavior of ERL 4221/HMPA system, 272*t*

reaction constant, 273

screen test of catalytic behavior of, 268–269, 272

solubility and catalytic latency, 269

thermal stability, 268

Metallophthalocyanines, transistor material, 6

Methacrylate monomers. *See* Thermally degradable photocrosslinking polymers

1-Methyl-1-(6-methyl-7-oxabicyclo[4.1.0]hept-3-yl)ethyl methacrylate (MOBH)

polymer characteristics, 240*t*

preparation of polymers, 239–240

structures, 241*f*

synthesis, 238–239

thermal degradation of homopolymer PMOBH, 243–246

thermal degradation of MOBH copolymers, 246–249

See also Thermally degradable photocrosslinking polymers

Methyl silsesquioxane (MSSQ)

adsorption/desorption isotherms, 218*f*

FT–IR comparison of MSSQ homo- and copolymers, 149*f*

gel permeation chromatography (GPC), 149, 150*f*

low molecular weight versions, 149

refractive index vs. porogen plots in MSSQ–SiO$_2$ copolymer, 157, 158*f*

relationship between refractive index and dielectric constant, 152*f*

scanning electron microscope (SEM), 167–168

SiOH end content, 149

small angle X-ray scattering (SAXS), 153, 154*f*

thermal and mechanical properties, 150, 151*f*

toluene and porosity, 217

transmission electron microscope (TEM) and field emission SEM (FE–SEM) micrographs, 152, 153*f*

See also Organosilicate materials

Metrology characterizing pores. *See* X-ray reflectivity

Microelectronics industry

advanced microprocessors, 188

demand for organic dielectrics, 200

low k dielectrics and, 210, 224

See also Polyarylene polymers; SiLK™ semiconductor dielectric

Microelectronic systems

evolution and requirements, 279–280

See also Benzocyclobutene (BCB)-based polymers

Micro-phase separation, block copolymers, 129–130

Microporous materials. *See* Nanoporous and microporous materials

Mobility, p-channel semiconductors, 6, 8

Models

experiments generating model line edge region, 110, 111*f*

film geometries studying photoresist roughness, 95*f*

Moisture absorption, divinyltetramethylsiloxane bisbenzocyclobutene (DVS–bisBCB) with elastomer, 289

Molecular brushes

atom transfer radical polymerization (ATRP) procedure, 201–202

ATRP of macromonomers, 204–205

characterization of macromonomers, 203*t*

characterization of polystyrene brushes, 205*t*

description, 200

evaluation of, as template, 205–207
experimental, 200–202
formation of crosslinked
   polyphenylene matrix, 206
synthesis of ATRP initiator with
   phenylethynyl functional group,
   200–201
synthesis of functional initiators, 202
synthesis of macromonomers, 203
transmission electron microscopy of
   porous films, 207
Molecular interactions
acid-base constants from C=O peak
   shifts, 258, 260
acid-base constants from OH peak
   shifts, 260–261
carbonyl peak shift in IR spectra,
   256–257
chemical structures of model
   molecules, 254f
concentration dependence of
   carbonyl peak, 260, 261f
cycloaliphatic epoxy resin, 253–254
determination of Drago constants,
   256
Drago constants determined in study,
   262t
Drago constants for model molecules
   from carbonyl peak shift, 259t
Drago constants for model molecules
   from hydroxyl peak shift, 262t
Drago constants for model molecules
   from literature, 260t
Drago's acid-base theory, 252
estimating heats of complexation,
   252–253
experimental, 253–256
Fourier transform–infrared (FT–IR)
   spectroscopy method, 255
future work, 261
heats of acid-base complex formation
   ΔH_{AB}, 252
hydroxyl peak shift due to interaction
   with cycloaliphatic epoxy, 258f
hydroxyl peak shift in IR spectra,
   257–258
hydroxyl shift results for model
   molecules, 259t
infrared (IR) peak shifts, 252–253
materials, 253–254
objective of IR study, 253
preparation of dilute solutions, 255
relation between shift and heat of
   complexation, 253
results for model molecules in dilute
   cyclohexane with Drago constants
   for acids, 257t
spectroscopic methods, 251–252
See also Infrared (IR) spectroscopy
Montmorillonite
nanocomposite by intercalation of
   poly(2-ethynylpyridine) (P2EPy)
   with, 301, 303–304
schematic, 296f
See also Polymer-clay
   nanocomposites
Morphology
block copolymers, 130
comparison of methyl
   silsesquioxane–SiO_2 films by field
   emission scanning electron
   microscope (FE–SEM), 157, 158f
See also Polyarylene polymers
Multilayer assemblies
capping step, 35
covalent siloxane-based self-
   assembly scheme, 35f
examples of acentric siloxane-based
   monolayers, 36f
film stability, 34–35
fine tuning properties, 31–32
incorporation of refractive index
   modifying metal oxide layers, 37f
layer-by-layer formation of siloxane-
   based multilayer, 34f
moisture effects, 34–35
nonlinear optical (NLO) responses,
   39
push-pull chromophore building
   blocks, 38
self-assembled siloxane-based, 34–
   39

streamlined assembly method, 36–37
two-step layer-by-layer sequence, 37*f*
two-step method and automation, 37
zirconium and hafnium phosphonate, 32–33

N

Nanocomposites. *See* Polymer-clay nanocomposites
Nanofilms. *See* Poly(*p*-phenylenevinylene) (PPV)
Nanoindentation, dielectric constants and modulus by, 182*f*
Nanopatterning, poly(*p*-phenylenevinylene) (PPV) and carbon, 27
Nanoporosity
  formation of controlled, 155–156
  formation using blending and sacrificial porogens, 148*f*
  See also Organosilicate materials
Nanoporous and microporous materials
  block copolymers, 225, 226*f*
  carbon dioxide (CO_2) foaming, 228, 231
  cell densities of porous poly(ether sulfone) (PES) morphologies vs. CO_2 concentration, 228, 230*f*
  copolymers polystyrene-*g*-poly(dimethylsiloxane) (PS-*g*-PDMS) and PS-*b*-PDMS, 225, 226*f*
  foaming behavior of poly(ether imide) (PEI) and PES films, 228, 231*t*
  foaming in CO_2 of hydrogen bonded materials, 231
  formation via sc CO_2 extraction, 224–225, 228
  mass density of PEI vs. dissolved CO_2 and foaming temperature, 228, 229*f*

potential benefits of supercritical carbon dioxide, 224
scanning electron microscopy (SEM) of PS after PDMS removal, 225, 227*f*
SEM image of foam from triurea compound in CO_2, 232*f*
supercritical CO_2 as development solvent for low k material formation, 225
supercritical CO_2 for oligomer extraction from polymer matrix, 225, 228
synthesis of urea compounds, 232*f*
transmission electron microscopy (TEM) of PDMS spheres in PS matrix, 225, 227*f*
Nanorods. *See* Poly(*p*-phenylenevinylene) (PPV)
Nanotechnology
  interest, 15–16
  See also Poly(*p*-phenylenevinylene) (PPV)
Nanotubes
  field emission properties of carbon, 26–27
  See also Poly(*p*-phenylenevinylene) (PPV)
Naphthalenetetracarboxylic diimide (NTCDI),n-channel semiconductors, 9
n-channel materials
  naphthalenetetracarboxylic diimide (NTCDI), 9
  organic semiconductors, 9, 10*t*
Near edge X-ray absorption fine structure (NEXAFS)
  advantages and disadvantages, 102
  carbon edge NEXAFS spectra for pure resist components, 103, 104*f*
  carbon K-edge electron yield NEXAFS spectrum for poly(*t*-butyloxy-carbonyloxy styrene) (PBOCSt), 101*f*, 102

carbon K-edge electron yield spectra as function of detector bias, 113, 114*f*, 115*f*
carbon K-edge fluorescence and electron yields for PBOCSt/PFOS (bis(*p-t*-butylphenyl)iodonium perfluorooctanesulfonate) film, 107, 109*f*
detector bias, 113, 116
differences in surface and bulk reaction rates, 107
dynamic secondary ion mass spectrometry (DSIMS), 102–103
electron and fluorescence yield for PBOCSt/PFOS films, 103, 105*f*, 106
ellipsometry, 103
experimental, 99–100
materials and methods for protected polymer PBOCSt, 99–100
measurement method, 100
mechanisms leading to incomplete surface deprotection, 107–108
post exposure delay (PED), 107, 108
pre- and post-edge jump normalized spectra, 110, 112*f*
principles, 100, 102
resist-air interface, 100–108
Rutherford back scattering (RBS), 102–103
schematic of experiments generating model line edge region, 110, 111*f*
schematic of principles, 101*f*
schematic of surface depth profiling, 109*f*
sensitivity of model photoresist to soft X-ray radiation, 106–107
surface depth profiling with, 108, 110, 113, 116
typical spectra, 101*f*
X-ray photoelectron spectroscopy (XPS), 102
Neutron reflectivity
experimental, 89
post-exposure bake (PEB) time, 91*f*
reaction front profile, 91*f*
See also Poly(butoxycarbonyl styrene) (PBOCSt)
Nitrogen adsorption isotherms. *See* Burning out sacrificial spacers (BOSS) resins; Ultra low k (ULK) dielectrics
No-flow underfill
process, 265
See also Metal acetylacetonates (AcAc)
Nonlinear optical (NLO)
active chromophores in poled-polymers, 30–31
responses of siloxane-based multilayers, 39
See also Electro-optical applications
Nucleation and growth (NG)
nanoporosity mechanism, 147
small angle X-ray scattering (SAXS), 154

O

4-Octylstyrene. *See* Block copolymers
Organic electronics
advantages of semiconducting materials, 1–2
dielectric materials, 9, 11
electrode materials, 3–4
field-effect transistor (FET) operation, 2–3
future outlook, 11–12
mobilities, 6, 8
n-channel organic semiconductors, 9, 10*t*
semiconducting materials, 4–9
soluble p-channel materials, 6, 8
transistor performance of n-channel organic semiconductors, 10*t*
transistor performance of organic semiconductors from solution, 7*t*
transistor performance of p-channel semiconductors by vacuum deposition, 4*t*, 5*t*

vacuum deposited p-channel
materials, 4, 6
See also Electronics
Organic polymers, self-assembled
electro-optical devices, 39–40
Organometallic compounds
latent catalysts, 265
See also Metal acetylacetonates
(AcAc)
Organosilicate materials
combining anionic and atom transfer
polymerization, 156–157
comparing transmission electron
microscope (TEM) and field
emission scanning electron
microscope (FE–SEM)
micrographs of porous MSSQ,
152, 153*f*
comparison of porogen foaming
efficiencies, 151*f*
controlled nanoporosity, 155–156
dielectric constants, 159
film morphologies by FE–SEM, 157,
158*f*
FT–IR comparison of various MSSQ
homo- and copolymers, 149*f*
functionalized star polymers, 156–
157
gel permeation chromatography
(GPC) of resins, 149, 150*f*
homopolymers and content of SiOH
ends, 149
low molecular weight methyl
silsesquioxane (MSSQ), 149
low molecular weight SSQ
derivatives, 149
nucleation and growth mechanism,
154
particle crosslinking, 155
pore size and structure of resin, 154
porogen templating vitrification of
matrix, 154
refractive index vs. porogen plots in
MSSQ–SiO$_2$ copolymer, 158*f*
relationship between refractive index
and dielectric constant for porous

MSSQ using poly(alkylene oxide)
porogen, 152*f*
representative silsesquioxane
structure, 148*f*
small angle X-ray scattering (SAXS),
153, 154*f*
TEM micrographs of porous MSSQ–
SiO$_2$ copolymer, 158*f*
thermal and mechanical properties of
blends, 150, 151*f*
thermosetting matrix resin, 148
variety of porogens, 150, 152
See also Semiconductor industry
Oxidative electrochemistry
oxidative solid-state crosslinking,
46–47
patterning, 49, 50*f*
Oxidative solid-state crosslinking.
See Intrinsically conducting
polymers (ICP)

P

Patterning
creation of closed structures, 137–
138
divinyltetramethylsiloxane
bisbenzocyclobutene (DVS–
bisBCB) thin film processes, 282,
286
hydrophilic and hydrophobic
interfaces, 136–137
hydrophilic-hydrophobic generation,
136–138
intrinsically conducting polymers
(ICP), 45
poly(dimethylsiloxane) (PDMS)
stamp, 136–138
scheme of surface patterning, 136*f*
solid-state oxidative
electrochemistry, 49, 50*f*
techniques, 45, 119
See also Block copolymers;
Photolabile terpolymers
p-channel materials

soluble, 6, 8
transistor performance, 4*t*, 5*t*, 7*t*
vacuum deposition, 4, 6
Performance, semiconductor chips, 146–147
Photoacid generator (PAG)
bis(*p-t*-butylphenyl) iodonium perfluorooctanesulfonate (PFOS), 99
electron yield near edge X-ray absorption fine structure (NEXAFS) spectra of PFOS, 103, 104*f*
term, 87
Photocrosslinking polymers
removal from substrates, 236–237
See also Thermally degradable photocrosslinking polymers
Photogenerated acid diffusion, side-wall or line-edge roughness (LER), 87
Photolabile terpolymers
amine polymers, 123, 125–127
approaches to thin functional films, 121
area selective UV imaging of protected amine terpolymer film, 126*f*
characterization methods, 119–120
confocal laser scanning fluorescence microscopy image of selectively deprotected amino polymer film, 126*f*
deprotection of protected amino polymer film, 125–126
diazosulfonate polymers, 121–123
experimental, 119–120
fluoresceine isothiocyanate, 126–127
formation of spatially defined gold metal layers, 123, 124*f*
growth of continuous metallic films, 123, 124
laser scanning microscopy of diazosulfonate terpolymer film

after ultraviolet (UV) irradiation, 124*f*
materials, 119
patterning technologies, 119
polymer design, 121
selective labeling of free amino groups, 126–127
stamps, 119
structure of diazosulfonate polymer, 122
structure of polymers with amino groups, 125
thin polymer films by spin-coating, 120
UV decomposition and imaging experiments, 120
UV imaging of diazosulfonate terpolymer film, 123*f*
Photolithography
1-[1-(*t*-butoxymethoxy)-2,2,2-trifluoro-1-trifluoro-methylethyl]-4-vinyl benzene (4-BOM) synthesis, 56–57
t-butyl[2,2,2-trifluoro-1-trifluoromethyl-1-(4-vinyl-phenyl)ethoxy]- acetate (4-*t*-BuAc) synthesis, 56
characterization methods, 57
decomposition temperature vs. blocking level, 62*f*
derivatives of poly(hydroxystyrene) (PHS), 54
etch conditions, 57–58
experimental, 55–58
4-hexafluoroisopropanol styrene (HFIP), 55
introduction of HFIP group, 59
lithography procedure, 57
lithography results, 61, 63–64
performance of fluorocarbon and hydrocarbon polymers during oxide plasma etching, 68*f*
performance of fluorocarbon and hydrocarbon polymers during poly plasma etching, 68*f*

photoresist formulation preparation, 55

plasma etch rates and selectivities, 67*t*

plasma etch studies, 64, 67–69

polymer absorbance at 157 nm, 60*f*

polymer properties, 58–61

polymer structures, 60*f*

polymer synthesis, 57

raw plasma etch rate, 67

relationship between glass transition temperature and blocking level, 62*f*

resolution capability of HFIP/acetal resists, 66*f*

resolution capability of HFIP/acrylate resists, 66*f*

resolution capability of hydrocarbon resists, 64, 65*f*

shorter wavelengths, 54

sizing dose with 157-nm exposure, 63*t*

summary of 157-nm exposure using phase shift mask, 63*t*

synthesis procedures, 55–57

thermal stability of partially blocked polymers, 65*f*

time to strip fluoroaromatic polymers and hydrocarbon resists, 67, 69

1-(2,2,2-trifluoro-1-methoxymethoxy-1-trifluoromethyl-ethyl)-4-vinyl benzene (4-MOM) synthesis, 56

use of 157-nm radiation, 55

See also Near edge X-ray absorption fine structure (NEXAFS)

Photoluminescence (PL)

films from poly(*p*-phenylenevinylene) (PPV) of varying thickness, 21

PPV films, nanotubes, and nanorods, 19–20

Photoresists

formulation preparation, 55

lithographic methods, 119

model film geometries studying roughness, 95*f*

See also Photolithography

Plasma enhanced chemical vapor deposition (PECVD), supercritical carbon dioxide, 225, 228

Plasma etching

hexafluoroisopropanol styrene (HFIP) copolymers, 64, 67–69

performance of fluorocarbon and hydrocarbon polymers, 68*f*

rates and selectivities, 67*t*

raw plasma etch rate, 67

See also Photolithography

Pockel effects, electro-optic, 30

Polyaniline, electrode materials, 3–4

Polyarylene polymers

effect of hot plate bake sequence on final pore morphology, 196*f*

effect of reaction time between matrix and functionalized porogen on pore morphology, 195*f*

effects of star porogen DPA functionality on pore morphology, 194*f*

experimental, 188–190

extent of grafting to star porogen, 194–195

film characterization, 196–197

interconnect architecture, 195–196

matrix during chemistry, 191

molecular weight data on star porogens, 193*t*

physical properties of SiLK™ polyarylene dielectric, 192*t*

pore morphology, 193–196

porogen molecular weight determination, 190

porogen synthesis, 192–193

preparation of nanoporous films, 188–189

reagents, 189

schematic of curing chemistry via Diels–Alder cycloaddition reaction, 191*f*

SiLK™, 188

synthesis of functionalized star
polystyrene polymer, 189–190
transmission electron microscopy
(TEM), 193–194
See also SiLK™ semiconductor
dielectric
Poly(butoxycarbonyl styrene)
(PBOCSt)
AFM (atomic force microscopy), 89–
90
AFM tapping mode vs. scan size, 89*f*
developed surface morphology, 92–
93
electron yield near edge X-ray
absorption fine structure
(NEXAFS) spectra, 103, 104*f*
Fourier transform of topographic
data, 93*f*
neutron reflectivity, 89
neutron reflectivity for varying post-
exposure bake (PEB) times, 91*f*
neutron reflectivity of reaction front
profile, 91*f*
PEB time, 87
reaction front profile, 91–92
reaction-front width and root-mean
squared (RMS) roughness, 94–95
schematic of deprotection reaction,
88*f*
tapping-mode AFM images, 92*f*
See also Near edge X-ray absorption
fine structure (NEXAFS);
Roughness
Poly(butoxycarbonyl styrene)
(PBOCSt)
AFM (atomic force microscopy), 89–
90
AFM tapping mode vs. scan size, 89*f*
developed surface morphology, 92–
93
electron yield near edge X-ray
absorption fine structure
(NEXAFS) spectra, 103, 104*f*
Fourier transform of topographic
data, 93*f*
neutron reflectivity, 89

neutron reflectivity for varying post-
exposure bake (PEB) times, 91*f*
neutron reflectivity of reaction front
profile, 91*f*
PEB time, 87
reaction front profile, 91–92
reaction-front width and root-mean
squared (RMS) roughness, 94–95
schematic of deprotection reaction,
88*f*
tapping-mode AFM images, 92*f*
See also Near edge X-ray absorption
fine structure (NEXAFS);
Roughness
Poly(dimethylsiloxane) (PDMS)
block copolymers forming
nanoporous materials, 225, 226*f*
transmission electron microscopy
(TEM) of, spheres in polystyrene
matrix, 225, 227*f*
Polyelectrolyte deposition, surface
modification, 142
Poly(ether imide) (PEI), foaming
behavior in CO_2, 228, 231*t*
Poly(ether sulfone) (PES), foaming
behavior in CO_2, 228, 231*t*
Poly(3,4-ethylene dioxythiophene)
(PEDOT), electrode materials, 3–4
Poly(4-hydroxystyrene) (PHS)
electron yield near edge X-ray
absorption fine structure
(NEXAFS) spectra, 103, 104*f*
neutron reflectivity, 89
preparation, 88
schematic of deprotection reaction,
88*f*
thermal stability, 61
See also Near edge X-ray absorption
fine structure (NEXAFS)
Polymer-clay nanocomposites
adsorption yield, area of aromatic
resonance, and d-spacing of 2-
ethynylpyridine (2EPy)
polymerization, 300*t*
basal d-spacing of in-situ
polymerized nanocomposites, 297*f*

chemical structure of 2EPy and
P2EPy, 300*f*
2EPy polymerization, in situ
intercalative, 295–296
experimental, 296–297
formation of lamellar, 295
intercalation of bulk polymer P2EPy
with montmorillonite, 301, 303–
304
materials, 296
nature of polymer environments,
300–301
nuclear magnetic resonance (NMR)
spectroscopy, 297
resurgence of research, 295
schematic of clay, intercalated, and
exfoliated, 296*f*
solid-state ^{13}C CP/MAS NMR
spectrum of nanocomposite, 298,
299*f*
solid state NMR spectra of
protonated and non-protonated
P2EPy, 301, 302*f*
solution-state ^{13}C NMR spectrum for
2EPy, 299*f*
synthesis of P2EPy/clay
nanocomposites, 296–297
tentative assignments of resonances,
298, 300
Poly(*N*-isopropylacrylamide), lower
critical solution temperature
(LCST), 140, 141*f*
Poly(4-octylstyrene)
contact angles, 135*t*
glass transition temperatures, 134*t*
See also Block copolymers
Poly(*p*-phenylenevinylene) (PPV)
atomic force microscopy (AFM)
images of patterned PPV, 26*f*
carbonization temperature
dependence of Raman spectra of
carbonized PPV films, 23*f*
comparison of electrical
conductivities of graphitic carbon
nanofilms, 25*f*

cross sectional image, transmission
electron micrograph (TEM) of
PPV, 21*f*
cross sectional TEM image of carbon
film, 24*f*
degree of graphitization of nano
products, 23–24
electrical conductivities of graphitic
films, 25–26
field emission properties of carbon
nanotubes, 26–27
field emission property of carbonized
PPV nanotubes, 26*f*
formation of graphitic structure by
carbonization of PPV, 25
Gilch–Wheelwright polymerization,
17
infusibility, 16
insolubility, 16
methods for preparation, 16–18
nanopatterning, 27
nanotechnology, 15–16
photoluminescence (PL), 19–21
PL spectra of PPV films of varying
thickness, 21
possible formation of carbon-silicon
bonds, 22
preparation and properties of
graphitic nano objects from, 23–27
preparation of nanofibers, 18–19
Raman spectra of carbonized films
from PPV films, 24–25
Raman spectra of carbon nanotubes
and nanorods, 22*f*
scanning electron micrograph (SEM)
of nanotubes, nanorods, and
nanofilms, 17*f*
synthesis in nano shapes, 18
TEM of nanotubes and nanorods, 18*f*
thickness dependence of Raman
spectra of carbonized PPV films,
23*f*
UV–vis absorption, 19–20
Wessling–Zimmerman method, 17
wide-angle X-ray diffractograms, 19

X-ray photoelectron spectra (XPS), 21*f*, 22
Polyphenylene, formation of crosslinked matrix, 205–206
Polystyrene
  block copolymers with poly(dimethylsiloxane) (PDMS) forming nanoporous materials, 225, 226*f*
  characterization of brushes, 205*t*
  scanning electron microscopy (SEM) of PS nanoporous film after PDMS removal, 225, 227*f*
  See also Molecular brushes; Poly(4-hydroxystyrene) (PHS); Poly(4-octylstyrene)
Polystyrene star porogens
  molecular weight data, 193*t*
  molecular weight determination, 190
  schematic of synthesis and functionalization, 192–193
  synthesis, 189–190
  See also Polyarylene polymers
Polythiophene deposition, surface modification, 140, 141*f*
Pore morphology
  effect of hot plate bake sequence on final polyarylene, 196*f*
  effect of reaction time between matrix and functionalized porogen, 195*f*
  effect of star porogen functionality on polyarylene, 194*f*
  transmission electron microscopy (TEM), 193–194
  See also Polyarylene polymers
Pore size
  approximate distribution, 220*f*, 221
  range of critical, 212–213
  resin structure, 154
Pore structure
  porous dielectrics, 169–170
  theory of open vs. closed pores, 183–185
  visualization approaches, 167–169
Porogens

compatibility with resin, 175–176
foaming efficiencies, 151*f*
variety in silsesquioxane derivative resins, 150, 152
vitrification of matrix, 154
See also Burning out sacrificial spacers (BOSS) resins; Hydrogen silsesquioxane (HSQ); Methyl silsesquioxane (MSSQ); Organosilicate materials; Polyarylene polymers
Porosimetry
  basics, 210–213
  See also X-ray reflectivity
Porous dielectrics
  back end of the line (BEOL) wiring structure, 163*f*
  characterization, 165, 167–170
  coefficient of thermal expansion, 164
  critical parameters in screening materials, 164
  ellipsometric porosimetry (EP), 169
  extendibility of low k technologies, 164–165
  low k dielectrics, 162, 164
  mechanical and thermal characterization, 165, 167
  mechanical properties, 166*t*
  pore characterization, 167–170
  pore characterization approach providing averaged information, 169–170
  pore structure, 169–170
  positron annihilation lifetime spectroscopy (PALS), 169
  scanning electron micrograph (SEM), 167–168
  SiLK™, 165
  small angle neutron scattering (SANS), 169
  small angle X-ray scattering (SAXS), 169–170
  transmission electron micrographs, 168
  visualization approaches, 167–169
  See also Dielectric materials

Positron annihilation lifetime
spectroscopy (PALS)
pores of burning out sacrificial
spacers (BOSS) resins, 181
porous dielectrics, 169
Post-exposure bake (PEB)
neutron reflectivity of reaction front
profile, 91*f*
See also Poly(butoxycarbonyl
styrene) (PBOCSt)
Post exposure delay (PED)
description, 107
incomplete surface deprotection
reaction, 107–108
t-topping, 108
PPV. *See* Poly(*p*-phenylenevinylene)
(PPV)
Printed wiring boards (PWB), organic
substrates, 265
Printing, soluble conjugated polymers,
8
Profiling, depth. *See* Surface depth
profiling
Protection and deprotection, amino
polymer film, 125–126
Push-pull modulators
chromophore building blocks,
38
electro-optic, 30–31

films from carbonization of (PPV),
24–25
thickness dependence of carbonized
films, 23*f*
Reaction front
neutron reflectivity of profile, 91*f*
profile, 91–92
width and root mean squared
roughness, 94–95
See also Poly(butoxycarbonyl
styrene) (PBOCSt)
Ring opening metathesis
polymerization (ROMP). *See*
Intrinsically conducting polymers
(ICP)
Ring-opening reactions
benzocyclobutene (BCB), 280–281
See also Benzocyclobutene (BCB)-
based polymers
Roughness
line-edge (LER), 87
model film geometries studying
photoresist, 95*f*
reaction-front width and root-mean
squared (RMS), 94–95
See also Poly(butoxycarbonyl
styrene) (PBOCSt); Surface
roughness

Q

Quartz crystal microbalance (QCM)
fluoropolymers, 80, 81*f*
method, 74

R

Raman spectroscopy
carbonization temperature
dependence, 23*f*
carbon nanotubes and nanorods from
poly(*p*-phenylenevinylene) (PPV),
22*f*

S

Scanning electron microscopy (SEM)
films, nanotubes, and nanorods of
poly(*p*-phenylenevinylene) (PPV),
17*f*
polystyrene nanoporous film after
poly(dimethylsiloxane) (PDMS)
removal, 225, 227*f*
porous dielectrics, 167–168
Self-assembled monolayers,
modifying metal electrode surface,
3
Self-assembly
foaming hydrogen bonded materials
in CO_2, 231, 232*f*

See also Electro-optical applications
Semiconducting materials
  advantages, 1–2
  n-channel organic semiconductors, 9,
    10*t*
  soluble p-channel, 6, 8
  transistor performance of n-channel
    organic semiconductors, 10*t*
  transistor performance of organic
    semiconductors from solution, 7*t*
  transistor performance of p-channel
    semiconductors by vacuum
    deposition, 4*t*, 5*t*
  vacuum deposited p-channel
    materials, 4, 6
Semiconductor industry
  back-end-of-the-line (BEOL), 145
  copper CMOS logic chip, 146 *f*
  formation of nanoporosity using
    blending and sacrificial porogens,
    148*f*
  macromolecular pore generator
    (porogen) approach, 147
  nucleation and growth mechanism,
    147
  organosilsesquioxanes (SSQ), 147–
    148
  performance issues, 146–147
  performance issues for advanced
    chips, 147*f*
  schematic depiction of back-end-of-
    the line wiring, 146 *f*
  silicon dioxide, 162
  See also Organosilicate materials
Separation, block copolymers, 129–
  130
Silicon content, transparency, 73
Silicon dioxide (SiO$_2$)
  dielectric insulator, 162
  material properties, 163*t*
SiLK™ semiconductor dielectric
  back-end-of-the-line (BEOL)
    interconnects, 162
  mechanical properties, 163*t*, 166*t*
  ultra low dielectric constant material,
    165

See also Molecular brushes;
  Polyarylene polymers; Porous
  dielectrics
Siloxane-based multilayers
  capping step, 35
  covalent scheme, 35*f*
  examples of acentric monolayers, 36*f*
  film stability, 34–35
  formation, 34–35
  incorporation of refractive index
    modifying metal oxide layers, 37*f*
  layer-by-layer formation, 34*f*
  moisture, 34
  nonlinear optical response, 39
  push-pull chromophore building
    blocks, 38
  streamlined assembly method, 36–37
  two-step layer-by-layer sequence,
    37*f*
  two-step method and automation, 37
Silsesquioxane (SSQ)
  methyl SSQ (MSSQ), 149
  preparation, 148
  structure, 148*f*
  See also Organosilicate materials
Size, devices, 145
Small angle neutron scattering
  (SANS), porous dielectrics, 169
Small angle X-ray scattering (SAXS)
  nucleation and growth mechanism,
    154
  porous dielectrics, 169–170
  porous organosilicates, 153, 154*f*
Soft lithography
  block copolymers, 130
  See also Block copolymers
Solid-state crosslinking
  chemical oxidants, 47, 49*f*
  oxidative electrochemistry, 46–47,
    48*f*
  See also Intrinsically conducting
    polymers (ICP)
Solution, p-channel semiconductors,
  6, 8
Spin-coating, soluble conjugated
  polymers, 8

Stamp, poly(dimethylsiloxane) (PDMS), 136–138

Star polymers, functionalized, combining anionic and atom transfer radical polymerization, 156–157

Star porogens. *See* Polystyrene star porogens

Stepwise assembly, organized arrays, 31–32

Supercritical carbon dioxide
CO_2 foaming, 228, 231
development solvent for low k material formation, 225
extracting oligomers from polymer matrix, 225, 228
foaming behavior of poly(ether imide) and poly(ether sulfone), 231*t*
foaming hydrogen bonded materials, 231, 232*f*
potential benefits for nanoporous materials, 224
See also Nanoporous and microporous materials

Surface depth profiling
near edge X-ray absorption fine structure (NEXAFS), 108, 110, 113, 116
NEXAFS as function of detector bias, 113, 114*f*, 115*f*, 116
pre- and post-edge jump normalized NEXAFS spectra, 110, 112*f*
schematic, 109*f*
schematic of experiments generating model line edge region, 110, 111*f*
See also Near edge X-ray absorption fine structure (NEXAFS)

Surface modification
copper deposition, 139
functional groups, 130
grafting via atom transfer radical polymerization (ATRP), 139–140
polyelectrolyte deposition, 142
poly(*N*-isopropylacrylamide), 140, 141*f*

polythiophene deposition, 140, 141*f*
titanium dioxide deposition, 138

Surface morphology, atomic force microscopy, 92–93

Surface plasmon resonance (SPR)
electrochemically modulated (EM–SPR), 33
film thickness, 135

Surface roughness
atomic force microscopy (AFM), 50
conducting polymers, 50–51
See also Roughness

Swelling, block copolymers, 135

T

Template
approach of thermoplastic materials, 200
evaluation of molecular brushes as, 205–207
See also Molecular brushes

Tensile properties, divinyltetramethylsiloxane bisbenzocyclobutene (DVS–bisBCB) with elastomer, 288*t*

Terpolymers. *See* Photolabile terpolymers

Theory, open vs. closed pores, 183–185

Thermal analysis, divinyltetramethylsiloxane bisbenzocyclobutene (DVS–bisBCB) with elastomer, 287, 288*f*

Thermally degradable photocrosslinking polymers
baking temperature, 247, 249
characteristics of polymers, 240*t*
concept, 237*f*
dissolution properties of crosslinked copolymers after baking, 249*f*
dissolution properties of crosslinked poly(1-methyl-1-(6-methyl-7-oxabicyclo[4.1.0]hept-3-yl)ethyl

methacrylate) (PMOBH) and
copolymers, 248*f*
dissolution properties of thermally
decomposed PMOBH films, 245*f*
effect of baking temperature on
dissolution properties, 247, 249*f*
experimental, 238–242
9-fluorenilideneimino *p*-
toluenesulfonate (FITS) as
photoacid generator, 238, 242
FT–IR spectra of PMOBH film with
3.6 mol% FITS, 244*f*
materials, 238
measurements, 240, 242
1-methyl-1-(4-methyl-cyclohex-3-
enyl)ethyl methacrylate
(MMCEM) preparation, 238
MOBH preparation, 238–239
photocrosslinking, 242, 243*f*
polymer structures, 241*f*
preparation of polymers, 239–240
removing networks from substrates,
236–237
thermal degradation of MOBH
copolymers, 246–249
thermal degradation of MOBH
homopolymer, 243–246
thermolysis of crosslinked copolymer
film by in situ FT–IR
spectroscopy, 247*f*
Thermogravimetric analysis
(TGA),hexafluoroisopropanol
styrene (HFIP) copolymers, 61, 65*f*
Thin films. *See* Burning out sacrificial
spacers (BOSS) resins; Photolabile
terpolymers
Titanium dioxide, deposition for
surface modification, 138
Transition metal ions, chelates, 272
Transmission electron microscopy
(TEM)
comparing morphologies for methyl
silsesquioxane (MSSQ)–SiO₂
films, 157, 158*f*
cross sectional TEM of carbon film,
24*f*

nanotubes and nanorods of poly(*p*-
phenylenevinylene) (PPV), 18*f*
poly(dimethylsiloxane) (PDMS)
spheres in polystyrene matrix, 225,
227*f*
polystyrene brush/SiLK™ block
copolymer, 207
pore morphology of polyarylene
polymers with star porogens, 193–
194
porous dielectrics, 168
Transparency, resist material
requirement, 73

U

Ultra low k (ULK) dielectrics
blending approach, 174–176
BOSS (burning out sacrificial
spacers) approach, 176–177
BOSS technology, 173
compability between resin and
porogen, 175–176
compatibility of BOSS with metal
barrier, 182
cure temperatures of BOSS resins,
180
dielectric constants and modulus by
nanoindentation of films from
BOSS resins, 182*f*
dielectric constants of cured films
from BOSS resins, 179*t*
dielectric constants of films from
BOSS resins, 180*f*
evaluation of BOSS powder samples,
177, 179
evaluation of BOSS thin films, 179–
180
N₂ adsorption data on cured
hydrogen silsesquioxane
(HSQ)/porogen blends, 176*t*
N₂ adsorption isotherm of
HSQ/triaconcene blend, 175*f*
N₂ adsorption isotherms at 77K for
cured HSQ/triaconcene, 176*f*

N_2 adsorption isotherms for cured
    BOSS resins, 178*f*
pore characteristics, 181
pores in regular hexagonal
    arrangement, 183*f*
studies of proto-type BOSS resin,
    181–182
synthesis of BOSS resins, 177
theory of open vs. closed pores, 183–
    185
thickness of polymer matrix between
    two pores for given diameter vs.
    porosity, 184*f*
thin film properties for proto-type
    BOSS resin, 181*t*
See also Burning out sacrificial
    spacers (BOSS) resins
Ultrathin polymer films. *See*
    Photolabile terpolymers
Ultraviolet (UV) imaging. *See*
    Photolabile terpolymers
Ultraviolet–visible spectroscopy
before and after chemical oxidative
    crosslinking, 47, 49*f*
divinyltetramethylsiloxane
    bisbenzocyclobutene (DVS–
    bisBCB), 285*f*
poly(*p*-phenylenevinylene) (PPV)
    films, nanotubes, and nanorods,
    19–20
Underfill process, no-flow. *See* Metal
acetylacetonates (AcAc)
Urea compounds
foaming behavior, 231
scanning electron microscopy (SEM)
    image of foam from triurea
    compound in CO_2, 232*f*
synthesis, 232*f*

V

Vacuum deposition
p-channel materials, 4, 6
transistor performance of p-channel
    semiconductors by, 4*t*, 5*t*

Vacuum ultraviolet (VUV) absorption
fluoropolymers, 78*f*, 79
measurement, 74
Visualization approach, pore
    characterization, 167–169

W

Wide-angle X-ray diffractograms,
    poly(*p*-phenylenevinylene) (PPV)
    on different substrates, 19
Wiring structure, back-end-of-the-line
    (BEOL), 163*f*

X

X-ray porosimetry. *See* X-ray
    reflectivity
X-ray reflectivity
adsorption/desorption curves by
    temperature (T) and pressure (P)
    variation techniques, 216*f*
adsorption/desorption isotherms for
    porous hydrogen silsesquioxane
    (HSQ) and methyl silsesquioxane
    (MSQ) films, 218*f*
approximate pore size distribution,
    220*f*, 221
basics of porosimetry, 210–213
capillary condensation in pores, 221
comparing low k dielectric films with
    different porosity levels, 216–
    219
critical angle θ_c equation, 213
critical radius for capillary
    condensation, r_c, 211–212
curves as function of toluene partial
    pressure for low k dielectric film,
    214*f*
data collection, 214–215
effects of varying pressure or
    temperature, 215–216
interpreting adsorption/desorption
    isotherms, 219–221

monitoring electron density in porous
    thin films, 213
pore blocking effect, 218–219
porosity equation, 217
range of critical pore sizes, 212–213
varying r_c in Kelvin equation, 212*f*

Z

Zirconium phosphonate, multilayer
    assemblies, 32–33